D1342080

AEOLIAN DUST AND
DUST DEPOSITS

ACADEMIC PRESS INC. (LONDON) LTD.
24–28 Oval Road, London NW1 7DX

United States Edition published by
ACADEMIC PRESS, INC.
Orlando, Florida 32887

British Library Cataloguing in Publication Data
Is available

ISBN 0-12-568690-0
ISBN 0-12-568691-9 (paperback)

Phototypeset by
Paston Press, Loddon, Norfolk

Printed in G
Universi

AEOLIAN DUST AND DUST DEPOSITS

KENNETH PYE

Department of Earth Sciences
University of Cambridge

1987

ACADEMIC PRESS

Harcourt Brace Jovanovich, Publishers

London · Orlando · San Diego · New York
Austin · Boston · Sydney · Tokyo · Toronto

PREFACE

The entrainment, dispersion and deposition of dust is a matter of increasing geological interest and wider environmental concern. This book draws together much of the scattered multi-disciplinary literature dealing with aeolian dust and dust deposits and attempts to provide an overview which it is hoped will stimulate further research. The book has been written primarily for research workers and advanced students in sedimentology, geomorphology and Quaternary studies, but it is also likely to be of value to soil scientists, meteorologists, planetary geologists, engineers and others concerned with environmental management.

Inevitably, a book of this length cannot be comprehensive. I have deliberately concentrated on transport and deposition of dust derived by deflation of surface sediments and soils, since geologically this is the most important aspect, and because other types of dust (cosmic, volcanic and industrial) have been adequately reviewed elsewhere. I have referred extensively to publications by other workers, but the book also draws heavily on my own research in Australia, North America, Europe, Asia, the Middle East and Africa. Many of the ideas presented have evolved during discussions with colleagues and friends, notably Haim Tsoar, David Krinsley, Ann Wintle, Andrew Goudie, Edward Derbyshire and Ian Smalley, to whom I extend my thanks.

K. Pye, Cambridge, September 1986

for
Diane and Caroline

CONTENTS

Chapter One

THE GENERAL NATURE AND SIGNIFICANCE OF WINDBORNE DUST

1.1 THE NATURE OF AEOLIAN DUST

Dust can be defined as a suspension of solid particles in a gas, or a deposit of such particles. Dust particles transported in suspension in the Earth's atmosphere are mostly smaller than 100 μm (Udden 1894, 1898). Grains larger than about 20 μm settle back to the surface quite quickly when the turbulence associated with strong winds decreases, but smaller particles can remain in suspension for days or even weeks unless washed out by rain. Material which is transported very long distances in the Earth's atmosphere is mostly smaller than 10 μm and much is smaller than 2 μm. In locations remote from the main areas of natural and industrial particulate emission, such as Antarctica, the Arctic and the Central Pacific, a high proportion of the deposited dust is sub-micron in size. The cumulative grain size frequency curves of a number of local and far-travelled dust samples from different parts of the world are illustrated in Fig. 1.1. As discussed more fully in later chapters, continental loess deposits are composed mainly of particles in the 10–50 μm size range which have not been transported great distances, while aeolian deposits in the oceans are composed mostly of far-travelled material finer than 10 μm.

The term 'aerosol' strictly applies to both the gas and particle phases in a system, but it is widely used to refer to the particle phase alone (Prospero *et al.* 1983). Dust is a type of aerosol, related to, but distinct from, smokes, mists, fumes and fogs. Smokes are generally composed of smaller particles than dusts (mostly <1 μm) and result from burning or condensation of supersaturated vapours. Mists and fogs are suspensions of liquid droplets formed by atomization or vapour condensation. Like smokes, mists consist of very high concentrations of small particles. If the particle concentration is high enough to

1

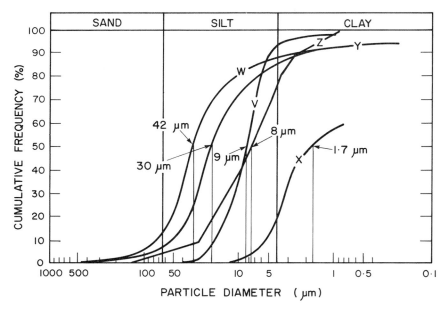

Figure 1.1. Cumulative grain size frequency curves of some atmospheric dust samples from different parts of the world: V – Saharan dust deposited in eastern England (Wheeler, 1985; W – local Kansas dust (Swineford and Frye, 1945); X – Saharan dust collected in Barbados (Prospero et al. (1970); Y – local Arizona dust (Péwé et al. (1981)); Z – Mongolian dust deposited at Beijing (Liu Tung-sheng et al. (1981)).

reduce visibility significantly, it is referred to as fog. Mixtures of smoke and fog, and by-products of their chemical interactions, are termed smog. Smokes, mist and smogs have been discussed in several existing texts on aerosols (Green and Lane, 1964; Friedlander, 1977; Reist, 1984; Hidy, 1984) and are not considered in detail in this book. Similarly, the reader is referred to texts by Cadle (1975), Allen (1981), Graf et al. (1976) and Perry and Young (1977) for information regarding methods of particulate sampling and analysis.

1.2 SOURCES OF DUST

Atmospheric dust particles originate from several different sources (Fig. 1.2). A number of attempts have been made to quantify the inputs from each of these sources, but accurate data are generally lacking.

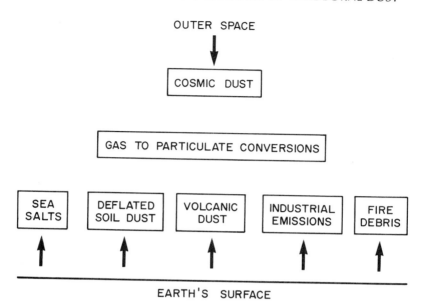

Figure 1.2. Sources of airborne particulates.

1.2.1 Cosmic dust

Cosmic dust is formed mainly by disintegration of meteorites entering the Earth's atmosphere (Whipple, 1964; Brownlow *et al.*, 1965; Fraudorf *et al.*, 1980). Cosmic dust in marine sediments occurs as black magnetic spherules, mostly 2–60 μm in size, which are similar in appearance and somewhat difficult to distinguish from industrial fly-ash particles (Handy and Davidson, 1953; Fisher *et al.*, 1976; Doyle *et al.*, 1976). The annual input of cosmic dust to the Earth's surface is small in comparison with the input from other sources, being of the order of $0.02–0.2 \times 10^6$ t (Hidy and Brock, 1971).

1.2.2 Volcanic dust

Volcanic eruptions provide an important, if periodic, supply of atmospheric dust (Lamb, 1970). The amount of dust produced varies with the type of eruption. Explosive eruptions tend to generate large amounts of dust, whereas effusive eruptions do not. Estimates of the average annual production of volcanic dust range from 4 to 25×10^6 t (Hidy and Brock, 1971; Peterson and Junge, 1971).

The main centres of present-day volcanic activity occur around the margins of the Pacific (Fisher and Schminke, 1984), but dust clouds following major volcanic eruptions have been observed to extend worldwide within three weeks or less (e.g. Robock and Matson, 1983). The development of dust plumes, and associated sedimentation patterns, following explosive volcanic eruptions are reviewed by Sparks (1986) and Carey and Sparks (1986). The physical properties and mineral composition of volcanic ash have been described and illustrated by Heiken (1974), Fisher and Schminke (1984) and Heiken and Wohletz (1985), while the environmental implications of volcanic eruptions have been treated in detail by Blong (1985).

1.2.3 Industrial emissions

Dust is produced by a wide range of industrial processes, and in mining and quarrying. Annual particle production was estimated by Hidy and Brock (1971) at $38-112 \times 10^6$ t. The major sources of industrial dust are smoke stacks, stockpiles, waste dumps, open-cast mines and quarries (e.g. Smit, 1980; Iversen and Jensen, 1981; Archibold, 1985; Nalpanis and Hunt, 1986). In addition to causing severe local pollution problems (Ledbetter, 1972; Perry and Young, 1977), industrial particulates may cause air quality deterioration at considerable distance from the source (e.g. Kerr, 1979; Rahn et al., 1981; Rahn, 1985; Barrie, 1986).

1.2.4 Gas to particulate conversions

Conversion of gases, such as SO_2 and NO_x, in the atmosphere produces substantial numbers of particles, although most are smaller than 1 μm (Fennelly, 1976). Conversion of industrial gases has been estimated to produce up to 284×10^6 t of particles each year (Peterson and Junge, 1971), while estimates of production by conversion of natural gases range from 478 to 2113×10^6 t yr^{-1} (Peterson and Junge, 1971; Hidy and Brock, 1971).

1.2.5 Fires

Fires produce large numbers of smoke particles; larger organic and inorganic particles may be raised high into the atmosphere by the associated buoyant thermal plumes. Peterson and Junge (1971) estimated an annual contribution of 5×10^6 t of airborne particles smaller than 5 μm from this source, while Hidy and Brock (1971) suggested a higher figure for particles of all sizes of 148×10^6 t yr^{-1}.

1.2.6 Sea salts

Breaking waves and bursting bubbles in the oceans are an important source of atmospheric aerosols (Heathershaw, 1974; Fairall *et al.*, 1983). Estimates of annual particle production range from 508 to 1113 $\times$ 10^6 t (Peterson and Junge, 1971; Hidy and Brock, 1971). Sea salts are particularly important in the atmosphere of oceanic and coastal areas, but their relative importance declines towards the continental interiors (Yaalon, 1964*a*, *b*; Hutton and Leslie, 1958; Hutton 1982).

1.2.7 Wind deflation of sediments and soils

Deflation of sediments and soils at the Earth's surface is the main source of atmospheric dust particles larger than 2 μm. Hidy and Brock (1971) estimated total annual dust production from this source to be 61–366 $\times$ 10^6 t. However, this figure is probably an underestimate since it does not take into account short-range, low-level dust transport. The main source areas of windblown mineral dust are the arid and seasonally-arid regions of the world (Prospero, 1981*a*; Péwé, 1981*b*; Coudé-Gaussen, 1984).

1.3 PREVIOUS STUDIES OF AEOLIAN DUST AND DUST DEPOSITS

Scientific observations on dust have been made for at least 200 years. Dobson (1781), in an account of the Harmattan wind between Cape Verde and Cape Lopez, reported conditions of poor visibility which he attributed to the dust content of the air. Darwin (1846) summarized earlier reports of dust falls on vessels off the West Coast of Africa and described a fall observed at first hand during the voyage of the Beagle:

> *On the 16th of January (1833), when the Beagle was ten miles off the N.W. end of St. Jago [Cape Verde Islands], some very fine dust was found adhering to the underside of the horizontal wind-vane at the mast head . . . During our stay of three weeks at St. Jago (to February 8th) the wind was N.E., as is always the case at this time of the year; the atmosphere was often hazy, and very fine dust was almost constantly falling, so that the astronomical instruments were roughened and a little injured. The dust collected on the Beagle was excessively fine-grained, and of a reddish brown colour; it does not effervesce with acids; it easily fuses under the blowpipe into a black or grey bead*
>
> *(1846, p. 27)*

Reports of falls of dust associated with rain and snow in Europe and elsewhere also extend back at least to the mid nineteenth century (Seignolis and

Arago, 1846; Ehrenberg, 1847; Mill and Lempfert, 1904). Early reports of dust storms and similar falls of 'red rain' in Australia and New Zealand were made by Brittlebank (1897), Liversidge (1902), Marshall (1903), Chapman and Grayson (1903), Marshall and Kidson (1929) Kidson and Gregory (1930) and Lindsay (1933). Many of the early references to dust falls were compiled by Fett (1958).

Some of the first laboratory and field investigations of sand and dust transport by wind were undertaken by Udden (1894, 1896, 1898, 1914). Free (1911) reviewed much of the early literature dealing with the mechanisms and control of soil deflation by wind. A popular account of sand and dust was compiled by Blacktin (1934). O'Brien and Rindlaub (1936) and Bagnold (1937, 1941) conducted detailed laboratory and field experiments on sediment mobilization by wind, but concentrated mainly on sand with a narrow range of grain sizes. Similar work on entrainment and dispersion of soil aggregates was initiated in the United States following the dust-bowl years of the 1930s, which has been described as 'the greatest man-made environmental problem the United States has ever seen' (Lockertz, 1978, p. 560). Major contributions to understanding of the mechanics and control of aeolian sediment transport were made by Chepil and his associates (Chepil, 1941, 1945; Chepil and Milne, 1941; Zingg and Chepil, 1950; Chepil and Woodruff, 1957, 1963). Investigations of wind action on soils were subsequently pursued in the United States by Clements *et al.* (1963) and by Gillette and his co-workers (Gillette and Goodwin, 1974; Gillette *et al.*, 1974, 1980, 1982). Further laboratory experimental work on sediment movement by wind was undertaken as part of the NASA research programmes during the 1970s and early 1980s when it was recognized that aeolian processes are also important on Mars and Venus (Iversen *et al.*, 1976; Iversen and White, 1982). Much of this work is summarized by Greeley and Iversen (1985).

The importance of additions of airborne salts and dust to soils only became widely recognized during the 1960s following investigations in Israel (Yaalon 1964*b*; Yaalon and Ganor, 1973, 1975), Australia, Hawaii and the Caribbean (Syers *et al.*, 1969; Walker and Costin, 1971; Jackson *et al.*, 1971, 1972, 1973; Bricker and Mackenzie, 1971). Subsequent work carried out in the 1970s and 1980s has confirmed that some soils consist almost entirely of weathered aeolian material (Macleod, 1980; Danin and Yaalon, 1982; Danin *et al.*, 1983; Rapp, 1984; Rapp and Nihlén, 1986).

Radczewski (1939), Rex and Goldberg (1958) and Bonatti and Aarhenius (1965) first identified the important contribution made by aeolian dust to deep ocean sediments. This has been confirmed by more recent work on sediment cores from the Atlantic and Pacific Oceans, although limited data are available from the Indian Ocean (Parkin and Padgham, 1975; Sarnthein *et al.*, 1981; Rea and Janecek, 1981, 1982; Rea *et al.*, 1985; Janecek and Rea, 1985; Lever and

McCave, 1983; Blank *et al.*, 1985; Tetzlaff and Peters, 1986; Rea and Bloomstine, 1986). Patterns of long-range dust transport have also been demonstrated by direct atmospheric sampling (Chester *et al.*, 1972; Aston *et al.*, 1973; Prospero and Nees, 1977; Windom and Chamberlain, 1978; Duce *et al.*, 1980; Uematsu *et al.*, 1983), and by satellite tracking (Bowden *et al.*, 1974; Vermillion, 1977; Iwasaka *et al.*, 1983; Chung, 1986; Takayama and Takashima, 1986).

Idso (1974b, 1976) drew attention to the frequency of contemporary dust storms in some parts of the world and suggested possible important consequences for climate modification. Goudie (1978, 1983a) re-emphasized the general importance of wind as a geomorphological agent and identified several major geomorphological consequences of dust storms. Global patterns of modern dust storms have been studied using meteorological records (Middleton, 1984, 1985, 1986a, 1986b; Middleton *et al.*, 1986). The relationship between recent dust-storm activity and environmental changes in Sahelian Africa and elsewhere was investigated by Prospero and Nees (1977, 1986), Middleton (1985) and McTainsh, (1985b, 1986), while the mechanics and geological implications of dust transport and deposition in deserts have been reviewed by Tsoar and Pye (1987) and Pye and Tsoar (1987). Volumes dealing with various aspects of dust have been edited by Morales (1979), Péwé (1981a) and Wasson (1982a).

The aeolian origin of loess became widely accepted following the publication of von Richthofen's works on China (von Richthofen, 1877–85, 1882), and there is now a vast international literature on the subject. Many of the early important contributions were drawn together by Smalley (1975), and Smalley (1980a) compiled a partial bibliography of loess. Other useful reviews, collections of papers and bibliographies include Scheidig (1934), Krieger (1965), Schultz and Frye (1968), Pecsi (1979a, 1984a), Pye (1984a), Lutenegger (1985) and Liu Tung-sheng *et al.* (1985a).

1.4 THE IMPORTANCE OF AEOLIAN DUST AND LOESS

The erosion, transport and deposition of dust has major environmental and economic consequences. Adverse effects include soil erosion (Kimberlin *et al.*, 1977; Lyles, 1977), undermining of structures and deflation of dirt roads (Hall, 1981), air pollution (Hagen and Woodruff, 1973), cattle suffocation, damage to young plants, visibility reduction leading to airport closures and road accidents (Houseman, 1961; Buritt and Hyers, 1981; Fig. 1.3), disruption to radio and satellite communications, damage to sensitive scientific and industrial equipment, contamination of food and drinking water (Clements *et al.*, 1963), causation and transmission of diseases (Bar-Ziv and Goldberg, 1974;

Figure 1.3. Dust storm warning sign, Highway 10, southeast of Phoenix, Arizona.

Bowes *et al.*, 1977; Hyers and Marcus, 1981; Leathers, 1981), corrosion of buildings and monuments (Goudie, 1977), and climate and weather modification (Bryson and Baerreis, 1967; Idso, 1981*a*, *b*). In battlefield situations serious consequences can arise from damage to engines and electronics (Hoidale *et al.*, 1967; Turner *et al.*, 1979), and dust has influenced the outcome of major political crises (Carter, 1979). There is also concern about dispersion and fallout of radioactive dust generated by nuclear weapons testing, nuclear accidents and possible nuclear war (Gabites, 1954; Dyer and Hicks, 1967; van der Westhuizen, 1969; Anspaugh *et al.*, 1973; Sehmel and Lloyd, 1976; Crutzen and Birks, 1982; Turco *et al.*, 1983; Covey *et al.*, 1984; Golitsyn and Ginsburg, 1985; Bach, 1986). It has even been suggested that the widespread extinctions which occurred at the end of the Cretaceous may have been related to marked global cooling and reduction in solar radiation caused by dense dust clouds generated by meteorite impact (Pollack *et al.*, 1983).

Geologically, dust deposition has been reported to play a role in case-hardening of rocks (Conca and Rossman, 1982), desert varnish formation (Potter and Rossman, 1977), duricrust development (Coque, 1955; Goudie, 1973; Watson, 1985), reddening and cementation of sediments (Waugh, 1970; Walker *et al.*, 1978; Pye and Tsoar, 1987). Aeolian processes are important in the transfer of elements from the continents to the oceans (Goldberg, 1971; Windom, 1975), and locally exert an important influence on seawater and marine sediment chemistry (Behairy *et al.*, 1975; Chester *et al.*, 1977; Chester *et al.*, 1979). In parts of the Pacific, more than half of the total sediment accumulated during the late Cenozoic is probably of aeolian origin (Rea *et al.*, 1985).

There are human and economic benefits associated with dust deposition, notably the addition of airborne nutrients to soils (Yaalon and Ganor, 1973; Bromfield, 1974; Beavington and Cawse, 1979), stabilization of sand dunes and other mobile surfaces in arid regions by formation of a cohesive dust-rich crust (Tsoar and Møller, 1986), and the accumulation of fertile loess soils which are of major agricultural importance (Chesworth, 1982). It has been estimated that 10% of the world's land surface area is covered by loess and loess-like deposits (Pecsi, 1968a). Loess regions in Central Asia and China were important centres for the development of early human cultures and civilizations (Ho Ping-ti, 1969, 1975; Ranov and Davis, 1979; Davis *et al.*, 1980). In many countries loess has provided an important raw material for the brick and cement industries (e.g. Ginzbourg, 1979), and in parts of China and North Africa houses are made from excavated loess.

Although naturally fertile, loess regions suffer some of the world's most serious soil erosion problems. In China, 80% of the sediment load of the Hwang Ho comprises material eroded from the Loess Plateau (Liu Tung-sheng *et al.*, 1985a). Major slope failures are common in the areas suffering erosion, and can have devastating human and economic consequences.

The record of dust deposition preserved in thick loess profiles, marine sediments and ice caps provides some of the most complete and detailed information regarding the effects of Quaternary climatic changes on the continents (Parkin and Shackleton, 1973; Parkin and Padgham, 1975; Fink and Kukla, 1977; Heller and Liu Tung-sheng, 1982; Thompson and Mosley-Thompson, 1981). In Soviet Central Asia, China, the Ukraine and Central Europe, the loess record extends back to at least 2 million yr BP. The oceanic dust record is even longer, and provides information about changes in global wind circulation since the Cretaceous.

Dust transport and deposition are also important on planets other than Earth, notably Mars and Venus, where they may be the dominant sediment transport processes (Sagan and Pollack, 1969; Arvidson, 1972; Cutts, 1973; Sagan, 1975; Hess, 1975; Greeley, 1979; Smalley and Krinsley, 1979; Greeley *et al.*, 1981; Zurek, 1982; Greeley and Iversen, 1985).

Some of these aspects are considered more fully in later chapters of this book.

Chapter Two

MECHANISMS OF
FINE-PARTICLE FORMATION

This chapter is concerned with the mechanisms of formation of fine particles found in dust deflated from soils and surface sediments. 'Fine' as used here refers to particles of silt and clay, which are defined as being 4–63 μm and <4 μm in size respectively (Wentworth, 1922).

It has been recognized for many years that inorganic clay-size particles are mainly, though not entirely, formed by chemical weathering, but the origin of silt particles, which form the bulk of typical loess, remains a matter of controversy. Several possible mechanisms of silt formation have been suggested (Table 2.1) but no clear picture regarding their relative importance has yet emerged.

Table 2.1
Some possible mechanisms of silt particle
formation suggested in the literature.

Mechanism	Best reference
Release from phyllosilicate parent rocks	Kuenen (1969)
Glacial grinding	Smalley (1966)
Frost weathering	Zeuner (1949)
Fluvial comminution	Moss *et al.* (1973)
Aeolian abrasion	Whalley *et al.* (1982)
Salt weathering	Goudie *et al.* (1979)
Chemical weathering	Pye (1983*b*)
Biological origin	Wilding *et al.* (1977)
Clay pellet aggregation	Dare-Edwards (1984)

2.1 RELEASE OF PARTICLES FROM FINE-GRAINED PARENT ROCKS

Kuenen (1969) suggested that release of particles from fine-grained metamorphic rocks such as schists and phyllites could provide an important source of silt. Outcrops of such rocks undoubtedly are important local sources of silt, but they cannot account for all the silt present in the geological record. Blatt (1967, 1970) emphasized the importance of granitoid rocks as sources of sand and silt, and pointed out that there is clear evidence of a reduction in the average size of detrital quartz in sediments compared with crystalline-source rocks.

2.2 GLACIAL GRINDING

The importance of glaciation in forming silt found in loess was recognized by Tutkovskii (1899, 1910) and Geikie (1898), but these views received little attention until the 1960s when Smalley (1966) and Smalley and Vita-Finzi (1968) claimed that glacial grinding is the only mechanism capable of generating large amounts of quartz silt. Smalley emphasized that virtually all of the great loess deposits of the world are located on the margins of areas which experienced Pleistocene glaciation (Smalley and Krinsley, 1978; Smalley, 1980b). Smalley's hypothesis was supported by Boulton (1978):

> *Vita-Finzi and Smalley (1970) have argued that a major proportion of the world's silt is produced by glaciers. I would support them in this, and go further to suggest that most of this is produced in the basal zone of traction which I believe to be a uniquely glacial environment in which large forces at non-inertial shear contacts produce fine-grained wear products.*
>
> *(1978, p. 796)*

Observations by Boulton and others (Vivian, 1975; Boulton, 1978, 1979) confirmed the presence of a debris-rich layer at the ice–bedrock interface beneath temperate glaciers, and showed that fines are produced preferentially in the zone of traction (Fig. 2.1). However, later work has questioned whether large amounts of quartz silt are produced in such environments. Haldorsen (1981, 1983) conducted a series of ball-mill experiments under wet and dry conditions designed to simulate the effects of crushing and abrasive particle–particle contact on samples of a Norwegian sandstone. At the end of the crushing experiments the rock fragments were reduced to a series of monomineralic particles whose size reflected that of the minerals in the parent material, suggesting that brittle fracture occurred mainly along intergranular boundaries. The abrasion experiments produced rock flour, most of which was finer than the mineral grains present in the parent rock. Feldspars and micas were found to be preferentially comminuted, with little change in the size of

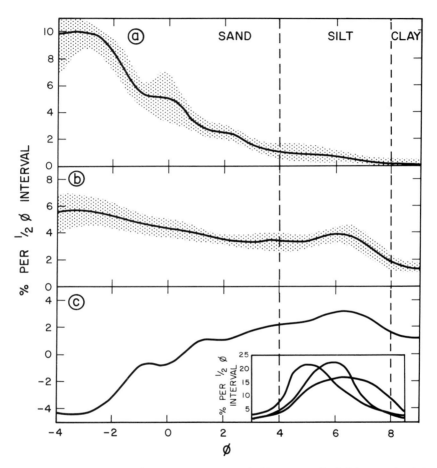

Figure 2.1. Mean weight percentages per half φ interval of debris in high-level transport: (a) in a medial moraine (23 samples); and (b) in the zone of traction (27 samples), at Breidamerkurjokull, Iceland; and (c) the difference between curves (a) and (b) indicating the deficiency in fines less than 1 φ (0·5 mm) in high-level transport. The inset in (c) shows the grain size distribution of rock flour from the contact between a clast in traction and bedrock and clearly indicates the source of much of the fines in the debris from the zone of traction. The shaded area gives the standard deviation at each half φ interval. (After Boulton (1978), Sedimentology **25**, 773–799; © International Association of Sedimentologists.)

quartz grains. Haldorsen related these results to observed grain-size distributions in tills of Astadalen, southeast Norway. Basal meltout tills in this area are characterized by a low silt content, which Haldorsen attributed to a predominance of sand-particle formation by crushing. Lodgement tills in the same area were found to contain significant amounts of silt but had only a small component of quartz. It was concluded that sub-glacial abrasion has little effect on quartz sand grains. This conclusion was supported by Sharp and Gomez (1985) on the basis of SEM surface textural evidence. These authors suggested that the limited abrasion experienced by quartz reflects its relative hardness, and that quartz is more likely to act as an abrasive tool which produces comminution of other mineral species. They point out that the 'edge grinding' on sub-glacial grains reported by Whalley (1979) is not true abrasion, but brittle fracture localized on grain edges caused by frequent particle-bed or interparticle contact in the debris-rich layer above the bedrock–ice interface. Such edge grinding appears not to produce large amounts of medium and coarse silt (cf. experimental results of Whalley, 1979). Sharp and Gomez (1985) argued that: 'the brittle fracture of gravel-sized clasts is the principal mechanism by which quartz sand grains are produced in the sub-glacial environment. Subsequent modification of these grains is limited to the localized collapse of sharp grain edges as a result of stresses imposed by interparticle and particle-bed contacts' (1985, p. 42).

Little weight was apparently given by these authors to the importance of microfractures and other lines of structural weakness as factors controlling the durability of quartz, although they acknowledged that surface textural microfeatures can be influenced by such weaknesses (Fig. 2.2). Several other studies have, however, concluded that the hardness of quartz is not the only, or even the major factor, affecting its durability. Nur and Simmons (1970), Moss *et al.* (1973) and Moss and Green (1975) pointed out that all minerals in plutonic and many metamorphic rocks contain defects and partially-healed fractures which cause some grains to disintegrate rapidly when exposed to any kind of stress. The spacing, nature and ease with which the cracks can be opened varies with the thermal and stress history of the rocks (Sprunt and Brace, 1974; Simmons and Richter, 1976; White, 1976). Many quartz-rich sandstones, however, are composed of relatively 'sound' grains with few fractures, the weaker grains having been selectively destroyed during transport and weathering prior to deposition. This may explain the apparent anomaly presented by Haldorsen's (1981, 1983) results. Sharp and Gomez (1985) gave no information about the lithologies from which their quartz grains were derived. A typical angular quartz silt grain, collected from the traction zone of a glacier at Storsbreen, Norway, is shown in Fig. 2.2.

Riezebos and Van der Waals (1974) and Nahon and Trompette (1982) pointed out that the glacial grinding mechanism cannot account for the

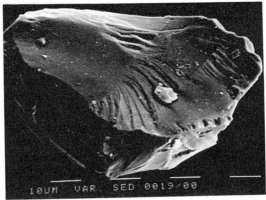

Figure 2.2. SEM micrograph of a quartz silt grain collected from a sub-glacial environment, Storsbreen, Norway, showing typical angular form, sharp edges, and conchoidal fracture surfaces. (Sample collected by J. L. Innes.) (Scale bar = 10 μm.)

formation of silt during periods of the Earth's history when glaciers did not exist on a large scale. Furthermore, Nahon and Trompette maintained that the abundance of silt in some fresh glacial sediments does not prove a glacial grinding origin, but may indicate only that large amounts of silt were available for entrainment by glacial processes. These authors argued that the major role of Quaternary and earlier glaciations was probably in reworking deep weathering profiles and in concentrating silts by fluvioglacial meltwaters. They point out that whereas ice sheets and glaciers are restricted in space and time, weathering is ubiquitous and continuous, although the relative intensities of individual processes vary.

2.3 FROST WEATHERING

The importance of frost action in forming silt on Mount Kenya was observed by Zeuner (1949). St Arnaud and Whiteside (1963) also suggested that frost weathering is important in increasing the silt content of Canadian soils. Laboratory experiments by Brockie (1973) and Lautridou and Ozouf (1982) showed that silt can be produced by frost action (Fig. 2.3). Moss *et al.* (1981) also demonstrated that simulated frost action combined with moisture absorption caused more damage to plutonic quartz grains than water absorption alone. Konischev (1982) argued that quartz is unstable under cryogenic conditions and that frost action is responsible for the dominant coarse silt and fine sand fractions found in soils developed on a range of sub-strata in the northern part of the European Soviet Union. Pye and Paine (1984) concluded

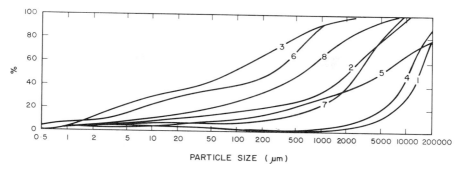

Figure 2.3. Grain size distribution curve of rock debris produced by frost shattering (with freezing to a temperature of −8° C): (1) lithographic limestone from Sonneville (Charentes), after 500 freeze–thaw cycles; (2) chalky limestone from Caen (Lower Normandy), after 500 cycles; (3) chalk from Tankerville (Normandy), after 500 cycles; (4) Precambrian schist/siltstone from Lengronne (Lower Normandy), after 1000 cycles; (5) Corsican schist, after 920 cycles; (6) argillite from Hyenville (Lower Normandy), after 1320 cycles; (7) weathered schist from Mesnil–Herman (Normandy), after 1320 cycles; and (8) weathered granite from Grand–Celland (Lower Normandy), after 1320 cycles. (After Lautridou and Ozouf (1982), Prog. Phys. Geog. 6, 215–232, © Edward Arnold.)

that silt and fine sand near the summit of Ben Arkle in northwest Scotland was formed by aeolian reworking of frost-weathered debris derived from quartzite. Smalley *et al.* (1978), Smalley (1980*b*) and Smalley and Smalley (1983) accepted that cold weathering processes complement glacial grinding in mountain regions and could have contributed significantly to the world's major Pleistocene loess deposits.

Several authors have drawn attention to the difficulty of separating the effects of moisture absorption and ice crystallization in cold environments (Hudec, 1973; White, 1976; Fahey, 1983; Walder and Hallet, 1986), but in the context of the amount of silt produced this problem is not critical. More important are the overall rates of rock breakdown and the nature of the weathering debris produced under such conditions. Laboratory experiments and field observations have shown that both are dependent on the rock and/or sediment properties, and on the prevailing environmental conditions (McGreevy and Whalley, 1982; McGreevy, 1981; Lautridou and Ozouf, 1982). The grain size, mineralogy, porosity, permeability and moisture content are important material properties (Hall, 1986), while rate and frequency of freezing are important environmental controls on rock breakdown. Although generalizations are difficult to make, the balance of evidence indicates that porous rocks such as chalk and limestone are more prone to frost damage and produce more fines than igneous rocks and quartz sandstones.

2.4 FLUVIAL COMMINUTION

Early tumbling-mill experiments using Brazilian quartz carried out by Kuenen (1959) suggested that large amounts of fine particles are not generated by fluvial abrasion processes. However, Moss (1966, 1972) noted that coarse plutonic quartz in southeast Australia disintegrated rapidly during fluvial transport. Rapid downstream reduction in the size of quartz and lithic clasts has also been noted by other authors (Blatt, 1967; Slatt and Eyles, 1981). Experiments by Moss et al. (1973) using natural plutonic quartz supported Moss's (1973) hypothesis that at least some granitic quartz grains experience fatigue due to many small impacts in water. Size analysis of the debris produced in Moss et al.s' experiments indicated a lower size threshold of about 20 μm. This was related to the presence of microfractures within the quartz. Moss and Green (1975) extended this work and concluded that the minimum size likely to be achieved during reduction of quartz is about 2 μm. However, the 'microcracks' observed by Moss and Green have been reinterpreted as Wallner lines formed during brittle fracture, and some of their 'quartz' grains appear to be feldspars (Whalley, 1979).

The amount of silt produced by fluvial and fluvio-glacial action is likely to depend on the energy of the stream and the size range of the sediment load. Highest rates of silt production might be expected in mountain areas with steep stream gradients and high seasonal discharges due to rainfall or snowmelt. Rivers in such areas typically carry a wide range of particle sizes ranging from cobbles to silt and clay, and the opportunity for crushing of smaller particles is great. The nature of interparticle contacts in water (a relatively high-viscosity fluid) is such that significant production of fines is unlikely in rivers which transport predominantly sand-sized sediment. The geographical distribution of silt supplied by fluvial comminution should therefore in part coincide with that supplied by glacial grinding and cold-weathering processes. Unfortunately, there are no quantitative field data available to test these predictions.

2.5 AEOLIAN ABRASION

Smalley and Vita-Finzi (1968) suggested that loess on the margins of some warm deserts, such as the Negev, may be composed partly of silt formed by aeolian abrasion during saltation of sand grains. Experimental investigations of aeolian abrasion were initiated by Knight (1924), Anderson (1926) and Marsland and Woodruff (1937), but the first detailed experiments were conducted by Kuenen (1960) using three different wind tunnels. The abrasion behaviour of limestone, feldspar and quartz, in the form of cubes, crushed crystals and natural grains, was compared. Abrasion was found to increase with particle size, wind velocity, angularity and surface roughness.

Polished medium-sized quartz grains suffered no abrasion. Abrasion was also slight for fine quartz sand grains and zero for grains <50 μm. In experiments with crushed quartz, only angular fragments with a minimum size of about 50 μm, and fine flour <2 μm, were produced. However, medium and quartz silt were produced during abrasion of crushed feldspar. This led Kuenen to conclude that the medium- and coarse-quartz silt found in loess cannot be of aeolian abrasion origin. In a later paper, Kuenen (1969) suggested that the small amount of coarse silt produced during aeolian abrasion of quartz is due to its hardness and the nature of interparticle impacts during saltation. Kuenen suggested that the force of interparticle collisions is mitigated by the spinning motion of saltating grains and by the fact that they are free to ricochet when they collide in air.

Whalley et al. (1982) arrived at different conclusions in simple aeolian abrasion experiments using crushed vein quartz. These workers found both coarse and fine silt particles in the abrasion debris. The exact quantities of attrition products were not measured, but after 60 h of attrition a mass reduction of 1–5% was observed; about 20% of the debris was finer than 5 μm. It was argued that: (1) abrasion takes place primarily by point impacts of (originally angular) edges; (2) grain spinning actually accentuates abrasion by delivering a grazing blow to the surface; and (3) the chips knocked off are predominantly of silt size and rather blocky in form. In a subsequent experiment using crushed Brazilian quartz (Whalley et al., 1987) it was shown that the proportion of fine silt increases with time as the initial grains become more rounded and the general quantity of silt produced decreases. Krinsley et al. (1981) and Krinsley and Greeley (1986) also observed in experiments that abrasion begins at grain edges and proceeds gradually to cover the entire grain surfaces, but these authors gave no details of changes in the sizes of particles produced over time.

It is not clear how applicable these experimental findings are to field conditions. Brazilian quartz differs from typical plutonic quartz in having few inherent crystallographic flaws. The grain–grain contacts in Whalley et al.s' experiments in no way resemble those during natural saltation, and the equivalent wind velocity in their experiments was not specified. It is also uncertain how much breakage resulted from grain contacts with the sides of the containment vessel. Field observations of silt formation by abrasion during saltation have not been made, but the common tendency for dune sands to become more rounded and quartz-rich with age and distance of transport (e.g. Mazzullo et al., 1986) suggests that abrasion is important. At present, however, little can be said about the rate of particle generation, or the size of the particles produced in nature.

The occurrence of yardangs (Blackwelder, 1934; McCauley *et al.*, 1977; Ward and Greeley, 1984) and ventifacts (Whitney and Splettstoesser, 1982) in deserts, together with the results of laboratory experiments (Dietrich, 1977; Suzuki and Takahashi, 1981), testify that abrasion of rocks by windblown sand and dust is also significant, but once again there are no published data concerning the amount or size of natural abrasion products.

2.6 SALT WEATHERING

Substantial experimental and field evidence accumulated during the past decade has demonstrated the effectiveness of salts in causing disintegration of rocks and sediments (Goudie, 1977, 1985, 1986; Cooke, 1981; Goudie and Day, 1980; McGreevy and Smith, 1982; Goudie and Watson, 1984; Goudie, 1984; Fahey, 1985). Experiments by Goudie *et al.* (1979) and Pye and Sperling (1983) showed that medium and coarse silt particles are formed by the action of sodium sulphate on quartz dune sand and other materials under simulated warm desert conditions. Pye and Sperling's study indicated that wetting and drying alone is much less effective in producing silt than combined wetting and drying and salt action. 'First cycle' sand grains in Cairngorm regolith were found to suffer more damage than multi-cyclic quartz dune sand grains from northeast Australia. Feldspars and micas were also found to be much more susceptible to salt damage than quartz on account of their well-developed cleavage. Examination of quartz grains in the scanning electron microscope (SEM) after 40 and 80 cycles of salt crystallization revealed numerous cracks and angular breakage features (Fig. 2.4). The fines produced in the experi-

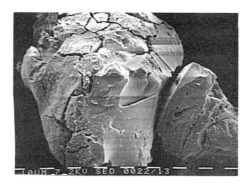

Figure 2.4. SEM micrograph showing damage to a quartz sand grain after 40 cycles of sodium sulphate 'weathering' under simulated hot desert conditions. (Experiments of Pye and Sperling (1983). (Scale bar = 10 μm.)

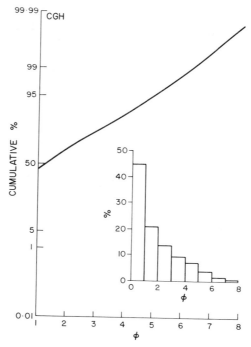

Figure 2.5. Grain size frequency distribution of particles formed from 1–2 mm-size grains of Cairngorm regolith sand after 40 cycles of experimental sodium sulphate weathering. (After Pye and Sperling (1983), Sedimentology 30, 49–62; © International Association of Sedimentologists.)

ments showed a continuum of sizes with a significant proportion in the coarse and medium silt range (Fig. 2.5). Quartz in the weathering products displayed dominant angular breakage features, though some surface textural features were clearly inherited from the parent grains (Fig. 2.6). Feldspar silt grains displayed a more blocky, equi-dimensional form, indicating cleavage-controlled fracture (Fig. 2.7).

Goudie (1986) subjected rectangular blocks of York Stone (a quartzose sandstone) to simulated capillary rise (the 'wick effect') using a variety of salt solutions. Analysis of the debris produced after 40 cycles indicated that silt comprised 25·3–80%, with a mean of 47·97%. Clay-size material averaged only 3·7%. In the debris produced after 60 cycles silt comprised an average of 43·7% and clay 3·5%.

Smith *et al.* (1986) examined sandstone blocks in the SEM and in thin section after 60 diurnal cycles of experimental salt weathering using sodium chloride, sodium sulphate and magnesium sulphate solutions. It was concluded that

Figure 2.6. SEM micrograph showing an angular quartz silt grain formed by simulated sodium sulphate weathering of quartz dune sand. (Scale bar = 10 μm.)

silt-size material is formed in two ways: (1) breaking away of silica cement and silica coatings from around sand grains; and (2) microfracturing of quartz sand grains in areas of point loading. It was inferred that crystallization and/or hydration expansion of salts within pores in the sandstone exerted sufficiently large stresses to cause fracturing of quartz grains under the confined conditions within the rock.

The laboratory studies cited above showed that sodium sulphate, magnesium sulphate, sodium carbonate and calcium chloride are the most destructive salts (Kwaad, 1970; Goudie et al., 1970; Goudie, 1974; Goudie, 1986). Sodium chloride and calcium sulphate, which are the most common salts

Figure 2.7. SEM micrograph showing a typical blocky feldspar silt grain formed by sodium sulphate attack on Cairngorm regolith sand. Fracture of the feldspar grains is strongly cleavage-controlled. (Scale bar = 10 μm.)

found in nature, are relatively less destructive, but there is substantial field evidence that they can also cause rock breakdown and generate fines (Goudie and Day, 1980; Chapman, 1980; Goudie and Watson, 1984).

2.7 COMBINED SALT AND FROST ACTION

In cold arid climates rocks and sediments are exposed to both salt and frost action. However, whether the net effect is an increase or a reduction in the rate of disintegration is still debated. Goudie (1974) suggested that a sharp fall in temperature could lower solubility and precipitate salts without evaporation. Hudec and Rigbey (1976) noted that sodium chloride enhances the water absorptive properties of rocks, thereby increasing the amount of moisture available for freezing. Experiments by Williams and Robinson (1981) suggested that presence of certain salts enhances the susceptibility of rocks to frost damage by lowering the freezing-point of the pore fluids and allowing larger, potentially more destructive ice crystals to develop. They noted that sodium sulphate precipitated very quickly from solution at the point of freezing, so the rock was subject to combined salt and ice crystallization pressures. However, experiments by McGreevy (1982), using salt solutions of different strengths, indicated that if salts are present in low concentrations they may reduce frost damage by preventing freezing taking place while not giving rise to salt crystal growth.

2.8 CHEMICAL WEATHERING

Many studies have documented a progressive reduction in particle size during soil development in humid areas. In part this is due to the chemical alteration of unstable silicates and formation of authigenic clay minerals, but a reduction in the average size of quartz has also been noted (Eswaran and Stoops, 1979). Disintegration of quartz can occur by dissolution and partial replacement by iron oxyhydroxides in ferricretes and by calcite in calcretes (Nahon and Trompette, 1982), or by simple dissolution of silica along lines of crystal weakness in podzol profiles (Pye, 1983a). In North Queensland, active coastal dune sands contain $<0.5\%$ silt but stabilized, podzolized dune sands contain up to 15% silt. Analysis indicated a continuum of particle sizes in the silt range (Fig. 2.8), suggesting *in-situ* formation by breakdown of sand grains. SEM examination showed preferential dissolution of silica along microfractures (Figs. 2.9 and 2.10). The silt grains produced were angular, although the surfaces and corners show rounding and pitting due to silica dissolution (Fig. 2.11). Silt is formed mainly in the A-horizons and is translocated to the

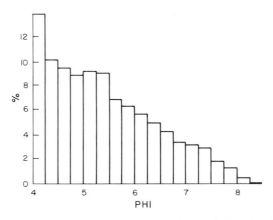

Figure 2.8. Grain size frequency histogram of the silt fraction from a stabilized sand dune at Cape Flattery, North Queensland. (After Pye (1983a), Sedim. Geol. *34*, 267–282; © Elsevier Publishing Co.)

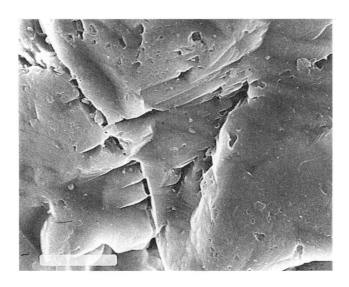

Figure 2.9. SEM micrograph showing preferential dissolution of silica along microfractures in a quartz sand grain from a podzolized sand dune, North Queensland. (Scale bar = 10 μm.)

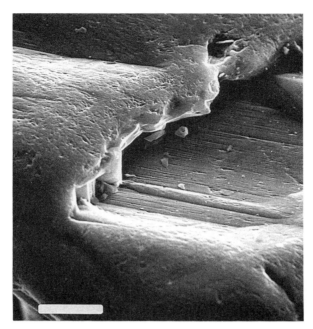

Figure 2.10. SEM micrograph showing crystallographic control of silica dissolution on a quartz sand grain from North Queensland. Note the formation of clay-size quartz particles beneath the 'overhang'. (Scale bar = 5 μm.)

Figure 2.11. SEM micrograph showing coarse and medium silt grains from the A_2 horizon of a podzol soil developed in a stabilized coastal dune, Cape Flattery, North Queensland. (Scale bar = 50 μm.)

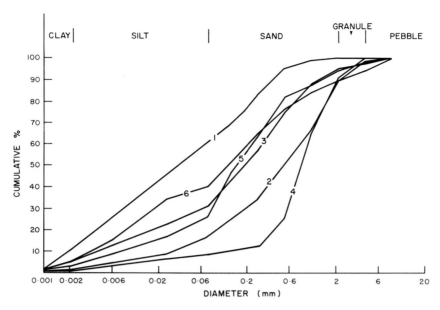

Figure 2.12. Cumulative grain size frequency curves of silty regolith developed on different rock types in the semi-arid Kora area of central Kenya: (1)–(4) granitoid gneiss; (5) granulite; and (6) gabbro. Silt comprises 10–60% of the regolith and clay comprises 1–10%.

B-horizons by percolating water. Although the solubility of quartz is normally low below pH 9 (Morey *et al.*, 1962; Siever, 1962; Iler, 1979), it is increased substantially by organic acids (van der Waals, 1967; Crook, 1968) of the type present in podzol soil profiles.

In the perennially wet tropics, soils formed by weathering of crystalline rocks typically have a low silt/clay ratio due to rapid chemical decomposition of silicate minerals to form clay. In the seasonally wet tropics, however, silicate decomposition is slower and the soils often have a higher silt/clay ratio. Boulet (1973) described 1–3 m thick soils on granite in Upper Volta which contained 15–20% silt. Pye *et al.* (1985) found that soils developed on a range of metamorphic and igneous lithologies in semi-arid central Kenya have a high silt/clay ratio (Fig. 2.12) due to the importance of granular disintegration relative to chemical decomposition (Pye, 1985).

Formation of silt by weathering of quartz and feldspar also occurs in temperate soils (e.g. Nornberg, 1980). In Scotland, many soils developed on granite contain 10–20% silt (Fig. 2.13), but the processes which produce these

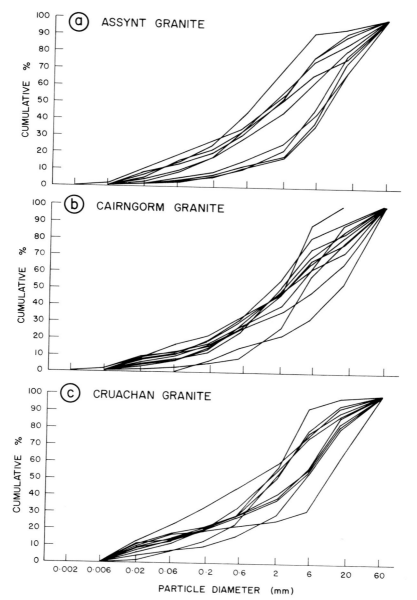

Figure 2.13. Cumulative grain-size frequency curves showing the particle size distribution of regolith developed on three Scottish granites: (a) Assynt granite; (b) Cairngorm granite; (c) Cruachan granite. Silt typically comprises 5–20% of the total material, clay less than 2%. (Samples collected by J. L. Innes.)

fines have not yet been investigated. Moisture absorption, frost and chemical action may all play a part.

Formation of silt-size authigenic quartz as a by-product of chemical weathering of other silicate phases in soils was reported by Robinson (1980).

2.9 INSOLATION WEATHERING

The role of diurnal and/or seasonal temperature changes, thermal gradients from the surface into the rock, and the different coefficients of thermal expansion of different minerals in causing rock breakdown is still disputed (Yaalon, 1974; Rice, 1976; Winkler, 1977; Smith, 1977). Early experimental work indicated that temperature changes alone are probably ineffective unless combined with wetting and drying (Blackwelder, 1925, 1933; Griggs, 1936). Although surface spalling and splitting of boulders in deserts have been attributed to insolation effects (Roth, 1965; Peel, 1974), there is no evidence that significant amounts of silt are produced by this mechanism.

2.10 FINE PARTICLES OF BIOLOGICAL ORIGIN

Small particles of biogeneic opal (phytoliths) are synthesized in the cellular tissues and walls of vascular plants, particularly grasses, sedges, horsetails and nettles. Grasses commonly contain 3–5% silica on a dry weight basis, and occasionally as much as 20% (Norgren, 1973). Conifer woods are generally low in silica (<1%), but some deciduous woods contain significant amounts (Geis, 1973). Biogenic opal is a minor but ubiquitous component of soils, ranging from <0·1–3% by dry weight (Wilding et al., 1977). The concentration is normally highest near the soil surface (up to 20%). Jones and Beavers (1963a) and Wilding and Drees (1971) found that 50–75% of the phytoliths they studied were <5 μm in size, but Yeck and Gray (1972) reported a majority of phytoliths in the 5–50 μm size-range. Phytoliths are readily recognized in dust by their distinctive morphology (Baker, 1959a, b; Folger et al., 1967; Jones and Hay, 1975).

Siliceous diatoms, radiolaria and echinoderm spicules are produced in large numbers in some aquatic environments and are sometimes found in soil (Jones et al., 1964; Jones and Beavers, 1963b; Wilding and Drees, 1968, 1971; Weaver and Wise, 1974). Most are of fine- or medium-silt size and may be entrained as dust from dry lake beds or exposed marine sediments.

Pollen, spores and other organic particles, which are frequently hollow and have a low density, are also readily transported by the wind in suspension (Horowitz et al., 1975; Stix, 1975; Lepple and Brine, 1976; Melia, 1984). Fungi and fragments of lichens have been reported to be major constituents of some dusts (Padu and Kelly, 1954; Donkin, 1981).

2.11 FORMATION OF CLAY PELLETS

Silt and sand-size pellets can form by aggregation of clay particles in playa lakes and similar desert environments. The composition of the pellets varies considerably, including palygorskite and sepiolite (Fig. 2.14), smectite, illite, kaolinite, carbonate, sulphates and chlorides. The pellets originate by wind erosion of mud curls and salt–mud efflorescences on dry lake beds, but the mechanism by which the pellets become rounded and compacted is not fully understood. Greeley (1979) and Krinsley and Leach (1981) suggested that electrostatic forces are important in forming silt aggregates on Mars. Electrical charges also build up on the surfaces of moving dust and sand particles on Earth due to abrasion (Rudge, 1914; Gill, 1948; Stow, 1969) but there is no evidence that they result in aggregates as suggested by Greeley. Most pellets on Earth appear to be held together by thin moisture films and soluble salts. The larger clay pellets normally accumulate as low dunes (lunettes) close to the source, but finer pellets can be dispersed more widely (Coffey, 1909; Hills, 1939; Huffman and Price, 1949; Bowler, 1973; Lancaster, 1978a, b; Coudé-Gaussen et al., 1984; Dare-Edwards, 1984).

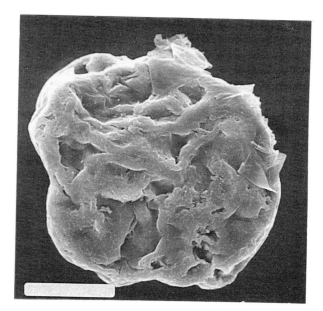

Figure 2.14. SEM micrograph showing a palygorskite pellet in North African dust. (Photo by M. Coudé-Gaussen. Scale bar 50 μm.)

2.12 THE RELATIVE IMPORTANCE OF DIFFERENT PARTICLE FORMATION MECHANISMS

The relative importance of different fine-particle formation mechanisms differs from area to area, reflecting the effects of climate, relief, lithology and geomorphic history. In cold climates, glacial abrasion, frost action and fluvioglacial abrasion are the dominant silt-forming processes, while in the wet tropics chemical weathering and possibly moisture absorption appear to be most important. In deserts, salt weathering is significant, but its role in relation to aeolian abrasion and other weathering processes remains to be established. Assessments of total silt production in different climatic zones are difficult to make due to the lack of quantitative data. Relief is a major complicating factor, since rates of weathering and erosion, and therefore of sediment production, are much greater in active mountain regions. Low-relief landscapes are characterized by landscape stability, low rates of surface stripping, and low rates of debris production.

The balance of distributional, stratigraphic, and sedimentological evidence presently available strongly suggests that most of the silt in the world's major loess deposits is of cold weathering and/or glacial origin. However, glacial and cold-weathering silt probably makes up only a relatively small proportion of the total silt present in the oceans and in the rock record.

Chapter Three

DUST ENTRAINMENT, TRANSPORT AND DEPOSITION

Aeolian dust transport involves three stages: (1) entrainment; (2) dispersion; and (3) deposition. The nature of wind sediment transport is controlled both by the nature of the airflow near the ground and by the properties of the ground surface over which the flow occurs.

3.1 PARTICLE FORCES

Several different forces act on a particle at rest on a bed and exposed to a turbulent windflow (Fig. 3.1). The forces include drag (D), lift (L), moment

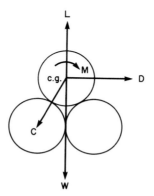

Figure 3.1. Forces acting on an erodible spherical particle in a windstream: L = lift; D = drag; W = weight; M = moment; C = interparticle cohesion; c.g. = centre of gravity.

(M), weight (W) and interparticle cohesion (C). The lift and drag forces and moment result from the flow of air over the particle. The weight and cohesion forces reflect properties of the particles on the bed, including size, density, nature of packing, chemical composition, surface charge, and moisture content. At the onset of movement the balance of lift and drag forces must outweigh the combined inertia forces of weight, cohesion and bed friction.

3.2 PARTICLE ENTRAINMENT

In theory, movement of particles at rest on a bed can be initiated by different mechanisms acting alone or in combination including (1) the drag exerted by the wind; (2) aerodynamic lift; (3) impacts by rolling or bouncing particles (including hail, rain and sediment grains); and (4) disturbance by pedestrian or vehicle traffic. However, despite intensive study, the precise manner in which particles are ejected from the bed is not yet fully understood.

3.2.1 Fluid drag

Wind passing over a stable bed is retarded at its base by friction and is characterized by turbulent eddies which move in all directions and at variable velocities. There is a very thin layer just above the bed in which the velocity is zero. The thickness of this layer, here referred to as the roughness height, z_0, is about 1/30 of the diameter of the particles on the bed surface (von Kármán, 1934; Prandtl, 1935; Bagnold, 1941). In the case of surfaces covered by tall vegetation or high densities of other roughness elements, the plane of zero velocity may be displaced upwards to some height determined by the height, density, flexibility and permeability of the roughness elements. This is referred to as the zero plane displacement height (d). Above the height of zero velocity the time-averaged forward velocity (U) of the wind in a neutral atmosphere turbulent boundary layer increases approximately logarithmically with height. The slope of the logarithmic wind profile is defined by the Prandtl–von Kármán equation:

$$u_* = \frac{kU}{\log_n (z/z_0)} \qquad [3.1]$$

where

k is the von Kármán constant which varies with temperature gradient but is usually taken to be 0·4
U is the wind velocity measured at a height z above the ground
z_0 is the height at which the velocity is zero
u_* is the drag velocity

The drag velocity is related to the shear stress (τ) exerted by the wind on the bed and to the density of air (ρ_a) by the expression:

$$u_* = \sqrt{\frac{\tau}{\rho_a}}$$ [3.2]

Both u_* and τ increase as the wind velocity (U) increases. At some critical point grains on the bed start to move. This point, referred to by Bagnold (1941) as the 'fluid threshold', can be expressed by

$$u_{*t} = A\sqrt{\frac{\rho_p - \rho_a \cdot gD}{\rho_a}}$$ [3.3]

where

u_{*t} is the fluid threshold velocity
ρ_p is the relative density of the grains (2·65 g cm^{-3} for quartz)
ρ_a is the relative density of air (1·22 × 10^{-3} gm cm^{-3})
g is the acceleration due to gravity (980 cm s^{-2})
D is the mean grain diameter (cm)
A is an empirical coefficient equal to 0·1 for particle friction Reynolds numbers (Re$_p$) greater than 3·5 (Re$_p$ = $u_* D/v$, where v is the kinematic viscosity of the air). Re$_p$ provides a measure of how turbulent the flow is around a particle.

The relationship between the fluid threshold velocity and mean grain diameter established by Bagnold is shown in Fig. 3.2; u_{*t} attains a minimum value when the mean grain size is approximately 80 μm. Bagnold (1941) suggested that with finer average particle sizes, the surface becomes aerodynamically 'smooth', and the air drag, instead of being carried by a few more exposed grains, is distributed more or less evenly across the whole surface. Consequently, a relatively greater drag is required to set the first grains in motion. As a result of the changes in the nature of the airflow, the value of the coefficient A starts to increase when the grain size falls below about 200 μm. In a simple experiment, Bagnold (1937) showed that when an airstream is passed over a scattered layer of fine Portland cement, no particle movement occurs even when u_* exceeds 100 cm s^{-1}; i.e. the wind is strong enough to move pebbles 4·6 mm in diameter.

These relationships between grain size and threshold velocity have generally been confirmed by many other workers (Horikawa and Shen, 1960; Chepil and Woodruff, 1963; Belly, 1964; Iversen *et al.*, 1976). However, Iversen *et al.* (1976), Sagan and Bagnold (1975) and Iversen and White (1982) suggested that the upturn of the fluid threshold curve for particles finer than 80 μm is due more to interparticle cohesive effects than to Reynolds number effects. These interparticle forces include moisture films, van der Waals forces and

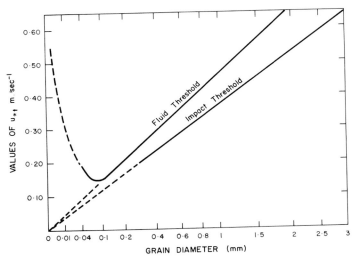

Figure 3.2. Relationship of the fluid threshold velocity (u_{*t}) and the impact threshold velocity (u_*) to median particle diameter as determined by Bagnold (1941). Values of u_{*t} for particles smaller than 60 μm are approximations only. (After Bagnold (1941), The Physics of Blown Sand and Desert Dunes, p. 88; © Chapman and Hall Ltd.)

electrostatic charges. Fletcher (1976b) obtained a semi-empirical expression which includes the effect of interparticle cohesive forces:

$$u_{*t} \propto \left(\frac{\rho_p - \rho_a}{\rho_a}\right)^{1/2}\left[0.13(gD)^{1/2} + 0.057\left(\frac{c}{\rho_p}\right)^{1/4}\left(\frac{v}{D}\right)^{1/2}\right] \qquad [3.4]$$

where

c is the particle cohesion
ρ_p is the density of the particle
ρ_a is the density of air
v is the kinematic viscosity

Both Fletcher (1976b) and Iversen and White (1982) obtained minimum values of friction velocity similar to Bagnold's and those obtained by Chepil (1945, 1951), (Fig. 3.3).

The presence of moisture and cementing agents can also significantly influence the threshold velocity for larger particles. Wind-tunnel experiments by Belly (1964) showed that 0.6% moisture (by volume) could double the value of u_{*t} compared with dry sand. On a natural beach in The Netherlands, Svasek and Terwindt (1974) found that a similar amount of moisture could treble the

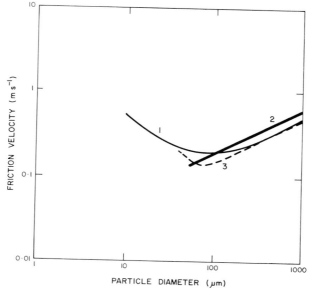

Figure 3.3. Relationship between threshold friction velocity and particle diameter: (1) as predicted theoretically by Iversen and White (1982); (2) as predicted theoretically by Fletcher (1976b), assuming no cohesion; and (3) as determined empirically by Bagnold (1941).

value of u_{*t}. Belly (1964) suggested the following modified relationship for entrainment of sand, taking into account moisture content:

$$u_{*t} = A \sqrt{\frac{\rho_p - \rho_a}{\rho_a} \cdot gD} (1 \cdot 8 + 0 \cdot 6 \log^{10} W) \qquad [3.5]$$

where W = % moisture content

Chepil (1956) showed that an equivalent moisture content of only 0·71% markedly reduced the amount of wind erosion from a silt loam soil. Erosion loss was very small with an equivalent moisture content of 1·03%. Similar findings were reported by Bisal and Hsieh (1966). Spraying with water is a technique widely used to suppress dust emission in industrial plants and mines (Walton and Woolcock, 1960).

Wind-tunnel experiments by Nickling and Ecclestone (1981) and Nickling (1984) demonstrated that even low concentrations of salts can effectively cement particles and significantly raise the threshold velocity (Fig. 3.4), although the effect is not as great as that of moisture. The bonding action of salts has also been observed in the field (Pye, 1980).

(a)   (b)

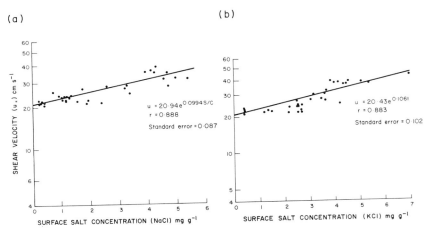

Figure 3.4. Effect of increasing surface salt concentration on threshold shear velocity: (a) NaCl; and (b) KCl. (After Nickling and Ecclestone (1981), Sedimentology **28**, 505–510; © International Association of Sedimentologists.)

Clay-rich crusts are formed on bare soils by raindrop impact (Chen *et al.*, 1980). Algae and fungi can also contribute to the formation of soil crusts (Fletcher and Martin, 1948; Foster and Nicolson, 1980; van den Ancker *et al.*, 1985). Smalley (1970) pointed out that the erodibility of a soil is dependent largely on its cohesiveness, and that a measure of the erodibility is indicated by the tensile strength, which is directly proportional to the packing density, to the coordination number of the particles, and to the interparticle bond strength. Gillette *et al.* (1980, 1982) subsequently used a portable wind tunnel to demonstrate that erodibility of various soils and sediments in the Mojave Desert is related to their modulus of rupture (Table 3.1). They found that in the case of undisturbed soils, even a weak crust (modulus of rupture <0·07 MPa) will protect the soil from wind erosion. Disturbed soils, on the other hand, are likely to be eroded unless the undisturbed crust has a modulus of rupture of >0·1 MPa.

The fluid threshold velocity is also affected by the bed roughness, which is controlled, for example, by the presence of non-erodible gravel particles or crop stubble on the ground surface (Bisal and Ferguson, 1970; Lyles and Allison, 1976). Chepil (1950) introduced a critical surface roughness constant to describe the importance of roughness elements. In experiments designed to simulate the effects of wind on a mixture of erodible and non-erodible elements, he found that when deflation ceased the height of the roughness elements (H) divided by the distance between them (d) was a constant. As

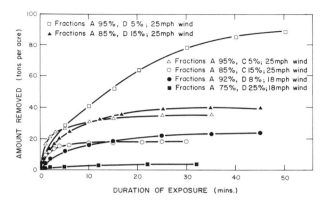

Figure 3.5. Rate of soil removal with duration of exposure in a wind tunnel. As non-erodible particles are exposed, the roughness increases and the rate of sediment removal decreases. (After Chepil (1950), Soil Science **69**, *149–162; © by William and Wilkins.)*

non-erodible fractions are exposed by deflation of the erodible material, the roughness increases and the rate of removal decreases (Fig. 3.5). Chepil and Woodruff (1963) used the reciprocal of this ratio and termed it the critical surface barrier ratio (CSBR). On cultivated land it was found to have values ranging from 4 to 20, depending on the friction velocity of the wind and the threshold velocity of the erodible particles. Further work by Lyles *et al.* (1974) and Lyles (1977) showed that as roughness increases, the total surface stress increases and a greater proportion is taken up by the non-erodible elements, leaving less stress to move the erodible material. Experiments by Logie (1982) confirmed that low densities of roughness elements (gravel particles and glass spheres) reduce the threshold velocity and activate erosion by scouring around the roughness elements. Dense covers, on the other hand, increase the threshold and limit erosion. For any size of roughness element, a certain cover density exists, called the inversion point, where its influence changes from activation to protection (Table 3.2). It was found that the shape of the roughness elements affects the value of the inversion point. Irregularly-shaped gravel produces more turbulence than glass spheres and consequently a higher density of particles is needed to stop deflation.

Pebble-covered surfaces which are apparently stable with respect to wind erosion occur widely in deserts. Sand and dust blow across such surfaces, but they experience little net deflation. Symmons and Hemming (1968) investigated the grain size distribution of soils beneath such surfaces in the southern Sahara (Table 3.3). They also cleared the surface stones from trial plots in order

Table 3.1
Threshold velocities for different surface types in the southwestern United States, determined using a portable wind tunnel.

Group	Number	Threshold velocities cm s^{-1}		Soil subgroup and family*
		Undisturbed	Disturbed	
I. Salt crusts	1	>250 (NR)	NR	
II. Desert pavements	1	271	66	
	2	278	66	
	3	154	59	Typic haplargids; coarse loamy, mixed, thermic
	4	163	42	Typic calciorthids; coarse loamy, mixed hyper-thermic
III. Crusted soils	1	>285 (NR)	182	
	2	>339 (NR)	158	
	3	>154	40	
	4	265	29	
	5	204	35	Udic pellustert; fine, montmorillonite, thermic
	6	>230 (NR)	36	
	7	—	35	
	8	>191 (NR)	35	
	9	>200 (NR)	27	
	10	>300 (NR)	51	
	11	>317 (NR)	101	
	12	121	33	
	13	>339 (NR)	88	
	14	>222 (NR)	19	
	15	261	83	

Table 3.1 (*continued*)

Geomorphological setting	Location	Comments
centre of playa	Lake Danby, Calif.	hard salt crust with moist soil lying below
alluvial stream deposit	Lake Danby, Calif.	fine desert pavements, no varnish, not mature
alluvial stream deposit	Lake Danby, Calif.	coarse desert pavement, no varnish, not mature
alluvial fan	Turtle Mountains near Parker, Ariz.	mature, varnished desert pavement, rounded cobbles
alluvial fan	Turtle Mountains near Parker, Ariz.	immature pavement, no varnish
centre of playa	Lake Danby, Calif.	cracked, curled clay crust
centre of playa	Lake Danby, Calif.	cracked, curled clay crust
edge of playa	Lake Danby, Calif.	silty crust
edge of playa	Lake Danby, Calif.	smooth crust
edge of playa	Hale County, Tex.	clay crust broken into 2–5 mm pellets
centre of playa	Battle Mountain, Nev.	thin peels of clay on thick flat crust
flat near playa	Battle Mountain, Nev.	silty soil near desert road
centre of playa	El Mirage Lake, Calif.	thick, hard clay crust; no cracks
edge of playa	El Mirage Lake, Calif.	silty crust, more easily broken than at centre of playa
centre of playa	Harper Dry Lake, Calif.	cracked clay crust
centre of playa	Emerson Dry Lake, Calif.	hard clay crust; narrow cracks
edge of playa	Emerson Dry Lake, Calif.	curled clay peels on hard clay crust
centre of playa	Lucerne Dry Lake, Calif.	hard clay crust; narrow cracks
centre of playa	Soggy Dry Lake, Calif.	hard clay crust; narrow cracks
prairie: flat	Pueblo, Colo.	thin clay crust; flat and soft

Table 3.1 (*continued*)

Geomorphological setting		Location		Comments
IV. Other soils	1	191	43	Typic torripsamments mixed, hyperthermic
	2	147	33	Typic camborthids coarse loamy, mixed, hyperthermic
	3	146	26	
	4	—	26	Arenic aridic paleustalt; loamy, mixed, thermic
	5	—	25	Aridic calciustoll; fine loamy (calcareous)
	6	40	28	
	7	59	34	
	8	31–56	44	Typic torripsamment; mixed, thermic
	9	134–237	89	Typic haplargid; coarse loamy, mixed, thermic

NR means threshold velocity not reached. Where a number is reported, it is simply the velocity of the wind tunnel for that surface which could not be increased. Some tests were done in several nearby locations to gain information for other BLM projects. Ranges are given for those tests. (After Gillette *et al.*, 1980, 1982, *J. Geophys. Res.* **C85**, 5621–5630; **87**, 9003–9015; © American Geophysical Union.)

Table 3.1 (*continued*)

Geomorphological setting	Location	Comments
aeolian deposit on a fan; thin coating of grus	Riverside Mountains, Calif.	near mountains; fine sand under thin layer of grus
aeolian deposit on alluvial fan; thin coating of grus	Riverside Mountains Calif.	lower on fan; fine sand under thin layer of grus
lower alluvial fan; near playa	Iron Mountain, Calif.	vesicular crust; sandy soil
flat, prairie	Plains, Tex.	loose, sandy soil
flat, prairie	Bronco, Tex.	loose, loamy fine sand
sand dune	Palm Springs, Calif.	sand dune with very soft crust
sand dune	Dale Dunes, Calif.	sand dune with very soft crust
alluvial fan	Emerson Fan, Calif.	gravel cover
alluvial fan	Stoddard Valley, Calif.	gravel cover

Table 3.2

The minimum and maximum measured value of the inversion point for various sizes of spheres and gravel.

Roughness elements	Diameter (mm)	Inversion point (%)
Gravel	2–3	3·0–6·3
	7	7·4–10·5
	8·7	7·0–10·2
	15	14·0–22·0
	17·5	10·5–24·5
Spheres	13·4	11·0–13·9
	16·9	15·0–20·4
	25	18·8–27·0

(After Logie, 1982, *Catena Supplement* **1**, 161–173; © Catena Verlag.)

Table 3.3

Particle size of soil samples taken from below a wind-stable stone pavement, Plain of Tamesna, southern Sahara.

	Depth of soil	
	0·01–0·1 m	0·1–0·6 m
stones (>2 mm), % of total soil	11	21
% of <2 mm fraction		
200 μm–2 mm	28	64
50 μm–200 μm	29	10
20 μm–50 μm	6	2
2 μm–20 μm	14	5
<2 μm	21	18

(After Symmons and Hemming, 1968, *Geog. J.* **134**, 60–64; © Royal Geographical Society.)

to measure the rate of accelerated deflation and to monitor the rate of regeneration of the protective stone cover. They observed a maximum of 7 mm surface lowering in 1 month, and calculated that deflation of 4·3 cm is required to restore the stone-mantled surface to its previous condition. They also noted that, in the absence of a stone mantle, the rate of surface deflation is strongly influenced by the degree of sediment compaction.

3.2.2 Aerodynamic lift

Bagnold (1941) concluded that sand grains at the threshold of movement on a bed exposed to turbulent flow first began to roll and then to bounce, presumably due to the combined effect of the velocity pressure exerted by the air on the upwind side of the grain and the negative viscosity pressure on the downwind side. Other workers have found no evidence of rolling and bouncing at the onset of motion, but observed that sand grains first begin to rock backwards and forwards and then are lifted directly into the flow, apparently due to instantaneous differences in air pressure near the ground (Bisal and Nielsen, 1962). A steady sheared flow near the ground decreases the static pressure at the top of a grain relative to the bottom (the Bernoulli effect), thereby creating lift (Chepil, 1945). Discontinuous lift forces also result from fluctuations in pressure above the bed due to turbulent eddy motion, and are probably partly responsible for the initial rocking movements (Lyles and Krauss, 1971). Short-term fluctuations in drag and lift forces are related and both contribute to entrainment of sand and silt particles (Chepil, 1959).

3.2.3 Entrainment by ballistic impact

Bagnold observed that when particle movement starts, bombardment of the bed initiates movement of new grains, so that sediment movement can be maintained at velocities lower than the fluid threshold. This lower threshold was referred to by Bagnold (1941) as the 'impact threshold', which is defined by

$$u_t = 680\sqrt{D} \cdot \log\left(\frac{30}{D}\right) \qquad [3.6]$$

Values of the impact threshold for different particle sizes determined by Bagnold (1941) are compared with the fluid threshold in Fig. 3.2.

Bagnold (1960) also noted that, although very high threshold velocities are required to erode fine powders and settled dusts, these particles can readily be ejected into the airflow by ballistic impacts of saltating grains larger than 100 μm. Settling of smaller grains in short-term suspension may have a lesser effect. The number and size of particles ejected from a sand bed by ballistic impact varies with the speed and trajectory of the impacting particles, and the nature of the impacts. At present, however, there are few available data concerning impacts on beds composed of sand/silt and silt/clay mixtures. Fletcher (1976*a, b*) found in wind-tunnel experiments that erosion along a continuous deposit of <50 μm limestone dust increased significantly downstream, possibly suggesting that grains impacting on the surface eject other grains. However, other wind-tunnel studies have shown that the aerodynamic shear increases with distance from the leading edge of a plane bed, and may itself be sufficient to set particles in motion at some critical point downwind if cohesion is not too great (Punjrath and Heldman, 1972).

In the case of natural crusted soils, abrasion by impacting particles has been observed to be important in breaking-up soil aggregates and releasing fine particles into the airstream (Chepil, 1945; Gillette *et al.*, 1974; Hagen, 1984). Chepil and Woodruff (1963) expressed the view that: 'wind erosion occurs only when soil grains capable of being moved in saltation are present in the soil . . . dust clouds are merely the result of movement in saltation' (1963, pp. 215–216). Aeolian action on many natural sediments and soils, other than well-sorted dune sands which contain few fines, involves simultaneous saltation and suspension transport (Clements *et al.*, 1963). The relative importance of each transport mode depends on the particle-size characteristics and structure of the deflated material and the nature of turbulent diffusion in the area concerned (Nalpanis and Hunt, 1986; Anderson and Hallet, 1986). In the case of wind-eroded soils, aggregates of silt and clay-size material are initially transported in saltation but may break up during interparticle or ground–particle contacts to release fines which are then carried away in suspension (Gillette *et al.*, 1972, 1974; Gillette and Walker, 1977; Hagen, 1984).

3.2.4 Disturbance by pedestrians and vehicles

Bagnold (1960) described raising of dust clouds by flocks of sheep on the Anatolian Plateau and pointed out that even deposits of very fine, uniform powders can be eroded by the wind where previous disturbance, as created by the passage of animals or vehicles, has caused sharp edges to protrude into the windstream. However, once the surface irregularities have been removed, no further movement takes place. In addition to physically ejecting particles from the surface and creating local turbulent eddies, disturbance may increase the susceptibility to wind erosion by breaking-up soil crusts. Large amounts of dust are raised into the atmosphere during ploughing and harrowing of dry soils, construction activities, military manoeuvres in desert areas, and by vehicles on dirt roads (Wilshire, 1980). It has been estimated that a 4-wheel vehicle travelling at 60 km h^{-1} along a dirt road containing 12% silt will raise

3·7 kg km^{-1} of dust (Hall, 1981). The amount of dust emitted increases with higher vehicle speeds and higher silt content of the road materials.

3.2.5 Average wind speed required for dust entrainment in nature

It is evident from the preceding discussion that the threshold shear velocity required to entrain dust particles varies according to the grain-size distribution, sorting, cohesion and roughness of the source material. Field and wind-tunnel observations indicate that average u_* of the order of 0·2–0·6 m s^{-1} are required. In the case of poorly sorted sediments, the ejection of dust particles into the windstream is dependent mainly on the threshold shear velocity required to initiate saltation of sand grains. In the Yukon Territory of northern Canada, Nickling (1983) observed that during dust storms originating from exposed areas of poorly sorted fluvial sediments the mean wind speed ranged from 3·97 m s^{-1} to 12·68 m s^{-1}, with mean shear velocities ranging from 0·26 to 0·63 m s^{-1}. Clements *et al.* (1963) reported that the minimum wind speed required to generate blowing dust on different desert surfaces ranged from 6 m s^{-1} on sand dunes to >16 m s^{-1} on crusted alluvial fans and mature desert pavements. Hall (1981) found that average wind speeds >11 m s^{-1}, with gust wind speeds >16 m s^{-1}, were required for blowing dust events with visibilities of 10 km or less at Winslow and Tucson, Arizona, and at Denver, Colorado.

3.3 DISPERSION OF DUST DEFLATED FROM THE EARTH'S SURFACE

Once dislodged from the bed, a particle may move by sliding, rolling, bouncing (saltation) or in suspension. Sliding and rolling together are known as 'surface creep'. The mode in which a particle is transported is dependent on its physical characteristics and on the velocity and turbulent structure of the wind. Wind erosion of soils and sediments which contain a wide range of grain sizes gives rise to all modes of particle transport.

When a grain is dislodged it can either return to the surface almost immediately or remain in suspension. If the particle settling velocity exceeds the vertical velocity component of the wind, the particle will return to the surface a short distance downwind of the ejection point. Conversely, if the vertical velocity component of the wind exceeds the settling velocity, it will remain in suspension (Kalinske, 1943). The settling velocity of a particle

Table 3.4
Relative densities of some common minerals found in atmospheric dust.

Mineral	Relative density
quartz	2·65
albite	2·62
labradorite	2·70
anorthite	2·75
orthoclase	2·56
microcline	2·56
tridymite	2·2–2·3
cristobalite	2·2–2·3
gypsum	2·32
halite	2·16
ice	0·92
aragonite	2·95
calcite	2·71
dolomite	2·85
augite	3·2–3·4
hornblende	3·0–3·45
ilmenite	4·7–4·78
magnetite	5·20
illite	2·64–2·69
biotite	2·80–3·20
kaolinite	2·60–2·63
palygorskite	2·20–2·36
montmorillonite	2·3
nontronite	1·7–1·8
sepiolite	2·0
vermiculite	2·3
chlorite	2·66–3·3

(Source: Frye, 1981: *The Encyclopedia of Mineralogy*.)

depends on its mass and shape. Relative densities of a number of different minerals commonly found in atmospheric dust are given in Table 3.4. The more spherical a particle, the higher is its settling velocity for any given mass. The settling velocities of quartz spheres in the size range 1–50 μm can be calculated approximately according to Stokes's Law as a function of the square of the grain diameter (Green and Lane, 1964):

$$U_f = KD^2 \qquad [3.7]$$

where

U_f is the settling velocity (cm s^{-1})
D is the grain diameter (cm)
$K = \rho_p g/18\mu$,

where

ρ_p is the particle density
g is the acceleration due to gravity
μ is the dynamic viscosity of the air
(K is taken to be $8\cdot1 \times 10^5$ cm^{-1}s^{-1} for air at 15°C at sea-level and for quartz spheres)

The settling velocities of different sizes of quartz spheres in still air are shown in Fig. 3.6.

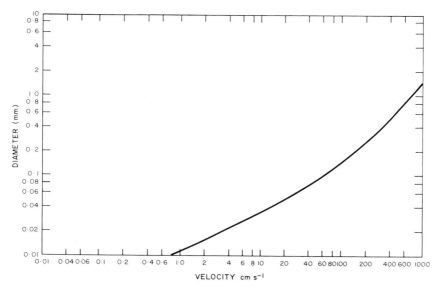

Figure 3.6. Settling velocity of quartz spheres in air. (Data of H. Rouse, in Malina (1941).)

Wind flow in the atmospheric boundary layer has both horizontal and vertical components due to turbulence. These forward horizontal and vertical velocity components can be designated u and w respectively. The mean velocities $\bar{u}$ and $\bar{w}$ can be defined for an interval of time. The vertical fluctuating velocity w', is given by $w - \bar{w}$. The value of w' depends on the stability conditions of the atmosphere at any given time. In a neutral atmosphere the distribution of vertical fluctuating velocity components (w') near the surface but above the saltating layer is approximately normal with a mean $(\bar{w}')$ of 0 since the upward and downward motions must be equal. The standard deviation of the vertical fluctuating velocity $(\sigma = \sqrt{(w'^2)}$ represents the force opposing the tendency of fine particles to settle. A particle should remain in suspension if $\sqrt{(w'^2)}$ is greater than U_f. The standard deviation of the fluctuating velocity is equal to Au_*, where A is a constant. The average value of A falls within the range 0.7–1.4_*, with a mean of 1; therefore $\sqrt{(w'^2)}/u_*$ is approximately equal to 1 (Lumley and Panofsky, 1964; Pasquill and Smith, 1983). Thus, a spherical quartz particle is likely to remain in suspension when U_f/u_* is <1.

Several workers have found the ratio U_f/u_* to be a suitable criterion for determining the degree of suspension (Rouse, 1937; Sundborg, 1955; Chepil and Woodruff, 1957; Gillette et al., 1974), but the upper limit of suspension at $U_f/u_* = 1$ is arbitrary since there is no sharp distinction between saltation and suspension. Pure saltation occurs when turbulent vertical velocity components have no significant effect on the particle trajectories. Pure suspension occurs when the particle's settling velocity is very small relative to the friction velocity. Between pure saltation and pure suspension both the inertia and settling velocity of a particle influence its trajectory. Particles transported in this mode, which has been called 'modified saltation' (Nalpanis, 1985; Hunt and Nalpanis, 1985), move with a random trajectory through the flow. The transition line between pure and modified saltation, as determined theoretically by Nalpanis (1985), is shown in Fig. 3.7. Following Gillette et al. (1974) the upper limit of pure suspension is taken to be $U_f/u_* = 0.7$.

To remain in suspension for a considerable period of time, particles must have a ratio of upward to downward movements >1. Using the ratio U_f/u_*, Gillette (1974, 1977) calculated the likely ratio of upward to downward motions for different particle sizes in air having a normal vertical velocity distribution (Fig. 3.8). He showed that, for grains having a $U_f = 0.4u_*$, the ratio of upward to downward movements is 0.5, and there is a high probability that these grains will settle back to the surface within a short time. However, particles finer than 20 μm have a U_f/u_* ratio of <0.1 for common storm winds, and are likely to remain in suspension for a long time. Tsoar and Pye (1987) used a U_f/u_* ratio of 0.1 to distinguish between 'settling' grains in short-term suspension and 'non-settling' grains in long-term suspension (Fig. 3.7). Under

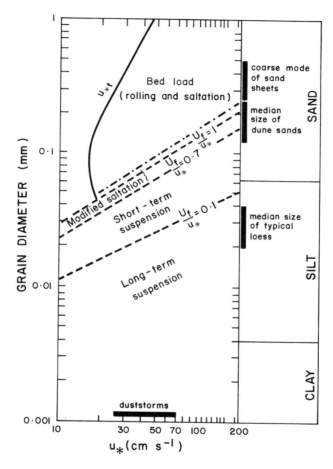

Figure. 3.7. Modes of transport of quartz spheres at different wind shear velocities (u_{*t}) (After Bagnold (1941), showing limits of modified saltation as defined by Nalpanis (1985) and arbitrary boundary between short-term suspension and long-term suspension.) (After Tsoar and Pye (1987), Sedimentology **34**, Fig. 1; © International Association of Sedimentologists.)

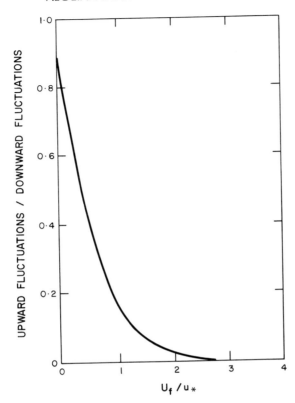

Figure 3.8. Distribution of upward to downward motions for a particle having a fall velocity (U_f) in air, with vertical velocity fluctuations with a mean $= 0$, and a standard deviation $= u_*$. (After Gillette (1977), Trans Am. Soc. Agric. Engnrs **20**, 890–897; © American Society of Agricultural Engineers.)

typical windstorm conditions, with u_* in the range $0{\cdot}2$–$0{\cdot}6$ m s^{-1}, medium and coarse silt grains, which comprise the bulk of typical loess, are transported mainly in short-term suspension or modified saltation (Fig. 3.9).

The degree of vertical air mixing is indicated by the coefficient of turbulent exchange (ε), which can be expressed approximately in the form:

$$\varepsilon = \sqrt{(\overline{w'^2})}l \qquad\qquad [3.8]$$

where l, the mixing length (Prandtl, 1935), is defined by:

$$l = \alpha kz \qquad\qquad [3.9]$$

where

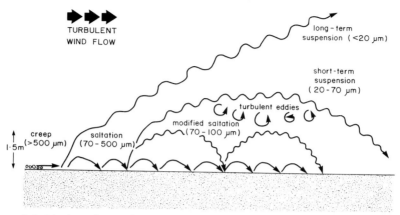

Figure 3.9. Modes of particle transport by wind. Indicated particle-size ranges in different transport modes are those typically found during moderate windstorms ($\epsilon = 10^4–10^5$ cm^2 s^{-1}).

$\alpha = 1$ for neutral atmospheres, $\alpha > 1$ for unstable atmospheres (where convective effects are significant), and $\alpha < 1$ for stable atmospheres)

k = the von Kármán constant = $0\cdot4$

z = the height above the surface within the lowermost boundary layer

Thus:

$$\varepsilon = \alpha\sqrt{\overline{(w'^2)}}kz \qquad [3.10]$$

For long-term suspension U_f should be much smaller than $\sqrt{\overline{(w'^2)}}$. If $U_f/\sqrt{\overline{(w'^2)}} = 0\cdot06$ there is a high probability that a particle will be carried to a level well above the ground. The coefficient of turbulent exchange needed to carry particles of different sizes to a particular height (z) within the boundary layer can be estimated by substituting for $\sqrt{\overline{(w'^2)}}$ in Eq. [3.10]:

$$\varepsilon = \frac{\alpha U_f kz}{0\cdot06} \qquad [3.11]$$

Figure 3.10 shows the approximate values of ε required to transport particles up to 100 μm in size to heights of 20 and 100 m in a neutral atmosphere. 'Settling' particles larger than 20 μm require values of ε in excess of 10^5 cm^2 s^{-1} in order to reach 100 m. Such values are attained during intense cyclonic storms and haboobs (Farquharson, 1937; el Fandy, 1953).

The average vertical extent (h) which dust particles will reach in an arbitrary time (t) is related to ε by the expression (Tsoar and Pye, 1987):

$$h = (2\varepsilon t)^{1/2} \qquad [3.12]$$

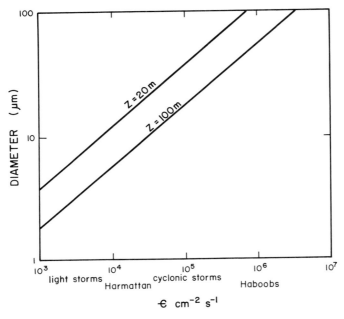

*Figure 3.10. The approximate coefficient of turbulent exchange (ε) needed to carry a quartz sphere to heights of 20 and 100 m under neutral atmosphere conditions. (After Tsoar and Pye (1987), Sedimentology **34**, Fig. 2; © International Association of Sedimentologists.)*

An estimate of the time a particle remains in suspension can be obtained by substituting $h = U_f t$ in Eq. [3.12]:

$$t = 2\varepsilon/U_f^2 \qquad [3.13]$$

The lower the rate of settling and the greater the turbulent exchange, the wider is the dispersion of particles. From Eqs. [3.7] and [3.13] it can be shown that the maximum distance and time travelled by a dust particle (that obeys Stokes's Law) is inversely proportional to the fourth power of the particle diameter:

$$L = \overline{U}t = \frac{\overline{U}2\varepsilon}{K^2 D^4} \qquad [3.14]$$

where

L is the distance travelled by a suspended dust particle
$\overline{U}$ is the mean wind velocity

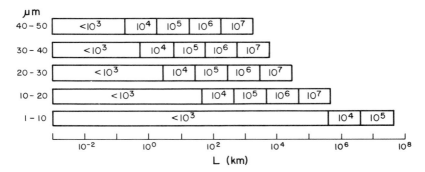

Figure 3.11. The maximum distances likely to be travelled by different-size classes of quartz spheres when $\bar{U} = 15$ m s^{-1} and ϵ varies from 10^3 to 10^7 cm^2 s^{-1}. (After Tsoar and Pye (1987), Sedimentology **34**, Fig. 3; © International Association of Sedimentologists.)

Figure 3.11, based on Eq. [3.14], shows how far different sizes of dust will be dispersed with an assumed $\bar{U} = 15$ m s^{-1} and values of ε ranging from 10^4 to 10^7 cm^2 s^{-1}. During moderate (neutral atmosphere) wind storms grains larger than 20 μm are unlikely to travel more than about 30 km from the source, while particles finer than 10 μm could be transported thousands of kilometres. Particles of sizes 10–20 μm could be dispersed over a distance of approximately 500 km when $\varepsilon = 10^4$ cm^2 s^{-1}. Under extreme wind-storm conditions, when $\varepsilon = 10^6$ cm^2 s^{-1}, 20–30 μm grains could be transported up to 3000 km.

The relative concentration of suspended dust of a particular size at a given height above the surface is given (Rouse, 1937) by:

$$C/C_a = [(h - y/y) \cdot a/(h - a)]^Z \qquad [3.15]$$

where

C/C_a is the relative concentration (number of particles per cm^{-3}) at any level, y, referred to some arbitrary elevation, a, above the surface (where $a < y$), and h is the height above which no particles are in suspension. The value of Z is given by:

$$Z = U_f/ku_* = 2 \cdot 5(U_f/u_*) \qquad [3.16]$$

Considering the lower zone of the atmospheric boundary layer, y is much smaller than h, so Eq. [3.15] can be simplified:

$$C/C_a = (a/y)^Z \qquad [3.17]$$

The change in relative concentration indicates the height range within which dust particles of particular size are mainly transported. For small particles

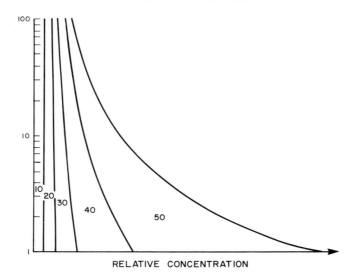

Fig. 3.12. Predicted vertical change in the relative concentration of 10, 20, 30, 40 and 50 μm-size particles (quartz spheres) under severe wind storm conditions ($u^* = 70$ cm s^{-1}). (After Tsoar and Pye (1987), Sedimentology 34, Fig. 4; © International Association of Sedimentologists.)

which obey Stokes's Law the following expression is obtained by substituting Eq. [3.7] in Eq. [3.17]:

$$C/C_a = \exp [2 \cdot 5 (KD^2/u_*) \log_n (a/y)] \qquad [3.18]$$

For $y = 100$ m and $u_* = 0 \cdot 7$ m s^{-1} (equivalent to a severe wind storm), the change in relative concentration with height of various dust sizes is shown in Fig. 3.12.

Under these conditions the majority of particles >20 μm are carried only a few metres above the ground, while particles <20 μm are almost evenly distributed with height up to 100 m. The predicted relative concentration of grains >30 μm at a height of 100 m is approximately 26% of the concentration at 0·5 m. When $u_* = 0 \cdot 3$ m s^{-1}, the respective figure is approximately 5% (Tsoar and Pye, 1987).

An exponential decrease in the mean diameter of suspended dust with height has been observed by several authors (Sundborg, 1955; Chepil and Woodruff, 1957; Gillette *et al.*, 1974; Nickling, 1983). Goossens (1985*a*) found dust clouds derived from a torn-up road in Belgium to be granulometrically stratified, with coarse silt moving mainly near the bottom of the dust cloud and fine silt moving both at the bottom and at the top (Fig. 3.13). In the Yukon,

(a)   (b)

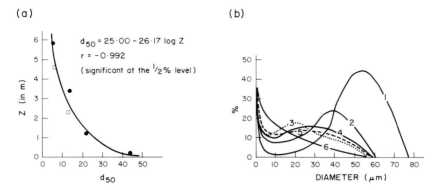

Fig. 3.13. (a) Median diameter of dust derived from a torn-up road as a function of height: ● = undisturbed sample, □ = disturbed sample. (b) Grain-size distribution of dust collected in buckets at different heights above the ground: (1) = 0·2 m; (2) = 1·2 m; (3) = 2·3 m; (4) = 3·4 m; (5) = 4·6 m; and (6) = 5·8 m. (After Goossens (1985a), Earth Surf. Proc. Landforms **10**, 353–362; © John Wiley & Sons Ltd.)

Nickling (1983) found that the decrease in grain size with height up to 12 m above the ground was somewhat irregular (Fig. 3.14), probably due to turbulent mixing.

It should be noted that the preceding discussion of dust dispersion relates only to neutral atmosphere wind-storm conditions. This atmospheric condition is said to exist when the temperature gradient above the surface is very close to 0 (0–0·01°C m), such that a parcel of air pushed up by turbulence or a topographic obstacle is cooled by expansion to exactly the same temperature as its surroundings; it is therefore in a state of neutral equilibrium, and will tend neither to rise nor sink due to density differences. If the air temperature decreases rapidly with height, the atmosphere is said to be 'unstable' since a parcel of air pushed up by turbulence will continue to rise because it is warmer, lighter and more buoyant than its surroundings. Conversely, if average air temperature increases with height above the ground an air parcel pushed up by turbulence will be colder and denser than its surroundings, so will tend not to rise. Such an atmospheric condition is described as 'stable'. In unstable air, turbulence is enhanced by buoyancy forces and in stable air it is suppressed (Thom, 1976). Under unstable atmospheric conditions, with strong convective activity, dust can be raised to higher levels than under neutral conditions and may be dispersed over a wider area, particularly if it becomes incorporated within a jetstream. Conversely, where air aloft is subsiding, and where inversions are well developed, dust may become trapped in the lower levels of the atmosphere, resulting in limited dispersion. The nature of different

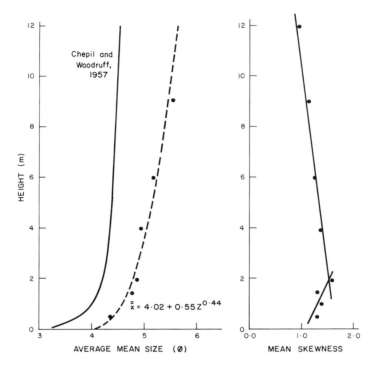

*Figure 3.14. Variation of average mean size and skewness of suspended sedi-
ment with height for 15 sampled dust storms in the Slims River Valley, Yukon
Territory. (After Nickling (1983), J. Sedim. Petrol. **53**, 1011–1024; © Society of
Economic Paleontologists and Mineralogists.)*

dust-transporting wind systems is considered in Chapter 5. However, the
pattern of dust dispersion under neutral atmosphere conditions predicted in
the model described by Tsoar and Pye (1987) agrees well with the observed
geographical extent, thickness and grain-size variations observed in Quaternary loess deposits. The grain size of modern dust-storm sediments also
indicates that coarse and medium silt, which forms the bulk of typical loess, is
mainly confined to low-level, relatively short-range transport (see Chapter 5).
The dispersion of non-settling dust particles and other aerosols smaller than 10
μm is considered in detail by Friedlander (1977), Pasquill and Smith (1983)
and Hidy (1984).

Dust will be deposited closer to the source under conditions of variable wind
direction than under a unidirectional wind regime. Handy (1976) showed that,
if winds blowing transverse to a linear source carry dust to a maximum distance
designated X_m, winds blowing at an angle α to the source will distribute their

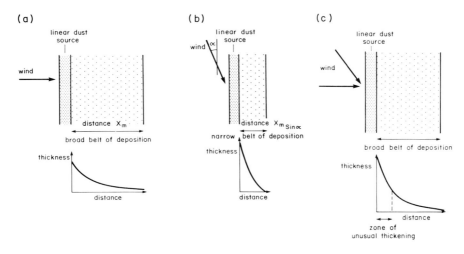

Figure 3.15. Schematic diagram showing the nature of dust dispersion from a linear source (such as a braided river channel) under different wind regimes. (a) With winds perpendicular to the source, dust is deposited in a broad belt and the thickness of deposited dust decreases gradually with distance downwind. (b) When winds blow at an oblique angle to the source, the belt of dust deposition is much narrower and the thickness of deposited dust decreases much more rapidly in a direction perpendicular to the source. (c) Under conditions of variable winds, the result is a broad belt of deposition with a zone of unusual thickening adjacent to the source. (Modified after Handy (1976).)

load over a reduced distance, $X_{m_{\sin \alpha}}$, leading to a proportionate increase in thickness by a factor of $1/\sin \alpha$. Thus, where a is small, with winds blowing almost parallel to the source, deposition should be limited to a narrow corridor adjacent to the source, leading to thick deposits in this area (Fig. 3.15).

The nature of the surface over which a dust cloud travels exercises an important control on the pattern of dispersion. Surfaces formed of bare rock or dry, featureless playa sediments can be regarded as 'reflective' in the sense that the majority of settling dust particles will be rapidly re-suspended if the windflow is maintained. Such surfaces provide only a limited input of additional deflated dust, so that the particle concentration in a passing dust cloud maintains a state of equilibrium. During passage of a dust cloud over an erodible surface containing fine material, such as a braided outwash plain, the number of particles leaving the surface may exceed the number of particles settling on it, leading to an increase in the concentration of suspended dust. On the other hand, water bodies and marshes are highly retentive with regard to settling dust, and do not act as sources of new deflated material, with the result

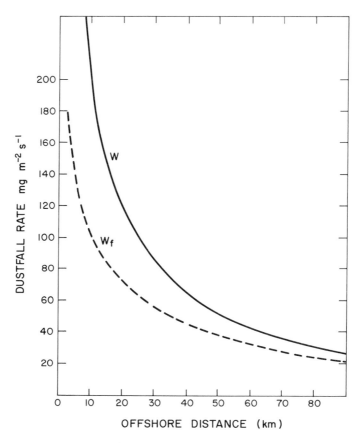

Figure 3.16. Variation in the total dustfall rate (w) and the dustfall due to fall velocity alone (w_f) as a function of offshore distance (x) associated with the passage of a dust cloud over the sea. (After Foda et al. (1985), Sedimentology *32*, 595–603; © International Association of Sedimentologists.)

that the dust concentration and dust sedimentation rate decrease rapidly downwind.

Foda (1983) and Foda *et al.* (1985) presented a theoretical model which assumes that dust dispersion and fallout rate over the sea are dependent not only on the gravitational characteristics of the particles, but also on retention and absorption processes in a thin air layer just above the sea surface. The model indicates that the total dustfall and dustfall rate are higher over the sea than over a 'passive' surface where particle deposition is controlled only by gravitational characteristics (Fig. 3.16).

3.4 DISPERSION OF DUST FROM VOLCANIC ERUPTIONS

The dispersion of volcanic dust during eruptions differs from the dispersion of dust deflated during windstorms because particles are initially carried to a greater height by explosive activity and because dispersion takes place from a point source. The dispersion characteristics of volcanic ash depend on: (1) the nature of the eruption which determines the ejecta size, shape and density, and the turbulent energy of the eruption cloud; and (2) wind conditions at different levels in the atmosphere. An eruption column can be divided into three parts (Fig. 3.17): (1) a gas thrust region in the basal part of the plume which is dominated by the momentum of the flow and in which buoyancy effects are negligible; (2) a convective region in which buoyancy effects are dominant (Sparks and Wilson, 1976); and (3) an upper umbrella region of predominantly lateral dispersion. The convective thrust commonly comprises up to 90% of the total column height. In a stratified atmosphere the plume eventually reaches a level where it has the same density as the surrounding air and starts to spread out laterally (Morton *et al.*, 1956). Initial vertical velocities in the gas thrust region range from about 100–600 m s^{-1} (Sparks, 1986). The rate of rise of the convective thrust is dependent partly on the initial eruption temperature and partly on the concentration and size distribution of particles in the column. The smaller the fragments, the more rapid the heat exchange with the surrounding air and the higher the column (Fisher and Schminke, 1984, p.

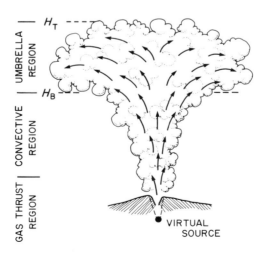

*Fig. 3.17. Structure of a volcanic plume. (After Sparks (1986), Bull. Volcanol. **48**, 3–15; © Springer Verlag.)*

64). The maximum rate of rise of the central column in the convective thrust region is of the order of $10–200 \, \mathrm{m \, s^{-1}}$ and decreases with increasing height. The rate of radial spreading of the umbrella region at the top of the column is related to the mass discharge rate of magma and the column height, rates of radial spreading increasing with column height (Sparks, 1986). Observed rates of spreading are of the order of $10–50 \, \mathrm{m \, s^{-1}}$. The temperature inversion above the tropopause does not appear to have a major effect on the rate of radial spreading, but strong upper-level winds can cause asymmetry in the umbrella. The implications of plume-spreading characteristics for deposition of tephra during explosive eruptions are considered by Carey and Sparks (1986). In the simplest case, with unidirectional winds throughout the height of the plume, the largest volume of tephra will be deposited in an elongate, symmetrical oval pattern, with an axis of maximum thickness coinciding with the axis of the umbrella region of the plume. Theoretical calculations and field observations indicate that the thickness and mean grain size of deposited tephra decrease exponentially downwind. The height of the plume and atmospheric wind strength determine the actual distance travelled by any particular particle size. Transport distances are, in general, greater than those for surface-derived dust particles of equivalent size.

3.5 DUST DEPOSITION

Deposition of dust can occur in one of four ways: (1) if there is a reduction in wind velocity and turbulence such that $\sqrt{(\overline{w'^2})} < U_f$; (2) the particles are 'captured' by collision with rough, moist or electrically charged surfaces; (3) the particles become charged and form aggregates which settle back to the ground; and (4) the particles are washed out of atmospheric suspension by precipitation.

Small particles which come into contact with a smooth surface (such as the floor, roof or walls of a building) during turbulent airflow may accumulate if they become immersed in the laminar sub-layer close to the surface (Owen, 1960). On a larger scale, deposition can occur where there is a local or regional reduction in wind velocity due either to meteorological or topographical factors. Rapid deposition of dust in short-term suspension frequently occurs as a dust cloud crosses a roughness boundary, e.g. between bare ground and a vegetated surface or an urban area. At the roughness boundary the roughness height (z_0) is increased and may be displaced upwards by a distance d (the zero plane displacement). For many types of tall vegetation, d is approximately equal to two-thirds of the height of the stand (Oke, 1978). The wind velocity gradient above the vegetation canopy is reduced (Fig. 3.18), and u_* may fall below the critical value for resuspension of settling grains. Forest

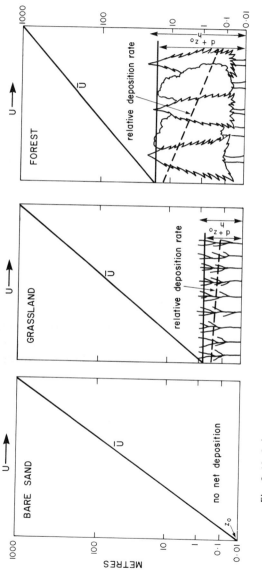

Fig. 3.18. Schematic diagram showing the effect of a change in surface roughness on the slope of the wind-velocity profile and height of effective zero velocity. Rapid deposition of dust close to the roughness boundary occurs as if dust-laden winds cross from a bare surface to one covered by tall vegetation. (After Tsoar and Pye (1987), Sedimentology **34**, Fig. 5; © International Association of Sedimentologists.)

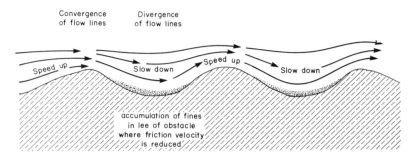

Figure 3.19. Deposition of dust in the lee of topographic obstacles due to flow divergence and reduction of u_*. Dust deposition is prevented on windward slopes where flow convergence and speed-up occur.

vegetation is more efficient in trapping dust than steppe or tundra vegetation due to its greater roughness. In the case of bare sand surfaces, values of z_0 and d are $0·0003$ m and 0 respectively. For surfaces covered by grass $0·25–1$ m high, typically $z_0 = 0·04–0·1$ m and $d = <0·66$ m, while in the case of forests, $z_0 = 1–6$ m and $d = <30$ m (Thom, 1976; Oke, 1978).

Above the level of $d + z_0$ dust particles may still be trapped if they come into contact with leaves and plant stems. The processes of air and fine particle flow through vegetation are reviewed by Chamberlain (1975).

In those parts of the world where loess accumulated under a cover of forest vegetation during the Pleistocene, as in the lower Mississippi Valley, the thickness of loess typically decreases very rapidly with distance from the roughness boundary (see Chapter 9). Loess which accumulated under steppe vegetation, as in much of Europe and the Soviet Union, tends to thin more gradually with distance from the source due to less rapid deposition close to the roughness boundary.

Settling grains can also be deposited and avoid resuspension in the lee of topographic obstacles where wind velocity and turbulence near the ground are reduced (Fig. 3.19). Accumulation of dust rarely occurs on the upwind sides of bare hills where an increase in wind velocity and turbulence is usually observed (Jackson and Hunt, 1975).

As noted in Section 3.3, moist surfaces of any kind are capable of permanently trapping dust which comes into contact with them. In some desert basins, the water table is maintained at sufficiently shallow depth to keep the ground surface permanently moist. As dust accumulates it reduces the rate of evaporation, allowing the groundwater table to rise, and leading to further dust accumulation (Cegla, 1969; Fig. 3.20).

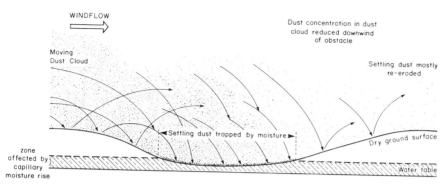

Figure. 3.20. Accumulation of dust in broad, flat-floored depressions where settling dust is trapped by capillary moisture.

Goossens (1985b) pointed out that in dust clouds containing very high particle concentrations the grains do not settle individually but instead behave collectively as a coherent mass. He showed that during the fall of experimental dust clouds an 'explosion point' is reached where lateral diffusion reduces the particle concentration to a level where the grains follow individual settling behaviour. However, the models developed by Goossens are not applicable to natural dust storms in which the particle concentration is relatively low.

Coarse- and medium-silt grains travelling at relatively low levels in the atmosphere either settle back to the surface or are trapped by obstacles transverse to the flow. Part of the fine silt and clay, on the other hand, which travels at higher levels, is less affected by changes in surface roughness and may remain suspended for very long periods. Such particles are deposited only if aggregation occurs or if they are washed out by rain or other precipitation. Collision and aggregation of fine particles can occur simply as a result of Brownian motion, laminar shear or turbulent motions (Friedlander, 1977, pp. 175–208; Suck et al., 1986), or through the build-up of bipolar electrostatic charges (Greeley and Leach, 1979; Marshall et al., 1981). The latter mechanism has been suggested to be responsible for secondary thickening observed towards the distal parts of some volcanic ash deposits (Brazier et al., 1983). The most important mechanism of fine particle deposition, however, is washout by rain or snow (Itagi and Koeunuma, 1962; Graedel and Franey, 1975; Knutson et al., 1977). In Israel, Ganor (1975) found that washed-out dust contained 50–65% clay, while dry-deposited dust contained less than 20% clay. Most falls of Saharan dust in Central and northern Europe are associated with precipitation and are widely referred to as 'red rain' or 'blood rain'. Two mechanisms of washout have been identified. In the first, particles are collected by rain, hail or snow falling through a cloud of material, while the second involves capture of very small particles by cloud droplets and the subsequent deposition of these droplets as rain (Pasquill and Smith, 1983, p. 259). This latter mechanism may also apply to the capture and precipitation of soluble gases and vapours.

Chapter Four

DUST SOURCES, SINKS AND RATES OF DEPOSITION

4.1 GLOBAL DUST SOURCES

The major sources of present-day dust emissions are: (1) the sub-tropical desert regions which occur in a broad belt from West Africa to Central Asia; and (2) semi-arid and sub-humid regions where dry, ploughed soils are exposed to severe winds at certain times of the year (Fig. 4.1). The Sahara Desert is believed to be the world's most important source of dust (Junge, 1979; Goudie, 1983a; Coudé-Gaussen, 1982, 1984; Middleton *et al.*, 1986).

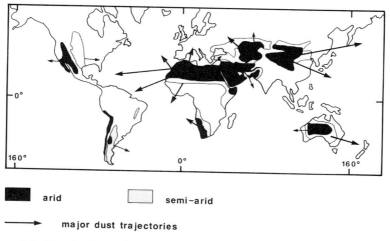

arid semi−arid

major dust trajectories

Figure 4.1. Distribution of areas with high dust-storm activity and major dust trajectories. (Modified after Coudé-Gaussen (1984).)

Table 4.1
Major global dust source areas with key station dust-storm day
frequency (days (D) with visibility < 1000 m).

Source area	Station	D	Data period	Number of years
Australia				
Central Australia	Alice Springs	15·6	1942–82	41
China				
Takla Makan	Hetian	32·9	1953–80	28
			1953–80	28
Kansu Corridor	Minqin	37·3	1953–80	28
Soviet Union				
Turkmenia	Repetek	65·5	1936–60	25
Kara Kum	Nebit Dag	60·0	1936–60	25
Rostov	Zevetnoe	23·3	1936–60	25
Altai	Rubtsovsk	25·1	1936–60	25
Alma Alta	Bakanas	47·7	1936–60	25
Kazakhstan	Dzhambeiti	45·9	1936–60	25
Southwest Asia				
Thar desert	Fort Abbas (Pakistan)	17·8	1951–58	8
Upper Indus plains	Jhelum (Pakistan)	18·9	1951–58	8
Afghan Turkestan plains	Chardarrah	46.7	1974–80	7
Seistan Basin	Zabol (Iran)	80·7	1967–73	7
Makran coast	Jask (Iran)	27·3	1970–73	4
Middle East				
Lower Mesopotamia	Kuwait Int. Airport	27·0	1962–84	23
N. Saudi/Jordan/Syria	Abou Kamal (Syria)	14·9	1959–79	21
North Africa				
Bodélé Depression	Maidurguri (Nigeria)	22·5	1955–79	25
S. Mauritania/N. Mali/C. Algeria	Nouakchott (Mauritania)	27·4	1960–84	25
Libya and Egypt	Sirte (Libya)	17.8	1956–77	22

(After Middleton *et al.* (1986), in W. G. Nickling (ed.) *Aeolian Geomorphology*, pp. 237–259; © Allen & Unwin.)

Table 4.2
Estimates of dust production.

Location	Source	Annual quantity (million tons)
World	Schütz (1980)	up to 5000
World	Peterson and Junge (1971)	500
World	Joseph *et al*. (1973)	128 ± 64
Sahara	Schütz (1980)	300
Aral Sea area (Soviet Union)	Grigoryev and Kondratyev (1980)	75
Southern Mediterranean	Yaalon and Ganor (1975)	20–30
Southwestern United States	Gillette (1980)	4
Mediterranean	Joseph *et al*. (1973)	3·2 ± 1·6

(After Goudie (1983*a*), *Prog. Phys. Geog.* **7**, 502–530; © Edward Arnold.)

However, dust emissions from other deserts have been less intensively studied, and no accurate estimates are available for the dust flux from the Gobi, Karakum and other deserts of Central Asia.

4.2 TYPES OF TERRAIN FAVOURABLE FOR DUST DEFLATION

Silt and clay-size particles occur widely in soils and sediments, but the susceptibility of different surfaces to dust entrainment varies greatly. Portable wind tunnels have been employed in the field to determine threshold velocities for dust entrainment (Malina, 1941; Zingg, 1951; Clements *et al.*, 1963; Gillette, 1978*a*, *b*; Gillette *et al.* 1980). Other techniques which have been used to estimate dust production potential from natural and disturbed surfaces include air photograph interpretation, analysis of particle-size distributions, moduli of rupture, mineral composition and variations in soluble salt content (Clements *et al.*, 1963; Gillette *et al.*, 1982; Gerson *et al.*, 1985).

A high silt and clay content by itself does not ensure large-scale dust production. If a sediment or soil containing fines is compacted, cemented, crusted or armoured, dust generation is likely to be minimal even during severe windstorms. Areas of bare, loose and mobile sediments containing substantial amounts of sand and silt but little clay provide the most favourable surfaces for dust production. Such sediments tend to be most common in geomorphologically active landscapes, where tectonic movements, climatic changes or human disturbance are responsible for rapid exposure, incision and reworking of sediment formations containing fines.

Coudé-Gaussen (1984) recognized a number of terrain types in the Sahara which are potentially important dust sources:

(1) *Les cuvettes de sebkhas* – the beds of dry salt lakes where the surface sediment is disrupted by curling of mudflakes and growth of salt crystals.
(2) *Les épandages d'oueds* – wadi sediments containing silt.
(3) *Les surfaces de fech-fech* – powdery surface sediments formed by weathering of ancient lake muds or argillaceous rocks.
(4) *Les takyrs* – clay soils with polygonal desiccation cracks.
(5) *Les affleurements rocheux meubles* – outcrops of unconsolidated Neogene fine-grained sediments.

On the global scale there are at least eleven natural terrain types which act as major sources of dust. All except the first occur mainly in arid and semi-arid areas: (1) glacial outwash plains and braided fluvioglacial channels; (2) dry wadi beds; (3) dry lake beds; (4) the surfaces of coastal sabkhas; (5) alluvial fans; (6) stony deserts with high weathering rates, such as the Gobi; (7) areas of exposed argillaceous bedrock; (8) areas of loess where the vegetation cover has been reduced by climatic change and/or cultivation; (9) areas of deeply weathered regolith where the vegetation cover has been reduced by climatic change and/or human activities; (10) alluvial floodplain sediments, particularly those which have been cultivated; and (11) areas of formerly stabilized dunes which are reactivated.

4.2.1 Glacial outwash

Outwash plain (sandur) deposits are extensive in ice-marginal areas of Iceland, Greenland, northern Canada and Antarctica. More localized outwash occurs in areas of montane glaciation (e.g. the Himalayas, the Andes and New Zealand). Sandur sediments are better-sorted than glacial sediments, containing fewer large clasts and less clay than till, but more poorly sorted than fluvial sediments. They are composed of mixtures of fine gravel, sand and silt, and characteristically show great variability of grain size and sorting between adjoining bed sets (Hjulstrom, 1952; Krigstrom, 1962; Augustinus and Riezebos, 1971; Church, 1972; Blück, 1974; Ruegg, 1977). Sandurs can be regarded as low-angle alluvial fans, typified by major seasonal fluctuations in discharge and channel instability. Complex cut-and-fill structures are abundant. The silt content typically ranges from 1 to 15%, but a combination of a high sedimentation rate and frequent surface instability means that sandurs

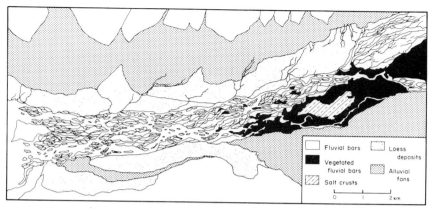

Figure 4.2. A geomorphological map of Lower Adventdalen, Spitzbergen, showing the distribution of loess deposits and other superficial sediments. (After
Bryant (1982), Polar Res. **2**, 93–103; © Norsk Polar Institut.)

and outwash trains confined within glacial valleys are important sources of
windblown dust (Tuck, 1938; Hobbs, 1943a, b; Péwé, 1951; Trainer, 1961;
Pissart et al., 1977, Nickling, 1978; Bryant, 1982; Fig. 4.2). Péwé (1951)
described a dust storm in August 1949 which was generated by 48 km h^{-1}
winds blowing across the braided outwash plain of the Delta River, Alaska.
The outwash plain was 1·5–3 km wide, with braided channels and numerous
silt-covered bars. During the wind storm, dust was carried to a height of
1200 m and dispersed over an area of 480 km^{-2}.

4.2.2 Wadi sediments

Flows in desert wadis take the form of flashfloods which are capable of
transporting very large amounts of sediment for short periods of time.
Deposition occurs very quickly during the phase of declining discharge due to
the high sediment load and rapid loss of water volume by lateral and vertical
seepage. The resulting sediments are generally poorly sorted, though the
grain-size distribution reflects the relief and lithology in the source area. Wadi
channels are normally braided and frequently migrate laterally, unless confined
(Glennie, 1970; Williams, 1971; Karcz, 1972; Fig. 4.3). Gerson et al. (1985)
reported that the non-gravel fractions in coarse alluvium in Israel and Sinai
typically contain 80–90% sand, 10–15% silt and 1–5% clay, although some of
this represents added aeolian dust.

Figure 4.3. Incised, bare alluvial fan sediments of Wadi Zeelim, western shore of the Dead Sea, Isreal.

Figure 4.4. Gravelly alluvial fan sediments grading into evaporite facies in an intermontane desert depression, Owen's Lake, California. Salt weathering of clasts on the lower parts of the alluvial fans produces substantial quantities of silt.

4.2.3 Lake and playa sediments

Dry lake and playa sediments vary enormously in composition, reflecting the balance between clastic, biogenic and chemical sedimentation, but they are usually rich in fine-grained material. Salt weathering around the margins of playas and lakes produces loose fine material (Fig. 4.4). Salt crusts on some playas restrict deflation, but on mud playas without continuous salt crusts deflation rates can be high. The exposed sediments of former deep-water lakes, either brackish or saline, are also susceptible to deflation unless they are cemented. Deflation of fine material is the major process responsible for the enlargement of pans (Goudie and Thomas, 1985), and for the formation of silt and clay dunes (lunettes) on their downwind margins (Hills, 1939; Bowler, 1973). Young and Evans (1986) reported airborne deposition of 2 kg m^{-2} yr^{-1} downwind of a playa in Nevada, and calculated that mud dunes on its margin contained a total of $24\cdot3 \times 10^6$ m^{-3} of deposited aeolian material.

4.2.4 Coastal sabkha sediments

Supratidal sabkha deposits on arid coasts such as those of the Red Sea and Arabian Gulf are in many respects similar to inland playas. The sediments commonly consist of a mixture of terrigenous and authigenic carbonate (aragonite, calcite and dolomite) mud, together with crystals of authigenic gypsum and anhydrite. Silt and clay typically comprise >70% of the sediment. Development of desiccation cracks, mud curls and growth of halite efflorescences loosens aggregates of salt and mud which are then deflated. The sand-size aggregates are transported in saltation, forming adhesion ripples on moist parts of the sabkha surface, but finer aggregates are blown further inland in suspension. Thin accumulations of carbonate- and gypsum-rich dust blown from sabkhas in Bahrain were described by Doornkamp et al. (1980).

4.2.5 Alluvial fans

Alluvial fans can act as net sources or dust sinks depending on the rate of supply of new alluvial sediment and the rate of weathering of the surface fan material. Fan complexes undergoing entrenchment do not provide major sources of dust since the rapid formation of a coarse lag deposit on the fan surface prevents deflation of fines from the underlying material. Airborne dust deposited on stable fan surfaces is infiltrated into the fan by rain and runoff (Gerson et al., 1985; Amit and Gerson, 1986). Some fine material is produced in situ, principally by salt weathering (Goudie and Day, 1980), but the presence of a coarse surface lag again limits the amount of material deflated. Aggrading fans, on the other hand, which are characterized by frequent sheetfloods and lateral channel migration, are more important dust sources since new supplies of fine material are continually being exposed to the wind and a permanent surface lag deposit is not formed. The dynamics and sediments of alluvial fans are discussed by Hooke (1967), Bull (1964) and Mayer *et al.* (1984).

4.2.6 Stony deserts

Stony (reg) deserts also generally provide limited amounts of dust due to surface armouring unless: (1) the rate of weathering of clasts in the armoured layer is exceptionally high; or (2) the surface is highly unstable due to rapid tectonic uplift, faulting and fluvial incision. Stony deserts (Gobi) are extensive in northern China and Mongolia (Chao and Xing, 1982; Walker, 1982):

> On the depositional Gobi, gravels are mainly transported and deposited by rivers or floods, they are rather elliptic or circular in shape. The wind acts on these erosional materials by deflating, sifting and sculpturing and blows selectively finer grains away so that eventually the area forms a barren 'gravel surface'. For example, on the diluvial gravel Gobi near Yumen, Hosi Corridor, are strewn everywhere gravels more than 10–20 cm in diameter; for the sake of gathering sands and other finer materials for engineering and agricultural uses, the local residents have to dig pits on the gravel surface so as to trap these fine grain materials which are saltating and creeping along with the wind stream.
>
> (Chao and Xing, 1982, p. 85)

Sand and dust production in the Gobi of northwest China and Mongolia may be higher than in some other stony deserts due to the arid, windy and periglacial nature of the winter climate. Frost action, salt weathering and wind abrasion are all viable mechanisms for production of fines in the Gobi, but at present the amount of fines produced cannot be quantified.

4.2.7 Weathered argillaceous bedrock

Outcrops of poorly consolidated shales, silts and marls, which are easily disaggregated by weathering or abraded by the wind, occur in some deserts. Yardangs, such as those of Lop Nor in the Takla Makan described by Hedin (1896, 1903), or those in Borku (Chad) investigated by Mainguet (1968), are best developed in soft fine-grained lake or alluvial sediments. The highest frequencies of pans and lunettes, in southern Africa, Texas, Australia and elsewhere, are also found on argillacous rocks and sediments (Goudie and Thomas, 1985).

4.2.8 Deeply weathered regolith and palaeosols

In parts of the Sudan, Egypt and elsewhere, deeply weathered regolith and palaeosols formed by weathering during periods of more humid climate are experiencing erosion under present semi-arid and arid climatic conditions.

Several lines of evidence indicate that the climate of the southern Sahara was wetter than present during the early Holocene and at certain times in the Pleistocene (Rognon and Williams, 1977; Nicholson and Flohn, 1980; Maley, 1982). In parts of Nubia red soils formed in the early Holocene are presently being eroded by gullying and deflation (Butzer and Hansen, 1968). Much of the red, kaolinite-rich dust which falls in the Atlantic and in Europe is derived from similar degraded soils in North and northwest Africa.

4.2.9 Loess regions

Loess lands act as an important source of dust where the natural vegetation cover has been destroyed by cultivation, grazing or a natural climatic trend towards aridity. Erosion of dust from fields is a serious problem in parts of the Chinese Central Loess Plateau (Liu Tung-sheng *et al.*, 1985a). A natural reduction in vegetation cover on loess during the late Holocene appears to have occurred in parts of Uzbekistan and Tajikistan, although grazing by sheep and goats has also played a part (Fig. 4.5).

Figure 4.5. Thinly vegetated surface of mid Pleistocene loess deposits in the desert fringe area southwest of Tashkent, Uzbekistan. The gully systems have developed partly by collapse of sub-surface pipes. At present the surface is subject to deflation during strong wind storms.

4.2.10 Alluvial floodplain sediments

Alluvial floodplain sediments such as those of the Tigris–Euphrates and Indus contain large amounts of fine material. Under natural conditions deflation takes place from point bars and other unvegetated channel areas at times of low flow, but more important at present is deflation from fields during, or immediately following, tilling. Deflation from undisturbed overbank flood-plain deposits is limited by the cohesive nature of the deposits and by vegetation cover. Khalaf *et al.* (1985) concluded that most of the dust fallout in Kuwait is derived from dry Mesopotamian floodplain deposits in Iraq. The importance of alluvial soils in the floodplain of the Ganges as a source area for dust storms in northern India was also stressed by Middleton (1986*b*). In Rajasthan (Gupta *et al.*, 1981; Goudie *et al.*, 1973) and many other parts of the world, including the area south of Tibesti (Grove, 1960), desiccation and reduction of vegetation cover during the later Holocene has allowed extensive aeolian reworking of sandy alluvial deposits.

4.2.11 Dunefields

Although some authors (e.g. Whalley *et al.*, 1982, 1987) have suggested that substantial amounts of dust may be generated by abrasion during saltation of sand grains in active dunefields, this mechanism appears to be less important than the release of fines during reactivation of formerly stabilized dunes. These fines are partly formed by *in-situ* weathering of the stabilized dune sands and partly represent airborne dust infiltrated into the dunes by rain (Pye and Tsoar, 1987). Several studies of stabilized desert dunes have indicated silt and clay

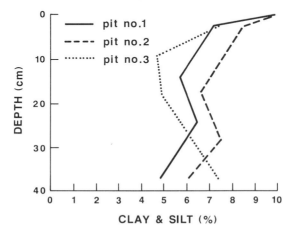

Figure 4.6. Variation in silt and clay content with depth in a stabilized linear sand dune, northern Negev. (After Tsoar and Møller (1986), in W. G. Nichling (ed.) Aeolian Geomorphology, pp. 75–95; © Allen & Unwin.)

contents of up to 15% (Goudie *et al.*, 1973; Tsoar and Møller, 1986; Fig. 4.6). Reactivation of stabilized dunes has occurred in many desert–marginal areas this century due to the combined effects of drought and human activity.

4.3 GLOBAL DUST SINKS

Long-term dust sinks occur both on the continents and in the oceans. On land, dust accumulates with other sediments in subsiding basins as loess, aeolian components in soils and sediments, as fluvially-reworked loess, or as lacustrine dust accumulations. Loess deposits also form blankets up to several hundred metres thick on valley sides, terraces and foothills in mountainous areas. The highest rates of continental dust deposition during the Quaternary occurred in vegetated areas relatively close to the sources of dust (see Chapter 9).

Some dust which is deposited initially on land ultimately finds its way to the oceans through fluvial erosion and transport. In the middle Hwang Ho loess lands, for example, erosion rates are of the order of $10\,000\,\mathrm{t\,km^{-2}\,yr^{-1}}$, reaching $34\,500\,\mathrm{t\,km^{-2}\,yr^{-1}}$ in areas of severely gullied terrain. Consequently, the Hwang Ho has one of the highest suspended sediment loads in the world (Derbyshire, 1978, 1983*a*).

The main oceanic dust sinks are located downwind of continental areas which provide, or have provided in the recent past, major sources of dust. The most important oceanic dust sinks occur in the eastern equatorial Atlantic and in the northwest Pacific. Smaller sinks occur in the eastern equatorial Pacific, the Tasman Sea, southeast Indian Ocean, the Mediterranean, the Arabian Gulf and the northwest Indian Ocean (see Chapter 8).

Fine dust is dispersed over a wide area. Deposition of Saharan dust has been recorded in Florida, Barbados (Delany *et al.*, 1967; Prospero *et al.*, 1970), Bermuda (Chester *et al.*, 1971) and northwest Europe (Pitty, 1968; Stevenson, 1969; Tullett, 1978, 1980, 1984; Bain and Tait, 1977; Vernon and Reville, 1983; Bucher and Lucas, 1984; Wheeler, 1985). Falls of dust derived from northern and Central China have been recorded by ships in the North Pacific more than 2500 km off the Chinese coast (Ing, 1969) and on the Hawaiian Islands (Shaw, 1980; Darzi and Winchester, 1982; Parrington *et al.*, 1983; Braaten and Cahill, 1986; see also Hirose and Sugimura, 1984; Uematsu *et al.*, 1985). Summer dust haze in the Arctic has also been attributed to Central Asian Desert sources (Rahn *et al.*, 1977). Australian dust sometimes reaches New Zealand (Kidson and Gregory, 1930; Glasby, 1971). In the United States, dust storms in the Great Plains and the Canadian Prairies have been observed to result in dust deposition more than 3000 km away (Alexander, 1934; Hand, 1934; van Heuklon, 1977). However, heavy dustfalls at great distance from the source are rare, and tend to be associated with unusual meteorological circumstances.

Available data concerning rates of dust erosion and deposition are limited and of uncertain reliability. Few meteorological stations have maintained long-term dustfall records, and even where such data exist their geological significance is uncertain because frequently no distinction is made between dust deposition and dust accretion. In arid continental areas a significant proportion of deposited dust is subsequently re-eroded by the wind; measurements of deposition may therefore considerably overestimate accretion. A further problem arises from the fact that many of the monitoring stations are located in urban areas, which themselves generate much local dust.

Measurements of atmospheric dust concentration give an indication of those parts of the world where dust transport is important but, on land at least, they do not discriminate between areas in which net erosion, net accretion or steady-state recycling of a limited dust pool is taking place. This is also true of records of dust storm frequency in different parts of the world. Estimates of longer-term dust accretion rates have been made on the basis of loess thickness and the amount of dust in ocean cores. However, high-resolution chronological control in such sequences is frequently lacking, and assumptions made about uniform sedimentation rates are almost certainly unfounded. As discussed in Chapters 8 and 9, there is strong evidence that aeolian dust sedimentation during the Quaternary has been episodic, partly in response to global climatic changes.

4.4 DUST CONCENTRATIONS IN THE AIR

A considerable number of measurements have been made of atmospheric dust concentrations, both over land and over the oceans (Table 4.3), but the short-term nature of many of the observations makes temporal variability difficult to assess. In continental arid areas where dust is mobilized, dust concentrations may reach 10^2–10^5 μg m^{-3} (Hagen and Woodruff, 1973; Jackson *et al.*, 1973; Orgill and Sehmel, 1976; Ganor and Mamane, 1982). Reported concentrations of dust in the oceanic atmosphere mostly lie in the range 0·01–10 μg m^{-3}, although higher figures have exceptionally been recorded (Prospero and Bonatti, 1969; Ferguson *et al.*, 1970; Chester and Johnson, 1971a, b; Chester *et al.*, 1972; Aston *et al.*, 1973). A mean concentration of 0·16 μg m^{-3} was reported for the equatorial mid Pacific from filter samples collected aboard the DV *Glomar Challenger* in November and December 1973 (Prospero, 1979). In the northwest Pacific, however, concentrations as high as 60 μg m^{-3} have been recorded (Kadowaki, 1979) over Japan in spring, when large-scale dust storms occur in northern China and Mongolia. Duce *et al.* (1980) also reported highest concentrations of terrestrial dust at Eniwetak atoll, as indicated by atmospheric aluminium concentrations, during April and May. This spring pattern was confirmed by data from other North

Table 4.3

Mineral aerosol concentrations in the atmosphere.

	Geometric Mean Conc.			Range		n
	C^b	$C(Al)^c$	$C(Fe)^d$			
North Atlantic						
Northern Norway (coast)		0·56	1·32	0·08	2·63	21
Lerwick, Shetland Island (coast)		0·84	1·84	0·28	4·21	11
Bermuda (island)		1·96	2·37	0·04	50·00	29–60
Central and Northern (ship)	0·36			⩽0·02	14·1	109
Eastern Tropical (ship)		0·70	1·05	0·17	2·90	8
Tropical & Equatorial (ship)	14·2			1·15	186·7	22
Tropical (Islands)	14·6			0·36	199·5	149
South Atlantic						
Gulf Guinea (ship)			3·16	2·11	4·47	9
Tropical & Central (ship)	0·69			0·04	7·52	35
Pacific						
Oahu, Hawaii (island)			0·24	0·03	1·32	56–119
28° N–40° S (ship)	0·35			0·05	2·34	24
Indian Bay of Bengal						
7° N–15° S (ship)	4·76			0·49	11·4	5
Mediterranean (ship)	4·29			2·76	9·50	13
MS, SC, and PS^e (ship)	1·09			0·24	3·89	6
Urban	22·45	22·45	44·74	4·77	126·34	4–10
South Pole	0·008	0·008	0·013	0·003	0·026	

[a] Units: 10^{-6} g/m^{-3} air (STP); [b] Bulk insol. aerosol concentration; [c] Concentration based on Al concentration; [d] Concentration based on Fe concentration; [e] Malacca Straits, South China and Philippine Seas (After Prospero (1981b), in C. Emiliani (ed.). *The Sea*, Vol. VII, 801–874; © John Wiley.)

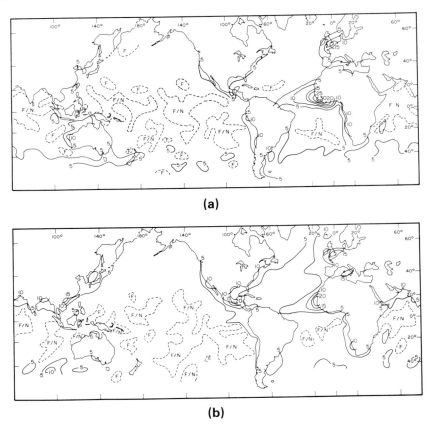

(a)

(b)

Figure 4.7. Frequency of haze at sea by season: (a) *December–February;*
(b) *March–May;* (c) *June–August;* (d) *September–November. F/N–Few or none.*
(Data of McDonald (1938), Atlas of Climatic Charts of the Oceans; © *USDA*
Weather Bureau, Washington DC.)

Pacific stations obtained by Uematsu *et al.* (1983, 1985). These authors
estimated the total annual dust input to the central North Pacific to be
$6–12 \times 10^6$ t yr^{-1}, with the highest deposition occurring between latitudes 25
and 40° N.

In the equatorial Atlantic dust concentrations show considerable spatial and
temporal variability, reflecting seasonal changes in dust emission and dispersion patterns. Maps showing the frequency of dust haze at sea prepared by
McDonald (1938) from ship data predating the 1930s showed that the highest
frequency occurs in the eastern equatorial Atlantic off the coast of Ghana,
Ivory Coast, Liberia and Sierra Leone during the Harmattan season between

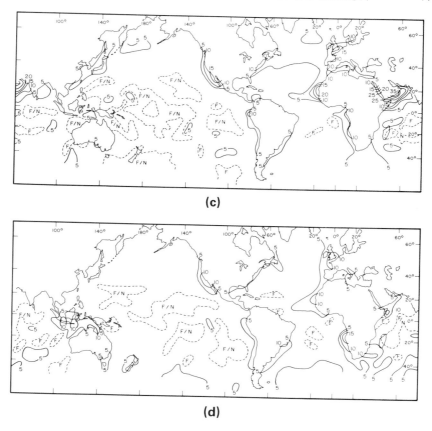

(c)

(d)

Figure 4.7 (continued)

December and February (Fig. 4.7). During the period March–August the frequency of haze is greatest off the coasts of Senegal and Mauritania.

Chester *et al.* (1972) compared the dust concentrations of the major wind systems in the eastern Atlantic between latitudes 27° N and 34° S. They reported average dust loadings of 57 μg m^{-3} for the Northeast Trades, 0.23 μg m^{-3} for variable winds of the Inter Tropical Convergence Zone (ITCZ), 1·14 μg m^{-3} for the Southeast Trades, and 0·07 μg m^{-3} for southerly winds blowing from the South Atlantic. They showed that the Sahara Desert in West Africa has a much greater effect on the dust content of the Northeast Trades than the Namib Desert has on the dust content of the Southeast Trades.

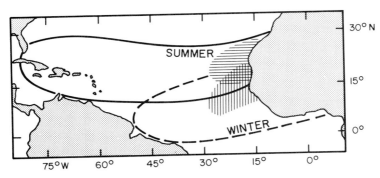

Figure 4.8. Areal extension of the Saharan dust plume and dust fallout areas over the Atlantic Ocean during summer and winter (northern hemisphere). Hatched areas close to the African coast represent the mean extension of dust fallout regions. (After Schütz (1980), Annals New York Acad. Sci. **338**, 515–532; © N.Y. Academy of Sciences.)

Atmospheric turbidity measurements have confirmed that a significant shift occurs in the location of the Saharan dust plume between winter and summer, associated with migration of the ITCZ (Fig. 4.8). During winter the dominant Harmattan Northeast Trade wind system transports dust across the equatorial Atlantic towards Brazil. Bertrand *et al.* (1974) reported dust concentrations of 900 μg m^{-3} at Abidjan, Ivory Coast, during heavy winter dust outbreaks. At the downwind end of the Harmattan dust plume, in French Guiana, Prospero *et al.* (1981) recorded maximum concentrations of about 21 μg m^{-3} during late winter and spring (Fig. 4.9). In summer, the axis of the main Saharan dust plume is located east–west close to Cape Verde, where dust concentrations of 20–200 μg m^{-3} have been reported (Jaenicke and Schütz, 1978; Schütz, 1980; Schütz *et al.*, 1981). Atmospheric dust concentrations at Barbados, close to the downwind limit of the summer dust plume, reach a maximum of about 20 μg m^{-3} (Prospero and Nees, 1977). During winter, average atmospheric dust concentrations at Barbados drop below 2 μg m^{-3} (Fig. 4.10).

Dust transport from the northern Sahara across the Mediterranean towards southern Europe occurs only in short-term outbreaks, mainly during spring. The hot, dusty surface winds associated with these outbreaks are known locally as the sirocco and garmsil. Few data are available concerning dust concentrations during such outbreaks, although Prodi and Fea (1979) reported a concentration of 154 μg m^{-3} in one major event over the Tyrrhenian Sea. Chester *et al.* (1977) reported a maximum dust loading of 23 μg m^{-3} in the lower troposphere of the southeastern Mediterranean, with the average being 13 μg m^{-3}.

Few atmospheric dust concentration data are available from the Indian Ocean. Aston *et al.* (1973) reported maximum dust concentrations of 1·9 μg

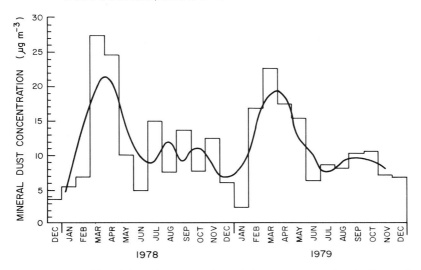

Figure 4.9. Monthly arithmetic mean mineral dust concentration in the surface-level air at Cayenne, French Guiana. The heavy line is the 3-month moving average. The means are computed from nominal 24 h daily samples. (After Prospero et al. (1981), Nature **289**, 570–572; © Macmillan Journals Ltd.)

m^{-3} in the Central Indian Ocean during the northeast monsoon, but the frequency of haze at sea (Fig. 4.7) suggests that atmospheric dust concentrations are much higher close to the Arabian peninsula and Makran coast.

Prospero and Bonatti (1969) reported dust concentrations ranging from 0·04 to 1·2 μg m^{-3} over the eastern equatorial Pacific during the spring of 1967. The main source of dust in the southern part of this area is provided by the deserts of northern Chile and Peru, while in the northern part of the area the main dust source was considered to be southwest Mexico.

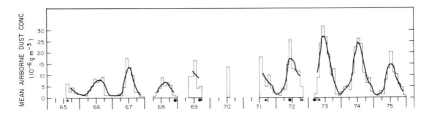

Figure 4.10. Arithmetic mean monthly mineral aerosol concentration at Barbados. Each monthly value is based on a minimum of 21 days of continuous sampling except for: (▲) 15–20 days; and (●) 10–15 days. Months for which less than 10 days of data are available are not plotted. The heavy line represents the 3-month moving average. (After Prospero and Nees (1977), Science **196**, 1196–1198; © American Association for the Advancement of Science.)

Table 4.4
Mineral aerosol deposition – estimates for various ocean regions.

Ocean region	Deposition Rate	
	$\times\ 10^{-6}\,\mathrm{g\,cm}^{-2}\,\mathrm{yr}^{-1}$	$\times\ 10^{12}\,\mathrm{g\,yr}^{-1}$
North Atlantic north of trades	82	12
North Atlantic trades	—	100–400
South Atlantic	85	18–37
Pacific	37	66
Indian Ocean	450	336
All oceans (minimum–maximum)		532–851

(After Prospero, 1981*b*, in C. Emiliani (ed.) *The Sea*, Vol. VII, 801–874; © John Wiley.)

Using such atmospheric dust concentration data, Prospero (1981*a*) concluded that average rates of dust deposition in the oceans range from 0·37 t km^{-2} yr^{-1} in the Pacific to 4·5 t km^{-2} yr^{-1} in the Indian Ocean (Table 4.4). The total amount of terrestrial dust deposited in the oceans each year is estimated to range from 532 to 851 $\times\ 10^6$ t, of which 19–47% is deposited in the North Atlantic trade wind belt and 39–63% is deposited in the Indian Ocean. It should be emphasized that these estimates are subject to large margins of error, particularly in the Indian Ocean, where very few data are available.

4.5 ANNUAL FREQUENCY OF MODERN DUST STORMS

Dust storms are defined by international convention as meteorological events in which visibility at eye level is reduced to less than 1000 m by dust actively raised from the surface by wind action, although some studies have adopted other visibility criteria (e.g. Oliver, 1945: <700 m; Péwé *et al.*, 1981: <800 m; Nickling and Brazel, 1984: <1600 m). A dust storm day is defined as a period of 24 h in which the visibility is reduced below 1000 m for all or part of the time. Blowing dust is defined by some authors as the condition where visibility is reduced to <11 km (Hinds and Hoidale, 1975; Orgill and Sehmel, 1976). Dust haze refers to a condition where visibility is reduced (sometimes, though not normally, below 1000 m) due to suspended particles in the air, and where the particles are not being actively entrained by wind at the time of observation.

The highest frequencies of dust storms occur in the arid and semi-arid regions of the world (Table 4.2). The highest reported frequency is 80·7 days in the Seistan Basin of Iran (Middleton, 1986*b*). In the Middle East, the highest frequencies of dust storms occur in northwest Saudi Arabia, Jordan and southeast Syria, and in the Mesopotamian Plain (Fig. 4.11). In the Soviet

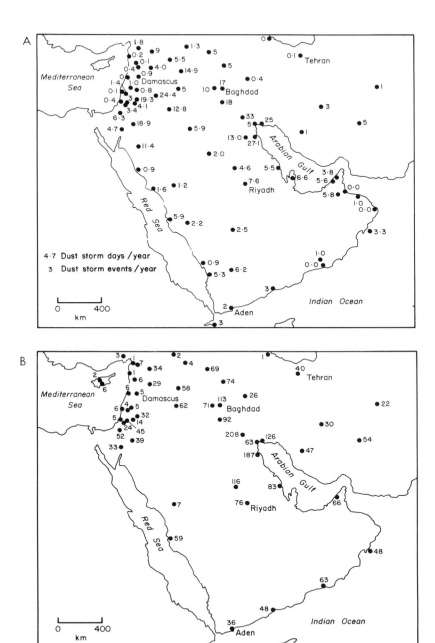

Figure 4.11. The annual frequency of: (a) dust storms (visibility < 1000 m); and
(b) blowing dust (visibility < 11 km) in the Middle East. (After Middleton (1986a),
J. Arid. Env. **10**, 83–96; © Academic Press.)

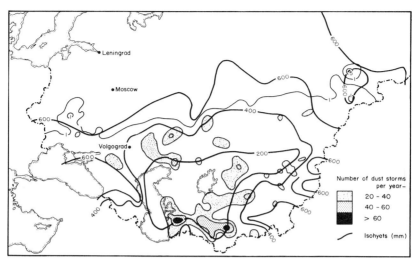

Figure 4.12. The annual frequency of dust storms (visibility <1000 m) in the Soviet Union in relation to mean annual rainfall (After Goudie (1983a), Prog. Phys. Geog. 7, 502–530; © Edward Arnold.)

Union the highest frequency of dust storms (>40/yr) occurs in a zone extending from the southern Caspian Sea, through Turkmenistan, Uzbekistan and into Tajikistan (Fig. 4.12). Other areas of high dust storm activity (>20/yr) occur in the Caucasus, southern Ukraine, and to the east of the Aral Sea in Kazakhstan (Zirkhov, 1964; Vasil'yev *et al.*, 1978; Klimenko and Moskaleva, 1979; Grigoryev and Kontratyev, 1980; Shikula, 1981). Dust storms are frequent (more than 30 per year) in parts of northwest China, particularly the Takla Makan and Gansu Corridor, but south of a line between Beijing and Xian fewer than three dust storms per year are recorded (Fig. 4.13). More than 30 dust storms per year occur in northwest Pakistan, southwest Afghanistan and eastern Iran (Fig. 4.14), but most of the Indian sub-continent experiences low frequencies (one to five per year). Frequencies are also relatively low in most of Australia, with only the Simpson Desert, part of western New South Wales, and the central coastal area of Western Australia experiencing more than five dust storms per year (Middleton, 1984; Fig. 4.15). In Africa, areas with moderate to high dust storm frequency (10–25 per year) include northern Nigeria, southern Mauritania, northern Mali and central Algeria, the Mediterranean coast of Libya and Egypt, and southern Sudan (Middleton *et al.*, 1986). In the United States, dust storms and blowing dust are most frequent in the southern Great Plains area of northwest Texas and eastern New Mexico, with secondary centres in the southwestern deserts of California and Arizona and in the northern Great Plains (Fig. 4.16).

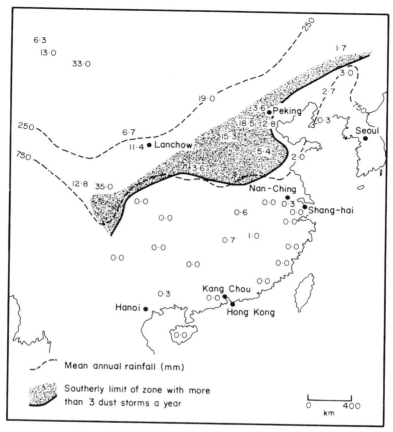

Figure 4.13. The annual frequency of dust storms (visibility <1000 m) in China. (After Goudie (1983a), Prog. Phys. Geog. 7, 502–530; © Edward Arnold.)

The frequency of dust storms generally shows a weak inverse relationship with mean annual precipitation, with markedly higher frequencies occurring in areas which receive 100–200 mm of rain (Goudie, 1978, 1983a; Fig. 4.17). In hyper-arid areas, receiving <100 mm of rain, dust storm frequency appears to decline. Goudie (1983a) suggested this may be because infrequent stream runoff limits the dust supply, or because strong winds associated with fronts and cyclonic disturbances are rare in such areas. Conversely, the markedly higher frequencies of dust storms in desert marginal areas may reflect greater fluvial activity, greater supply of dust, and more frequent occurrence of strong winds. However, it is more likely that it reflects the effects of recent cultivation of desert–marginal soils. The present-day pattern of dust storm activity is

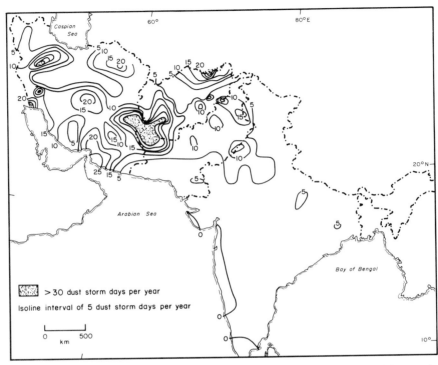

Figure 4.14. The annual frequency of dust storms (visibility <1000 m) in south-west Asia. (After Middleton (1986b), J. Climatol. *6*, *183–196*; © *John Wiley*.)

heavily influenced by agricultural activity, and does not provide a reliable basis on which to interpret patterns of dust transport and deposition earlier in the Quaternary. Estimates of dust production and deposition based on present-day measurements (Table 4.2) probably substantially overestimate the longer-term Holocene dust flux.

4.5.1 Seasonality of dust storms

In many parts of the world dust storms are highly seasonal in their occurrence, reflecting: (1) the occurrence of strong dust-transporting winds; and (2) times when dry, bare surfaces are exposed to the wind. In thinly vegetated desert areas the highest frequency of dust storms occurs in the months with the strongest winds (July–August in Arizona, May–July in Iraq, March–May in Saudi Arabia, December–April in Egypt, January–March in West Africa). In cultivated semi-arid and humid areas such as the Great Plains of the United

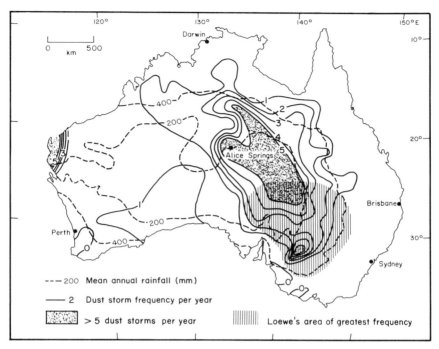

Figure 4.15. Average numbers of dust-storm days per year in Australia, 1957–82, also showing the high area of incidence according to Loewe (1943), and 200 mm and 400 mm annual isohyets. (After Middleton (1984), Search **15**, 46–47; © Australia and New Zealand Association for the Advancement of Science.)

States, the southern Soviet Union and Europe, the highest frequency of dust storms occurs in spring, when fields are ploughed and left bare until crops begin to germinate (e.g. Fullen, 1985; Fig. 4.18). Zhirkov (1964) reported that in the southern Soviet Union dust storms do not occur until the snow cover has melted and the soil dries out in late spring.

4.5.2 Longer-term variations in dust storm frequency

Dust storms are more frequent during periods of several successive drought years when the protective vegetation cover dies off and soil surfaces dry out. Prospero and Nees (1977) observed an increase in dust concentrations at Barbados during the period 1972–74, when severe droughts affected much of sub-Saharan Africa, and Middleton (1985) showed that dust storm frequencies in the Sudano–Sahel belt have increased dramatically since 1970 by a factor of 6 in Mauritania, and up to a factor of 5 in Sudan (Fig. 4.19). Particularly high

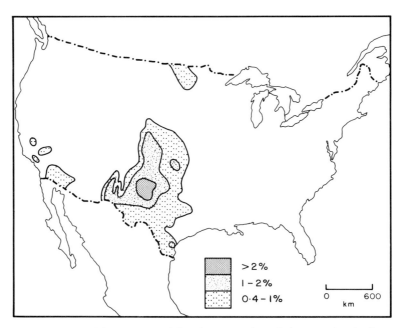

Figure 4.16. Annual frequency of dust hours with visibility <11 km in the United States. (After Orgill and Sehmel (1976), Atmos. Env. **10**, 813–825; © Pergamon Press.)

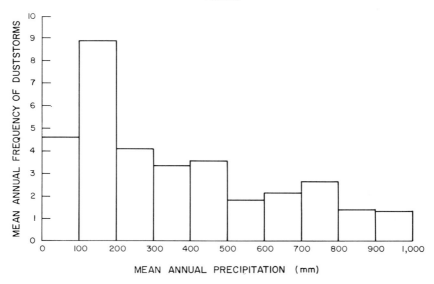

Figure 4.17. Relationship between mean annual frequency of dust storms and mean annual precipitation, based on data for several different parts of the world. (After Goudie (1983a), Prog. Phys. Geog. **7**, 502–530; © Edward Arnold.)

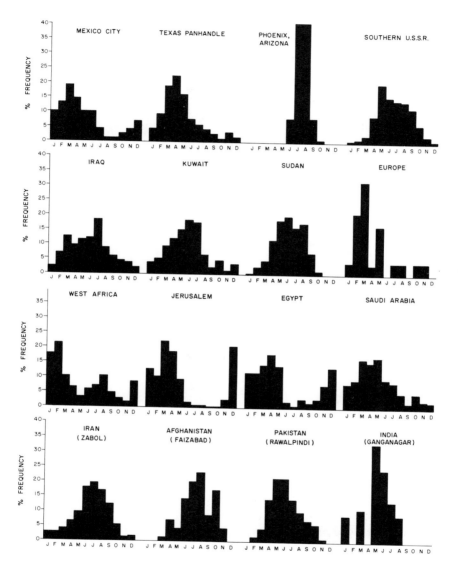

Figure 4.18. Percentage frequency of dust storms by month. (Drawn from data compiled by Goudie (1983a) and Middleton (1986a).)

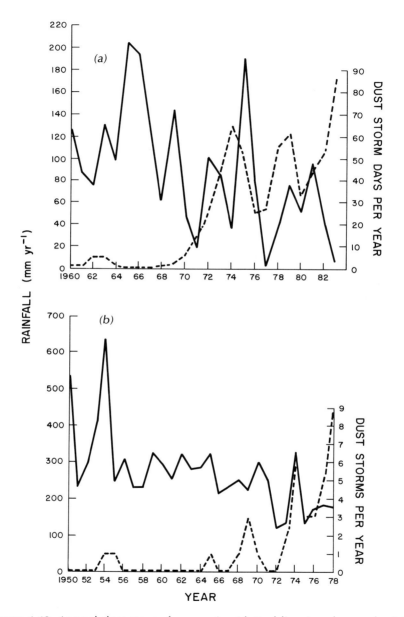

Figure 4.19. Annual dust-storm frequencies (dotted lines) and annual rainfall (solid line) at: (a) Nouakchott (18° 07′ N, 15° 56′ W); and (b) El Fasher (13° 32′ N, 25° 21′ E). Dust-storm data for Nouakchott are in the form of 'dust-storm days' (data from the Service Meteorologique, Nouakchott); mean annual number 25·3. Dust-storm data for El Fasher are the number of dust storms (visibility <1000 m) occurring (data from Sudan Meteorological Department Annual Meteorological Reports); mean annual number = 1·2. (After Middleton (1985), Nature **316**, 431–434; © Macmillan Journals Ltd.)

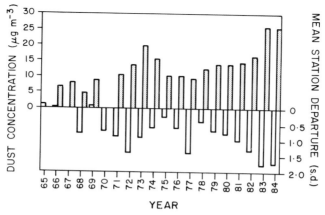

Figure 4.20. *Mineral aerosol concentrations at Barbados (stippled) and rainfall departures from the mean in the sub-Sahara for the period 1965–84. Dust concentrations are means for the period April to September, which encompasses a major portion of the dust-transport period observed at Barbados. Annual rainfall departures are expressed as standard deviations from the mean for the base period 1941–74. (After Prospero and Nees (1986),* Nature **320**, *735–738; © Macmillan Journals Ltd.)*

dust concentrations were recorded at Barbados in 1983 and 1984, when very severe drought affected sub-Saharan Africa (Prospero and Nees, 1986; Fig. 4.20). Similar increases in dust storm frequency were observed in the United States during the dry 'dust bowl' years of the 1930s (Martin, 1936; Lockertz, 1978). Changes in the frequency of dust storms at Dodge City, Kansas, related to variations in the climatic index devised by Chepil *et al.* (1963), are shown in Fig. 4.21.

Increased agricultural activities have also undoubtedly contributed to higher dust storm frequencies this century. For example, Klimenko and Moskaleva (1979) reported that following agricultural development of the Omsk region of the Soviet Union in the late 1950s dust storm frequency increased on average by a factor of 2·5 and in some areas by a factor of 5–6. Human activities have undoubtedly also contributed to the recent increased dust emissions from the Sahel. McTainsh (1985*b*, 1986) has suggested that careful monitoring of dust emissions can provide a sensitive indicator of desertification changes affecting northwest Africa.

Temporary periods of increased dust storm activity have also resulted from military activity in deserts. Oliver (1945) reported an increase in the annual number of dust storms in the western Desert of Egypt during the early years of the Second World War, presumably due to disturbance of the desert surface by tanks and shelling. The frequency of dust storms was 3–4 per year before the war; this increased to 40 in 1939 and 51 in 1941, before declining again to

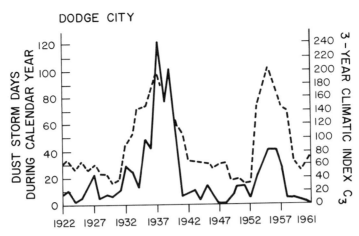

Figure 4.21. Frequency of dust storm days per year at Dodge City, Kansas, in relation to variations in the 3-year Climatic Index proposed by Chepil et al. (1963), Proc. Soil Sci. Soc. Am. **27**, 449–451; © Soil Science Society of America.

four per year in 1944. A similar increase in dust storm frequency due to tank manouevres occurred in Arizona during the Second World War when General Patton had his desert training camp in the Desert Center–Parker area (Clements *et al.*, 1963).

4.6 RATES OF DUST DEPOSITION ESTIMATED FROM OCEAN CORES

As discussed further in Chapter 8, recent studies of ocean sediments have provided much valuable information about changes in wind patterns and aeolian sediment transport during the late Cenozoic. At present, however, detailed dust deposition data are available only for the Pacific, where sediment traps have also been employed to measure directly the amount of material settling to the sea floor.

At Deep Sea Drilling Project (DSDP) site 463 in the Pacific (21° N, 174° E), the average rate of dust deposition over the last 700 000 years is about 0·15 t km^{-2} yr^{-1} (Janecek and Rea, 1983). In a core located 9° N, 104° W, the deposition rate in the surface sediment was determined to be 1·29 t km^{-2} yr^{-1} (Rea *et al.*, 1985). Sediment trap data from the same location gave a mineral flux to the sea floor of 1·1–1·26 t km^{-2} yr^{-1}. These values are only slightly higher than the dust deposition rates (0·15–0·38 t km^{-2} yr^{-1}) calculated from modern atmospheric dust concentrations in the Enewetak area (Duce *et al.*,

1980; Turekian and Cochran, 1981). Periods of more rapid aeolian dust deposition, up to a maximum of $4 \cdot 5$ t km^{-2} yr^{-1}, were identified in piston core KK75-02 (38° N, 179° E) (Janecek and Rea, 1985). The minimum rates of dust deposition in this core were < 1 t km^{-2} yr^{-1}. In the southeast Pacific, east of Tahiti, Rea and Bloomstine (1986) found that rates of aeolian dust deposition have been consistently much lower since the Oligocene ($0 \cdot 01$–$0 \cdot 04$ t km^{-2} yr^{-1}), reflecting the great distance from the Asian dust source.

4.7 RATES OF CONTINENTAL DUST DEPOSITION

Reliable data regarding rates of present-day continental dust deposition are sparse. The few data available refer only to very short time periods and may not be directly comparable due to differences in method of measurement. Reported rates range from less than 10 to about 200 t km^{-2} yr^{-1}. Maley (1980) reported a deposition rate of 42 μm yr^{-1} (equivalent to 109 t km^{-2} yr^{-1}, assuming an average particle density of $2 \cdot 6$) in 1966–67 at N'Djamene in Chad. McTainsh and Walker (1982) reported rates of 137–181 t km^{-2} yr^{-1} in northern Nigeria, in 1976–79. Yaalon and Ganor (1975) gave values of 57–217 t km^{-2} yr^{-1} for sites in southern Israel during the period 1968–73. Deposition rates were found to be 2–3 times higher in the arid Negev and its semi-arid fringe than in the sub-humid areas of central and northern Israel. Péwé et al. (1981) reported a figure of 54 t km^{-2} yr^{-1} in Arizona for the period 1972–73. On the basis of deposition from a single dust storm at Beijing in April 1980, Liu Tung-sheng et al. (1981) suggested an average annual dust deposition rate of about 240 t km^{-2} yr^{-1}. Khalaf and al-Hashash (1983) estimated an annual dustfall equivalent to 2600 t km^{-2} yr^{-1} in Kuwait, but the reliability of this figure is uncertain, since it is two orders of magnitude higher than that previously reported in Kuwait (55 t km^{-2} yr^{-1}) by Safar (1980). In Texas, Smith et al. (1970) obtained rates of 26–28 t km^{-2} yr^{-1}, while Chile and Grossman (1979) reported $9 \cdot 3$–$125 \cdot 8$ t km^{-2} yr^{-1} in New Mexico. Muhs (1983) obtained a dustfall rate of 24–31 t km^{-2} yr^{-1} based on one year's observations at San Clemente Island off the coast of California. In general, continental dust deposition rates are one or two orders of magnitude higher than those reported from the oceans, although no data are available for continental shelf areas where deposition rates may be expected to be intermediate between those of continental areas and mid oceanic areas.

Chapter Five

DUST-TRANSPORTING WIND SYSTEMS

Strong winds capable of entraining and transporting dust occur under many different meteorological conditions and at a variety of scales. At a very local scale, dust devils only a metre or two in diameter can raise dust for periods ranging from a few seconds to a few hours. At the other end of the spectrum, dust raised high into the atmosphere by turbulent mixing may be transported over distances of several thousand kilometres by transglobal wind systems, taking two or three weeks to distribute their sediment load. In this chapter a number of different dust-transporting wind systems are discussed by way of examples.

5.1 DUST DEVILS

Dust devils, also known as sand devils, dust whirls and diablos, are small-scale convective vortices which commonly develop where strong heating of the ground surface during the day forms a layer of superheated air just above the ground surface (Hallet and Hoffer, 1971; Sinclair, 1976). Contrary to popular belief, they do not occur only in deserts, having been described over bare ground in many temperate and even some sub-arctic areas (Grant, 1949). Dust devils are made visible by sand and dust drawn up from the surface, but the entrained dust has little effect on the dynamics of the vortex. Almost all dust devils have the shape of an elongate inverted cone, but their size and strength varies considerably. The height of the vortex typically ranges from 3 to 100 m, but in exceptional cases may exceed 1000 m (Sinclair, 1969; Idso, 1974*a*). The diameter of the vortex at ground level is generally 0·5–3 m. Ives (1947) referred to large dust devils, 800 m high, which lasted 7 h and travelled more than 60 km across the Bonneville Salt Flats in western Utah. Large dust devils can cause structural damage similar to a small tornado. On 2 June 1964, for example, a large dust devil destroyed a church under construction in Tucson, Arizona, and in May 1902 one reportedly demolished a livery stable in Phoenix (Ingram, 1973, cited in Idso, 1974).

On clear summer days the sun heats any bare surface to very high temperatures. In hot deserts the surface temperature may exceed 60° C in the early afternoon. This hot surface heats the air layer immediately in contact with it, creating an unstable situation with a layer of hot air overlain by a layer of much cooler and denser air. Since the air cannot turn over as a complete sheet, small thermal cells develop where upward convection is concentrated. The environment wind gives these cells an angular momentum which is intensified as the buoyant cells rise and the wind velocity increases with height. The direction of rotation may be either clockwise or anticlockwise, depending on the topography and flow pattern of the environmental wind (Ives, 1947; Sinclair, 1964, 1965). The energy required to maintain the dust devil is derived from the warm boundary-layer air which is continually fed into the visible vortex as it moves along. The strong up-draft around the vortex is balanced by a strong down-draught in the core (Kaimal and Businger, 1970).

Sinclair (1969) studied the spatial and temporal distribution of dust devils in two desert areas near Tucson. He found they most frequently developed on dry river beds in the lee of small hills. The great majority occurred during the early and mid afternoon when surface heating was most intense (Fig. 5.1). The duration lasted from a few seconds to 20 minutes, with most medium and large-sized dust devils lasting 2–3 min.

Ives (1947) classified dust devils into two types: stationary and migratory. He observed that in generally level terrain dust devils will sometimes migrate to, or originate over, a small topographic high, such as an anthill, and then remain fixed in position for an extended period, usually removing all loose dust and other light debris from the vicinity. In one case cited by Ives, a dust devil developed on a large railroad embankment under construction near Altar, Sonora, and proceeded to remove approximately 1 m^{-3} of material per hour for four hours. Erosion was eventually halted, and the dust devil broken up, by parking a bulldozer at the end of the embankment.

The trajectories of migratory dust devils vary considerably, some being essentially random and others taking the form of straight lines or regular spirals. Ives concluded that when the environmental wind velocity is low (<5 km h^{-1}), the local topography appears to control the trajectory, but at speeds >5 km h^{-1} the environmental wind direction appears to be the dominant control on dust-devil trajectory. The dust devils move downwind at approximately the same velocity as the environmental wind. In general, however, low environmental wind speeds favour dust-devil generation.

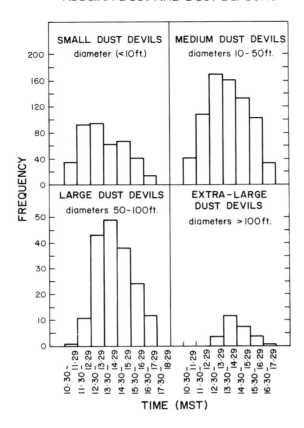

Figure 5.1. Diurnal frequency of dust devils of various sizes, Avra Valley, Arizona. (After Sinclair (1969), J. Applied Meterology **8**, *32–45;* © *American Meterological Society.)*

Ryan and Carroll (1970) studied the development of dust devils on an experimental 300 × 500 m plot of cleared ground in the Mojave Desert. In 30 days of observations, 11 of which were clear with low wind velocities, they recorded a total of 1119 dust devils, equivalent to about 250 dust devils $km^{-2} d^{-1}$.

In larger dust devils tangential velocities can reach 22 m s^{-1} (Schweisow and Cupp, 1976), with vertical velocities up to 3·5 m s^{-1} (Kaimal and Businger, 1970). Dust devils have been observed to carry quite large objects aloft, including kangaroo rats and rabbits. Ives (1947) determined the fall velocity of kangaroo rats by dropping several from the top of a control tower and timing the last 6 m of fall: 'This was found to vary from 25 to 30 mph (40–48 km h^{-1}), the animal being apparently unhurt after landing, although usually very angry.

This would indicate that the maximum upward component in a dust devil exceeds 25 miles per hour' (Ives, 1947, p. 172–173).

Measurements of airborne dust lifted from the surface at heights of 1·5 and 300 m indicated that dust devils near Tucson can raise 250 kg of dust km^{-2} on a hot summer day (Sinclair, unpublished, cited in Hall, 1981). Hall (1981) calculated that up to 250 $\mu g\ m^{-3}$ of dust might be introduced into the lower 1 km of the atmosphere of Pima County, Arizona, due to the action of dust devils on exposed dirt road surfaces; this is approximately one order of magnitude greater than the estimated contribution from vehicular traffic.

Idso (1974a) pointed out that in some instances there is no clear distinction between large dust devils and weak tornadoes. Large dust devils (or tornadoes) have been observed to form in association with advancing thunderstorm fronts, particularly where low-level density currents encounter topographic obstacles and form eddies in their wake.

5.2 DOWNDRAUGHT HABOOBS

The term 'haboob' (Arabic = violent wind) is used in Sudan to describe any dust storm raised by the wind without reference to its origin; the term 'downdraught haboob' is applied more specifically to those dust storms generated by outflow of cooled air from a cumulonimbus cloud and maintained by the resulting horizontal density gradients (Lawson, 1971). The leading edges of these storms have the appearance of solid walls of dust that conform to the shape of a density current head and rise to heights ranging from 300 to 3000 m. In the Khartoum area, downdrought haboobs travel at an average forward speed of approximately 50 $km\ h^{-1}$ and last on average between 30 minutes and one hour (Freeman, 1952). They occur about 24 times each year, particularly during summer, when warm monsoonal air and cooler more northerly air converge over the Sudan and form a zone of instability (Bhalotra, 1963). The haboobs occur mainly north of latitude 12–13° N and become less frequent north of latitude 15° as the depth of monsoon air becomes shallower and the conditions for thunderstorm generation become less favourable. In the Sudan, dust clouds associated with downdraught haboobs usually extend to heights of 1000–2000 m (Sutton, 1925, 1931; Farquharson, 1937; Lawson, 1971).

Similar dust storms associated with thunderstorm downdraughts occur less frequently in the southwestern United States (2–3 per year at Phoenix) where they are referred to as 'American haboobs' (Idso *et al.*, 1972; Idso, 1973a). In southern Arizona, as in Sudan, most of the storms develop during summer when warm sub-tropical maritime air from the Gulf of Mexico and Gulf of California converges over the region, creating potential instability. Thunderstorms appear to be triggered when the moist sub-tropical air is raised by

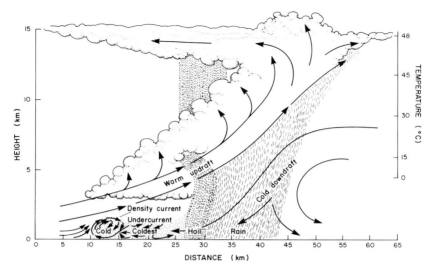

Figure 5.2. Schematic representation of a dust storm of the squall-line type. Rain and hail produce a downdraught of cold air that spreads out along the ground and moves forward to form a density current which picks-up loose dust and sand from the surface. The dense cold air pushes up warmer air which it meets, thereby reinforcing the warm updraft that creates new clouds and renews the cycle. (Modified after Idso (1976), Sci. Am. **235,** *108–114; © Scientific American, Inc.)*

canyon winds on east-facing slopes in the late morning and on strongly heated west-facing slopes in the afternoon. Storms which develop to the south and east of Tucson generally move up the Santa Cruz Valley as organized squall lines at an average rate of 48 km h^{-1}. Dust raised by individual thunderstorm cells along the squall line coalesces to form a solid wall up to 3000 m high. The haboobs arrive at Phoenix in the late afternoon or early evening; two-thirds of the dust storms are followed by rain within a period of two hours. When the dust storm arrives the surface air temperature drops by 3–13° C, with an average of 7°, and the visibility may drop to less than 400 m. Approximately 50% of all dust storms in the Phoenix area can be classified as downdraught haboobs (Idso *et al.,* 1972). The remainder are associated with the passage of fronts, rare tropical disturbances, and upper level cut-off lows (Brazel and Hsu, 1981; Brazel and Nickling, 1986).

The structure of a typical downdraught haboob is shown in Fig. 5.2. Some dust may be raised by winds associated with the warm updraft of the thunderstorm cell, but much more important in forming the dust 'walls' typical of most haboobs is the density current head of cold air which advances ahead of the main belt of hail and rain. The forward motion of a haboob over the ground is similar to that observed in laboratory investigations of gravity current heads (Simpson, 1969; Charba, 1974; Simpson and Britter, 1979). Dust is actively entrained by shear and pressure fluctuations associated with advancement of the 'nose' of the gravity current and is raised into the atmosphere by turbulent mixing in and behind the head. Parcels of warm air become trapped beneath the advancing cold air current, giving the head a billowed or lobed structure and giving rise to strong turbulent mixing (Lawson, 1971; Idso, 1976). Farquharson (1937) derived a value of 10^8 cm^{-2} s^{-1} for the coefficient of eddy diffusivity in one Sudan haboob, but this may be an overestimate. Dust which becomes incorporated into the convective updraught of thunderstorm cells may be carried to great heights (Joseph et al., 1980).

Lawson (1971) reported a maximum dust concentration of 40 μg m^{-3} in Sudan haboobs, and concluded that density differences due to the dust load play no part in their dynamic behaviour.

5.3 DUST STORMS ASSOCIATED WITH STEEP PRESSURE GRADIENTS AND LOW-LEVEL TROUGHS

5.3.1 The shamal in the Persian Gulf area

The shamal, derived from the Arabic word for north, is a persistent wind which blows during summer (late May to early July) over Iraq and the Arabian Gulf in response to the development of a monsoonal low over the northwest Indian sub-continent. The daytime surface winds are strong and are responsible for frequent dust storms in Iraq and Kuwait (al Najim, 1975; al Sayeh, 1976; Khalaf and Al-Hashash, 1983). Dust haze is carried across the Arabian Gulf to Bahrain, the Trucial States and Qatar (Houseman, 1961; Membery, 1983). Al Najim (1975) identified two main areas in Iraq which are seriously affected by dust storms, one in the Mesopotamian Plain and southern deserts west of Basrah, and the other centred on Baghdad. At Baghdad 53% of dust storms were found to be associated with northwesterly shamal winds. Fine dust entrained by the shamal, particularly during the afternoon when surface wind speeds are highest, is carried southeastwards, reaching Bahrain early the following day (Houseman, 1961).

The shamal occurs in response to the formation of an intense summer heat low over Pakistan and Afghanistan and development of a low-pressure trough to the lee of the Zagros Mountains (Fig. 5.3). A steep pressure gradient is created between this trough and a semi-permanent high-pressure cell located over northwest Saudi Arabia. The topographic effect created by the mountains of central Saudi Arabia and the Zagros also tends to enhance the low-level northwesterly wind flow over the Arabian Gulf.

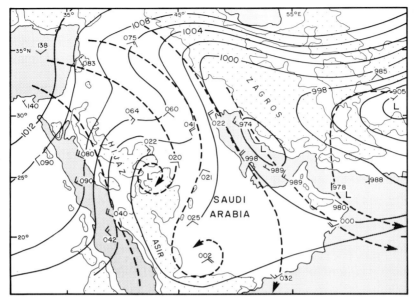

Figure 5.3. Map of the Arabian Gulf and surrounding area showing a typical shamal synoptic situation (1200 GMT, 9 June 1982) with surface isobars and 850 mb streamlines (broken). Shaded area represents land over 1000 m. (After Membery (1983), Weather **38**, 18–24; © Royal Meteorological Society.)

Most shamal dust storms do not exceed 1 km in height due to the fact that strong winds are restricted to low levels. During the strongest shamal on record, 11 June 1982, when visibility at Bahrain was reduced by haze to < 1500 m and in Kuwait to < 100 m by blowing dust, surface winds in excess of 41 km h^{-1} were recorded over the middle of the Gulf; maximum velocities of 80 km h^{-1} were recorded at a height of 350 m (Membery, 1983). Vertical dispersion of dust is also limited because at night a marked inversion commonly develops at a height of 400–450 m; under these conditions the vertical wind velocity gradient becomes very steep and a nocturnal jet sometimes develops at a height of 250–350 m (Membery, 1983; Figure 5.4).

5.3.2 The 'Santa Ana' winds of the western United States

The most common synoptic situation producing strong winds in the deserts of the American southwest occurs in late winter and spring when the Pacific High is located near its southernmost position and when a deep low-pressure cell is

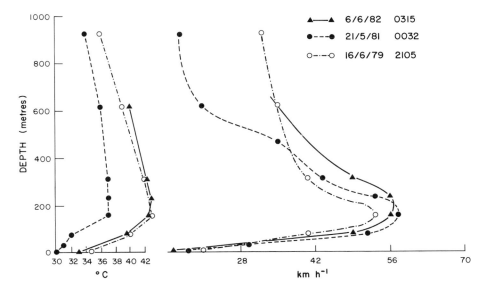

Figure 5.4. Examples of low-level wind profiles with marked temperature inversion present during the Gulf shamal. (After Membery (1983), Weather 38, 18–24; © Royal Meteorological Society.)

centred over Nevada. This situation produces a steep east–west pressure gradient and gives rise to strong southwesterly winds (Clements *et al.*, 1963). Less frequently, steep pressure gradients, strong northeasterly winds and dust storms also develop when a high-pressure centre lies over northern California and southern Oregon and a low is centred over northwest Sonora (Fig. 5.5). Strong, dry 'Santa Ana' winds originate over the deserts of Nevada and eastern California and blow across the Los Angeles Basin towards the Californian Channel Islands (Sergius, 1952; Sergius *et al.*, 1962). Nakata *et al.* (1976) reported dust entrainment from a dry lake bed near Edwards Air Force Base with Santa Ana surface wind velocities of 52 km h^{-1}. Dust deposition on the Channel Islands due to fallout from Santa Ana winds has been documented by Muhs (1983).

5.3.3 Easterly dust storms in Israel

A majority of dust storms in southern Israel are associated with southwesterly winds (Kastnelson, 1970; Ganor, 1975), but periodic easterly dust storms occur when a trough extends over the Red Sea towards the eastern Mediterra-

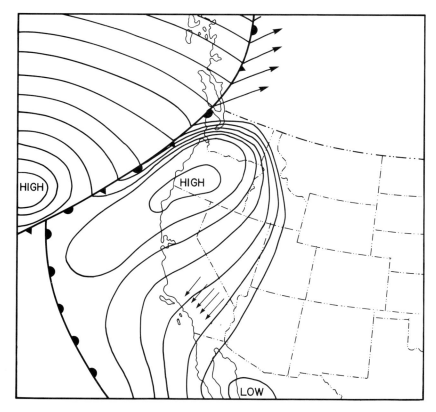

Figure 5.5. Typical 'Santa Ana' synoptic situation (surface winds, 4 November 1958). (After Clements et al. (1963), A Study of Windborne Sand and Dust in Desert Areas. U.S. Army National Laboratories Earth Science Division Technical Report ES8, Fig. 4.)

nean coast and a ridge of high pressure forms over Turkey, Syria and northern Saudi Arabia. This occurs mainly in winter when anticyclonic conditions exist over Central Europe and the Soviet Union, and movement of mid-latitude depressions across the southern Mediterranean is temporarily suppressed (Fig. 5.6).

Yaalon and Ginzbourg (1966) described the effects of an easterly dust storm which lasted for three days in November 1958. Maximum surface wind velocities reached 56 km h^{-1} at Beersheba and Lod, and 37 km h^{-1} in Jerusalem. Large amounts of sand and dust were transported in the northern Negev.

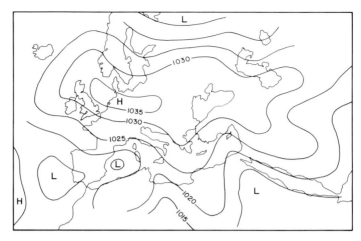

Figure 5.6. Synoptic situation associated with severe easterly dust storms which affected Israel and Egypt 21–23 November 1958. (After Kastnelson (1970), Israel J. Earth Sci. **19**, *69–76;* © *Weizmann Science Press.)*

5.4 DUST-TRANSPORTING WIND SYSTEMS IN WEST AFRICA

Three distinct wind systems are responsible for transporting dust across West Africa and neighbouring parts of the eastern Atlantic (Fig. 5.7). In the northwest Sahara, dust is transported in the shallow Trade Wind Layer from the Atlas Mountains and coastal plain in a direction almost parallel to the coast. These low-level winds only carry dust to a height of 500–1500 m and most of the dust load is deposited in a zone extending from the coast to the Canary Islands and Cape Verde Islands (Tetzlaaf and Wolter, 1980; Sarnthein *et al.*, 1981; Stein and Sarnthein, 1984).

During the northern hemisphere summer, dust storms are quite frequent in the southern Sahara and Sahel, and there are periodic outbreaks of dust-laden air over the eastern Atlantic. At this time the Inter Tropical Convergence Zone (ITCZ), which separates hot, dry Saharan air from moister, tropical air to the south, lies across the northern Sahel (Fig. 5.8). Dust is raised into the atmosphere by strong winds associated with squall lines which cross the Sahel from east to west, usually between latitudes 10° and 15° N, several degrees of latitude south of the monsoon front at the surface (Houze, 1977; Tetzlaaf and Wolter, 1980; Bolton, 1984; Tetzlaaf and Peters, 1986). Most of the squall lines are generated either east of 5° E or between 0 and 5° W and most of the

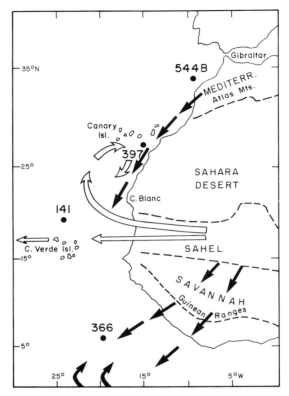

Figure 5.7. Airflow patterns and dust-transport directions in northwest Africa. Solid arrows: meridional trade winds; open arrows: mid-tropospheric zonal winds of Saharan Air Layer (SAL). Numbered dots refer to DSDP drilling sites. (After Stein and Sarnthein (1984), Paleoecol. Afr. and Surr. Is. **16**, 9–36; © A. A. Balkema.)

disturbances decay over the eastern Atlantic (Aspliden *et al.*, 1976). Many of the squall lines are not active for more than a day, and only some of them are able to reach the West African coast within 2–3 days (Tetzlaaf and Peters, 1986). The average speed of movement is 16 m s^{-1}, with a range of 10–22 m s^{-1}. The squall lines are typically 300–500 km in length, and are orientated north–south (Fig. 5.9). Strong easterly winds, having velocities of up to 30 m s^{-1} (Sommeria and Testud, 1984), form a gust front at the leading edge of the squalls and raise a wall of dust several hundred metres high, in a similar manner to the downdraught haboobs of the Sudan. Heavy rain and thunder follow the passage of the gust front, resulting in a sudden drop in temperature and

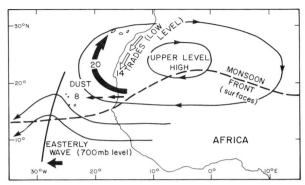

Figure 5.8. Lower mid tropospheric flow patterns and underlying trade winds associated with an easterly wave and a Saharan air outbreak in summer. Numbers indicate the transport velocities of the dust in m s^{-1}. (After Tetzlaaf and Peters (1986), Geol. Rund. 75, 71–79; © Geologische Vereinigung.)

increase in pressure. The wind velocity drops gradually and after a few hours returns to a steady southwesterly flow.

The dust lifted into the atmosphere tends to travel more slowly than the squall lines themselves and is pushed slightly further northward by a southwesterly flow component aloft. This allows the dust to become incorporated into the mid tropospheric easterly flow (the Saharan Air Layer) at a height of about 3 km (Fig. 5.9). This easterly flow is characterized by a series of large waves which propagate westward at an average velocity of 8 m s^{-1}, crossing the coast between latitudes 15 and 21° N (Reed *et al.*, 1977). In summer, three or four easterly waves per month pass over tropical West Africa towards the Atlantic. The outbreaks of dust-laden Saharan air travel above the surface Trade Wind Layer and generally take 5–6 days to cross the Atlantic (Prospero *et al.*, 1970; Carlson and Prospero, 1972; Ernst, 1974). Quite frequently, however, the flow at the 700-mb level develops a strong southerly component in the lee of an easterly wave, and after crossing the African coast the dust outbreak moves first northwest and then northwards, forming a hook-like trajectory between the Cape Verde Islands and the Canary Islands (Fig. 5.8).

During winter, when the ITCZ moves south to lie approximately over southern Nigeria, dry northeasterly surface winds (the Harmattan) dominate the surface flow pattern over much of West Africa. During this season dust haze frequently occurs in northern Nigeria 24–48 hours after sand and dust storms are reported in the Faya Largeau and Bilma areas of Chad and Niger (Samways, 1975, 1976; Adetungi *et al.*, 1979; McTainsh, 1985*a*; McTainsh and Walker, 1982). Visibility at Kano is frequently reduced to less than 400 m by haze which may persist for up to 48 hours.

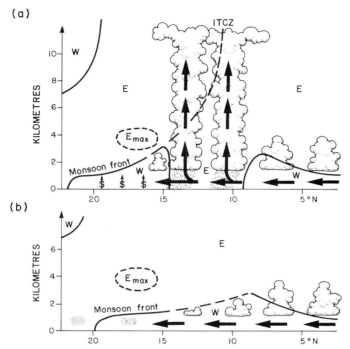

Figure 5.9. Vertical meridional section of the West African troposphere in summer at longitude 0°: (a) at the passage of a squall line; and (b) 12 h after the passage of a squall line. ITCZ = Inter Tropical Convergence Zone; E = easterly wind; W = westerly wind; S = dust raised by wind; stippled areas = dust haze (visibility below 8 km). (After Tetzlaaf and Peters (1986), Geol. Rund. 75, 71–79; © Geologische Vereinigung.)

Intense dust storms in the Faya-Largeau and Bilma areas occur mainly during periods when there is a low-level pressure surge associated with intensification of anticyclonic conditions over the northwestern and central Sahara (Samways, 1975; Kalu, 1979). Periodic intrusions of cold air from mid-latitudes give rise to high-level troughs during winter (Fig. 5.10). High-level convergence leads to subsidence, development of an anticyclone at

Figure 5.10. Typical meteorological conditions associated with the occurrence of dense Harmattan haze in northern Nigeria during winter: (a) an intrusion of an upper-level trough (200 mb, 9 February 1974); (b) low-level pressure surge indicated by isollobaric high-pressure centre (24 hour tendencies); (c) low-level 'jet' at 900 m on 31 January 1975 (isotachs in knots); and (d) strong surface winds responsible for generating sand and dust storms on 9 February 1974 (one full bar in the wind arrow equal to 10 knots (14 km h^{-1}; visibility in km). (After Kalu (1979), in C. Morales (ed.), Saharan Dust, pp. 95–118; © John Wiley.)

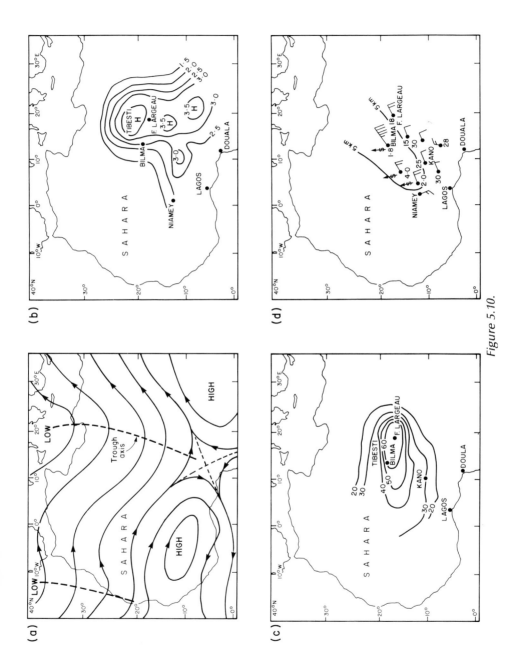

Figure 5.10.

approximately the 1500 m level, and low-level divergence. On the margins of these anticyclones winds may exceed 60 km h^{-1} for up to 48 h before the pressure surge decays. Dust haze normally reaches northern Nigeria when surface winds in the Faya-Largeau–Bilma source area exceed 42 km h^{-1}, since with such a wind speed there is sufficient turbulence to keep the dust particles airborne for a considerable period of time (Kalu, 1979). Samways (1975) reported the height of the haze at Kano to be 3500 m, which is approximately the height of the tradewind inversion. At Kano maximum wind shear and dust transport was reported by Kalu (1979) to occur at about the 900-mb level. On reaching the ITCZ over southern Nigeria the dust-laden Harmattan air is undercut by warm, moist tropical air and raised to a higher level (750–600 mb) over the equatorial Atlantic. Dust is then sometimes transported by upper-level winds as far as South America (Prospero et al., 1981). Dust is transferred downwards from the upper air layer to the surface air layer by gravity settling and turbulent mixing near the base of the upper air layer (Prospero and Carlson, 1972). However, the evidence from marine sediments (Chapter 8) suggests that southwesterly transport of Saharan dust over the Gulf of Guinea has been on a much smaller scale in the Quaternary than westerly and northwesterly dust transport over the eastern North Atlantic (Pastouret et al., 1978; Sarnthein et al., 1981).

5.5 DUST STORMS ASSOCIATED WITH DEPRESSIONS AND COLD FRONTS

The passage of frontal depressions is probably the most widespread cause of dust storms. Dust is frequently raised by winds either ahead or behind a cold front.

Ing (1969) described a major dust storm in northern China in April 1969 which was associated with the passage of three successive cold fronts across the Ordos Desert and Central Loess Plateau within a six-day period. The first two fronts were relatively weak and the third much stronger. Dust was entrained by northwesterly winds behind the first and second fronts but was raised only to a height of 1·5 km. Strong baroclinic conditions ahead of the third front gave rise to strong surface winds and pronounced vertical air motion, raising dust to heights of more than 3 km. On this basis, Ing suggested that dust storms generated by cold fronts can be classified into two types: (1) those affecting local regions, or limited regions downstream, where dust is lifted mechanically by winds behind a front and is generally confined below 1–1·5 km; and (2) those where dust is lifted higher, to about 3 km, by vertical motion ahead of a front and is transported large distances. A vigorous dry front, ahead of which is a zone of strong upward motion, is required for the second type.

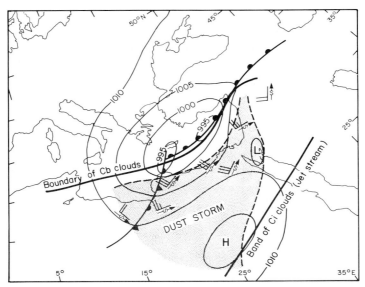

Figure 5.11. Synoptic situation associated with a typical southwesterly dust storm in northern Egypt and southern Isreal, 21 March 1967. (After Yaalon and Ganor (1979), in C. Morales (ed.) Saharan Dust, pp. 187–193; © John Wiley.)

Dust storms in southern Israel are most frequently associated with warm, dry, southwesterly winds ahead of a cold front moving slowly eastwards across North Africa (Fig. 5.11). The fronts may be associated either with low-pressure systems centred over Cyprus or with desert depressions which originate near the Atlas Mountains and move east along the North African coast (Kastnelson, 1970; Ganor, 1975; Yaalon and Ganor, 1979; Ganor and Mamane, 1982). The warm-sector winds are one of the main causes of 'sharav' (heat wave) conditions in Israel (Winstanley, 1972).

Frontal depressions are also responsible for hot, dry, dusty conditions known in Egypt as 'khamsin' (el Fandy, 1940). They occur mainly during spring and are classified according to whether the depression track lies over the southern Mediterranean, the North African coast, or the interior deserts (Sayed-Ahmed, 1949). All three types cause dust deposition at Cairo, but the type and rate of dust deposited varies with the type of storm (Abdel-Salam and Sowelim, 1967a, b). Mediterranean Sea-type depressions give rise to prolonged haze but have the lowest dustfall rates. However, because such conditions occur frequently in February and March they result in large aggregate deposits. North African coast-type khamsin depressions, which occur mainly in April, are very severe, with active dust storms and high dustfall rates in the Cairo area, but, being of short duration, these conditions are responsible for small-aggregate dust deposits. The African desert-type depressions occur mainly in May and June and are intermediate in duration and severity (at Cairo) between the other two types. They form or intensify mainly in southern Tunisia and western Libya, within a surface trough extending north from the ITCZ, and ahead of a shallow trough in the upper westerlies where positive vorticity advection is to be expected (Pedgley, 1972).

In the southern High Plains of Texas and New Mexico, dust storms are often generated by southwesterly winds associated with depressions centred to the east of the Rocky Mountains in Utah and Colorado (Warn and Cox, 1951; Fig. 5.12). McCauley *et al.*, (1981) described a particularly severe dust storm in the

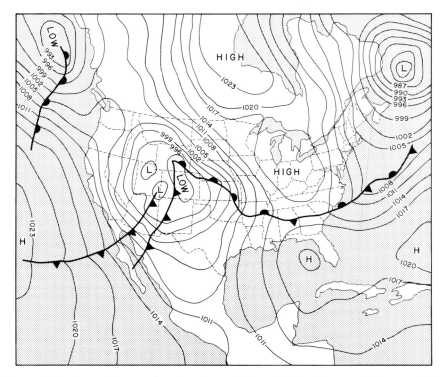

*Figure 5.12. Synoptic map showing the position of highs, lows and fronts on 25 March 1950, at the beginning of a severe dust storm which blew intermittently in the Southern High Plains for three days. A huge surface low was centred in Utah and Colorado with a shallow trough along the Central Rockies. (After Warn and Cox (1951), Am. J. Sci. **249**, 552–568; © American Journal of Science.)*

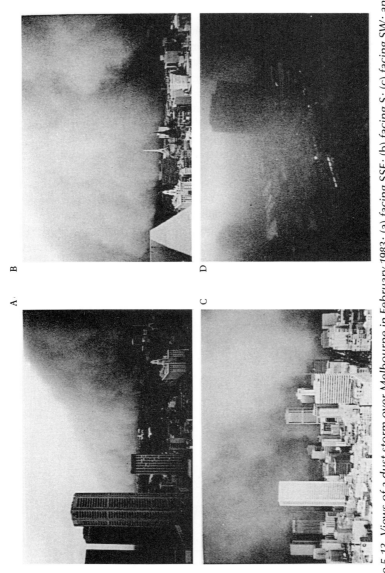

Figure 5.13. Views of a dust storm over Melbourne in February 1983: (a) facing SSE; (b) facing S; (c) facing SW; and (d) facing E, storm practically overhead. (After Lourensz and Abe 1983), Weather **38**, 272–275, © Royal Meteorological Society.)

High Plains in February 1977 which resulted from a strong eastward-moving frontal depression following a period of dry anticyclonic conditions. During this storm atmospheric dust concentrations in Oklahoma reached 5 g m^{-3} for several hours, and dust was raised to a height of 4000 m (Kessler *et al.*, 1978). In Texas, dust reached an altitude of 3000 m with the highest dust concentrations being found 1000–2500 m above the ground (McCauley *et al.*, 1981). The dust plume spread over much of the southeastern United States and reached the mid-Atlantic Ocean two days later (Windom and Chamberlain, 1978).

Many major dust storms in southern Australia are associated with non-precipitating cold fronts which extend up to 500 km inland from Melbourne (Loewe, 1943; Garratt, 1984). During a major dust storm which affected Melbourne on 8 February 1982 (Fig. 5.13), dust was raised by hot, dry northwesterly winds which averaged a speed of 33 km h^{-1} and gusted to 80 km h^{-1} as they blew ahead of a cold front across drought-affected grazing and wheat lands in northern Victoria and western New South Wales (Lourensz and Abe, 1983). The dust storm was up to 100 km wide and affected a zone 500 km long between Mildura and Melbourne. The height of the dust storm at Mildura was reported to be 3650 m but declined to 320 m at Melbourne. During the most intense phase of the storm, visibility in Melbourne was reduced to <100

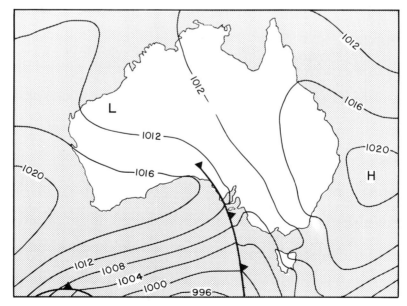

*Figure 5.14. Mean sea-level pressure chart, 00 GMT 8 February 1983, showing the strong northwesterly airstream over Victoria ahead of the approaching cold front and associated leading trough. (After Lourensz and Abe (1983), Weather **38**, 272–275; © Royal Meteorological Society.)*

m. An estimated $10 \cdot 6$ t km^{-2} of dust was deposited in the Melbourne suburbs. The cold front was associated with an intense low-pressure system which developed to the south of western Australia and moved east on 7–8 February (Fig. 5.14). The strong northwesterly dust-raising winds were channelled towards Melbourne by a pre–frontal trough which reached Melbourne at 3 pm on 8 February. As the pre-frontal trough passed the temperature dropped 8° C in 15 min. and the wind veered from northwest to southwest, carrying dust towards the eastern suburbs.

5.6 LONG-RANGE TRANSPORT OF DUST BY JET STREAMS

On a number of occasions heavy dustfalls have been recorded at great distance from the source only a short time after the dust-entraining event. For example, severe dust storms associated with movement of a deep trough of low pressure across southeast Australia on 5–6 October 1982 were followed by heavy thunderstorm-related dustfalls in the South Island of New Zealand later on 6 October. Healy (1970) calculated that the dust travelled 2700 km across the Tasman Sea within 17–22 h, i.e. at a speed of 120–160 km h^{-1}, and pointed out that such wind speeds would only occur in the mid to upper troposphere associated with a jet stream. He suggested that dust in southeast Australia was initially raised to great heights by strong vertical wind velocities ahead of a cold front, and then incorporated in a mid- to high-latitude jet stream at about the 300–500 mb level (Fig. 5.15). Healy further suggested that dust deposition in New Zealand may have been accomplished either by the dust being 'sucked down' from the mid troposphere by strong downdraughts within thunderstorm cumulonimbus clouds, or by vertical motions of the jet stream itself. The latter mechanism was considered to provide a better explanation for falls of dust without rain in the North Island of New Zealand on 8 October.

Jackson et al. (1973) concluded that long-range dust transport in the southwest United States and elsewhere is favoured by the following conditions: (1) strong heating at the ground surface which forms a deep mixed layer; (2) an upper-level momentum source; (3) an increasing horizontal pressure gradient; and (4) a large-scale convergence that produces ascending trajectories. In the American southwest and High Plains these conditions occur most frequently in spring when solar heating produces high surface temperatures, and when, upstream of these regions, vigorous cyclonic storm growth generates strong upper-level jet streams. Major dust storms are produced when the upper level jet and strong surface heating interact. Downward transfer of momentum from the jet stream causes strong surface gusts capable of raising dust which is then lifted by compensating ascending plumes of air. The strongest winds commonly occur ahead of a developing cold front, where low-level convergence contributes to the lifting of dust to higher elevations.

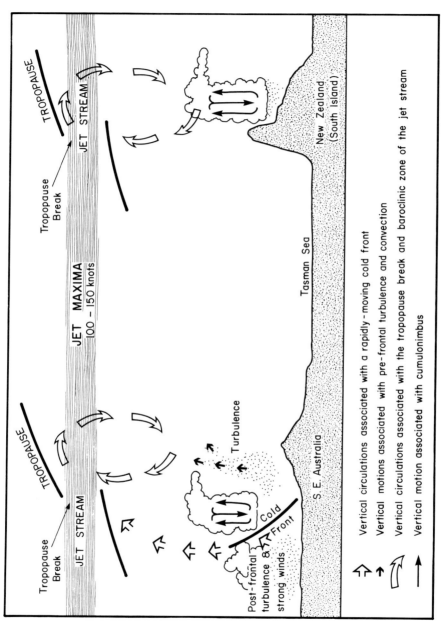

Figure 5.15. Schematic representation of a mechanism for transport of dust from Australia to New Zealand. (After Healy (1970), Earth Sci. J. **4**, 106–116; © Waikato Geological Society.)

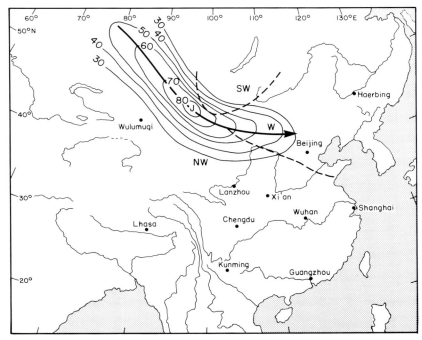

Figure 5.16. Position of the jet stream axis (arrow) and 300 mb isotachs associated with a severe dust storm in northern China, 12 GMT, 17 April 1980. Dust raised by strong turbulent mixing ahead of an advancing cold front in the deserts of northwest China was carried rapidly eastwards by the jet, leading to heavy dustfalls around Beijing. The dashed line denotes the boundary of wind directions. (© After Liu Tung-sheng et al. (1981), Geol. Soc. Am. Spec. Pap. 186, pp. 149–158.)

The importance of an upper-level jet in long-range dust transport was also identified in northern China by Liu Tung-sheng *et al.*, (1981). A deep eastward-moving depression which developed to the south of Lake Baikal produced major sand and dust storms in the northwestern Kansu corridor and western Inner Mongolia on 17 April. Dust was initially raised by warm-sector winds blowing ahead of a cold front. A short-wave trough with strong baroclinicity developed southeastwards from western Siberia, and a strong ascending current appeared in front of the trough. The increase in pressure gradient on the south side of the depression increased the surface wind-speed. In addition, surface wind-speeds and turbulence were intensified by a downward transfer of momentum from the upper westerlies. Strong vertical air motion ahead of the pre-frontal trough caused dust to be incorporated in the jet stream and to be rapidly transported eastward (Fig. 5.16). The velocity of the jet stream at 500–300 mb was 144–180 km h^{-1}, with the result that dust haze reached the northern China Plain, 1500 km east of the source, in 10 h. As the Mongolian depression moved eastward and began to weaken late on 18 April, the ascending current ahead of the upper-level trough moved into northeastern China. Subsidence of the weakening flow over east-central China resulted in dust haze conditions which lasted until 20 April. Liu Tung-sheng *et al.* (1981) estimated the dustfall rate at Beijing to be 1 t km^{-2} h^{-1}. Visibility reductions due to dust haze were reported as far away as Korea and Japan (Liu Tung-sheng *et al.*, 1985b; Chung, 1986). Periods of Asian dust haze are quite common in Japan, where they are known as 'kosa' (literally = yellow sands) (Iwasaka *et al.*, 1983; Hirose and Sugimura, 1984). The Asian dust plumes also frequently reach the Hawaiian Islands about 10 days after passing over Japan (Shaw, 1980; Chung, 1986; Braaten and Cahill, 1986).

5.7 TRANSPORT OF DUST OVER THE BRITISH ISLES AND CONTINENTAL EUROPE

Significant falls of dust in the British Isles are relatively rare, having been recorded this century in 1903 (Mill and Lempfert, 1904) and 1968 (Pitty, 1968; Stevenson, 1969). Smaller and more localized falls are more frequent, having been recorded in 1977 (Tullett, 1978; Bain and Tait, 1977), 1979 (Tullett, 1980; Pringle and Bain, 1981; Vernon and Reville, 1983), 1981 (George, 1981), 1983 (Tullett, 1984), and 1984 (Wheeler, 1985).

The main source of dust is the northwest Sahara. Fine dust raised into the atmosphere by sand and dust storms is carried northwestwards over the Atlantic at times when high-pressure centres become established over southern Europe and the western Mediterranean (Fig. 5.17). Strong winds on the western side of the anticyclone are reinforced when a depression moves in from the Atlantic. In March 1977 an outbreak of dust-laden air from the Ahaggar–Niger area initially moved northwestwards over the Canary Islands and Madeira before turning north and then east to pass over Ireland and the Western Isles of Scotland (Tullett, 1978). Dust deposition occurred as the air became involved in weak frontal activity. In September 1983 a dust cloud from Mali and Mauritania also moved out into the eastern Atlantic before curving northeastwards to pass over Ireland. The trajectories on these occasions lay much further to the west than in 1968, when dust-laden air from Morocco passed north of the Canary Islands and skirted the northern Iberian peninsula before reaching southern Britain. In this case the dust was deposited by intense thunderstorms which developed as the warm southerly airstream converged on a cold, moist Atlantic airstream from the west (Stevenson, 1969). General conditions favouring heavy dustfalls in Britain are: (1) strong surface winds

(a)

(b)

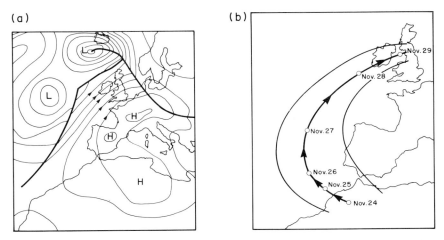

Figure 5.17. (a) The meteorological conditions for the week 21–28 November 1979, leading to falls of Saharan dust in Ireland and western Britain on 29 November. (b) Trajectory of the air mass carrying the dust (plotted on the 850 mb surface) from 24–29 November. (After Vernon and Reville (1983), J. Earth Sci. Roy. Dublin Soc. **5**, 135–144; © Royal Dublin Society.)

and vertical air motions in the Saharan source regions; (2) a steep pressure gradient associated with an anticyclone over Central and southern Europe, giving rise to strong southerly winds which have a short trajectory before reaching Britain; (3) small dust loss due to washout during transport; and (4) pronounced frontal and thunderstorm activity leading to almost total washout of the dust over the British Isles.

The synoptic situations giving rise to heavy dustfalls in Britain are substantially different to those responsible for Saharan dustfalls in Italy and Central Europe. In the latter case dust is entrained ahead of cold fronts associated with depressions in the western Mediterranean. The dust is then transported northeastwards across the Tyrrhenian Sea by warm-sector winds and deposited by orographic rains over the Appennines and the Alps (Haeberli, 1977; Prodi and Fea, 1979; Chester *et al.*, 1984*a*; Löye-Pilot *et al.*, 1986; Fig. 5.18).

Not all falls of dust in Europe are derived from North Africa. In Scandinavia and the Baltic Socialist Republics, for example, dustfalls are sometimes associated with southeasterly winds which transport dust from the Black Sea region of the Soviet Union (Lundquist and Bentsson, 1970). Such winds are most frequent in winter and spring when a steep pressure gradient develops between the Siberian anticyclone and temporary low-pressure centres over Central Europe.

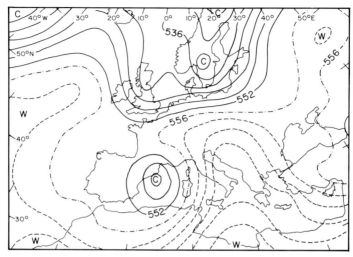

Figure 5.18. Synoptic situation associated with heavy falls of North African dust in Italy, 20 May 1977. (After Prodi and Fea (1979), J. Geophys. Res. **C84**, 6951–6960; © American Geophysical Union.)

5.8 FOHN AND VALLEY WINDS

Fohn and valley winds are sometimes responsible for generating dust storms in mountainous areas such as Soviet Central Asia and western Argentina. Fohn winds develop on the lee side of mountain ranges when air is forced to flow over them by the regional pressure gradient. During the ascent over the mountain range there is often a loss of moisture due to precipitation. The air, having cooled at the saturated adiabatic lapse rate above the condensation level, subsequently warms rapidly at the greater dry adiabatic lapse rate during descent on the lee side, leading to a reduction in humidity. Dust storms due to fohn-type winds occur in Turkmenstan during winter and spring when large depressions over the Near East and Saudi Arabia cause easterly airflow to spill over the mountains of northern Iran and down the north-facing slopes of Kopet-Dag (Petrov, 1968; Nalivkin, 1982). During a fohn-induced dust storm in January 1968 surface winds of 60–108 km h^{-1} occurred in the Ashkabad area, raising dust to a height of 8000–9000 m (Petrov, 1968).

Mountain ranges transverse to the windflow commonly give rise to wave motion in the atmosphere. Strong surface winds may be generated on the lee side of the range below wave troughs. Hallett (1969) reported low-level duststreams in Nevada which were generated by wave-generated winds blowing from the foot of the Pah Pah Range across the neighbouring valley floor, a distance of about 10 km. At other times Hallett also observed ascending, slowly rotating columns of dust rising almost vertically towards the base of the roll cloud in the lee of the mountain range.

Valley winds, sometimes produced only by patterns of diurnal heating, but more commonly reinforced by topographic funnelling of regional winds, are responsible for local dust storms in many mountain ranges including the Hindu Kush and Karakoram (Middleton *et al.*, 1986). Trainer (1961) reported that down-valley glacier winds are responsible for some dust blowing in the Matanuska and Kwik River valleys of Alaska, but confirmed that the major dust storms are caused by topographic funnelling of pressure-gradient winds.

Chapter Six

GRAIN SIZE, MINERALOGY AND CHEMICAL COMPOSITION OF AEOLIAN DUST

The composition of aeolian dust is determined by three principal factors: (1) the nature of the source material; (2) the speed and turbulence of the eroding wind, which determines whether selective entrainment of grains takes place; and (3) the vertical and horizontal distance over which transport takes place. During transport grains are sorted on the basis of size, shape and density. Consequently, mineral and chemical fractionation occurs with increasing distance of transport.

6.1 GRAIN SIZE OF SUSPENDED DUST ABOVE A WIND-ERODED SURFACE

The grain-size distribution of sediment transported above a wind-eroded surface is strongly influenced by the grain size and structure of the surface material and by the wind strength. Gillette and Walker (1977) investigated wind erosion of two different textured soils (fine sand and loamy fine sand) in Yoakum County, Texas, and found in both cases that the grain-size distribution of the sediment transported 0–1·3 cm above the ground was very similar to that of particles <0.4 mm in the parent soil, with a modal size of 90–105 μm and few particles <30 μm. They suggested that the size distribution close to the surface reflects the availability of particles in the parent soil. At a height of 1 m the transported sediment above the fine sand soil showed a bimodal distribution, with the finer mode composed of 1–10 μm particles and the coarser mode composed of 30–80 μm grains. At a height of 1·5 m above the

118

loamy fine sand soil the grain-size distribution was also bimodal at low wind speeds (modes composed of 1–5 μm and 10–30 μm grains respectively). More fine material was produced from the loamy sand soil than from the fine sandy soil even with lower wind speeds. In the case of the loamy soil the amount of fine material ($<10\,\mu$m) produced was found to increase at higher wind speeds, and with a wind velocity of 0·45 m s^{-1} the sediment transported 1·5 m above the surface became unimodal, in part due to breakage of soil aggregates by sandblasting. Gillette and Walker concluded that:

> *It is probable that the particles of the smaller size mode are derived from the exposed soils by sandblasting during wind erosion. Clay is removed from the soil as individual platelets, as coatings of platelets on quartz grains, and as coarse aggregates of platelets. Continued sandblasting by saltation removes the clay platelets from the surfaces of quartz grains and separates the aggregates. The finer-grained soil has a higher percentage of clay than the coarse-grained soil and yields higher percentages of fine airborne particulates for an equal amount of horizontal soil movement*
>
> (*1977, p. 109*)

By contrast, the moderately poorly sorted coarse silt and fine sand sediments of the Slims River Valley (Canada), which provided the source of dust investigated by Nickling (1983), contained little clay-size material. Nickling found that the material transported in creep and saltation was significantly coarser than the surface parent material, but that suspended sediment at all heights above the surface was significantly finer than the surface sediment (Fig. 6.1). Aggregates of silt-size particles were found to be a significant component of the creep load but formed only a small proportion of the saltation load. During 15 dust storms with differing wind speeds and turbulence the mean size of the suspended load at 0·5 m above the surface ranged from 40 to 65 μm with an average of about 53 μm. At a height of 12 m above the ground the mean size ranged from 8 to 30 μm with an average of about 20 μm. Nickling also observed that the sorting (as indicated by the standard deviation) of the suspended sediment decreased with height above 1·5 m. According to Folk and Ward (1957) fine sediments generally become more poorly sorted with decreasing mean size because as the mean size decreases the transporting fluid is better able to transport particles of all sizes in the sediment load. Nickling suggested additionally that the tendency for grain-size distributions to become more poorly sorted with height may be enhanced by the mixing of fine particles by turbulence. He argued that a reduction in skewness above 2 m, indicating that the suspended sediment distribution becomes more symmetrical with height, supports the suggestion that the sorting above 1·5 m becomes poorer as the size distribution becomes more uniform due to the decrease in mean particle size and more effective turbulent mixing. Below 1·5 m the mean grain-size was observed on average to decrease rapidly with height above the

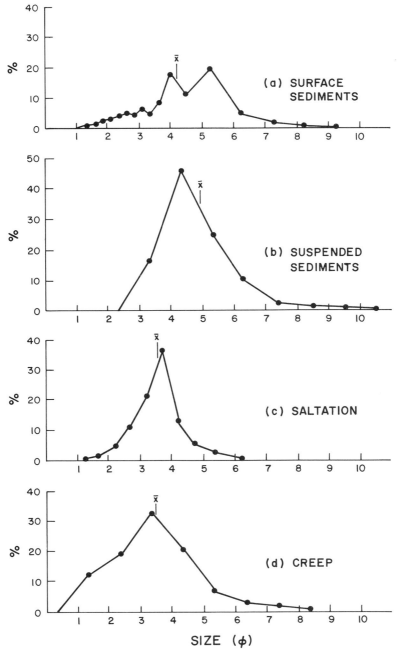

Figure 6.1. Average grain-size frequency polygons for the sediments collected from the surface and in creep, saltation and suspension samplers during dust storms, Slims River Valley, Yukon Territory. (After Nickling (1983), J. Sedim. Petrol. *53*, 1011–1024; © Society of Economic Paleontologists and Mineralogists.)

surface, while both the degree of sorting and positive skewness increased with height. Nickling attributed this to the fact that some saltating sand grains were trapped in the suspended sediment samplers close to the surface, adding a coarse tail to the suspended sediment grain-size distribution.

6.2 GRAIN SIZE OF AEOLIAN DUST TRANSPORTED OVER RELATIVELY SHORT DISTANCES

As discussed in Chapter 3, theoretical calculations indicate that the mean size of suspended dust should decrease downwind from a point or linear source (e.g. a braided river channel, road or dry lake bed). While no systematic field sampling studies have been carried out which demonstrate this with respect to individual wind storms, a substantial amount of evidence indicates that 'local' dust is on average much coarser than far-travelled dust or dust which remains suspended in the troposphere for long periods. A general decrease in the mean size of Harmattan dust in the downwind direction was demonstrated by McTainsh and Walker (1982). Dust which has been transported more than a few hundred kilometres generally contains few grains larger than 16 μm and contains a high proportion of clay-size material (Table 6.1). Dust sampled at continental sites frequently contains a mixture of local and far-travelled material, sometimes resulting in a markedly bimodal grain-size distribution.

In areas very close to a dust source particles are transported by a combination of suspension, saltation and modified saltation. 'Dust' accumulations in such areas typically show a wide range of particle sizes. For example, the 'dust' trapped in fissures on granitic highlands in Sinai (Coudé-Gaussen *et al.*, 1984) contained up to 50% sand grains, some as large as 1 mm, which must have been transported in saltation and modified saltation (Fig. 6.2).

Swineford and Frye (1945) analysed the grain-size distribution of dust deposited on a hotel balcony during a local Kansas dust storm in September 1939. The found that the dust (treated with acid to remove carbonates) contained about 10% fine sand, 77% silt, 13% clay, and had a median size of 33 μm. The size distribution of the dust showed considerable similarity to the local Sanborn loess, from which it may have been largely derived, except that the latter contained less fine sand (2–5%).

Yaalon and Ginzbourg (1966) reported that during the easterly dust storm which affected the northern Negev in November 1958, dust deposited at Yeroham, close to the major source, was much coarser (median 70 μm, mode 72 μm) than at Jerusalem (median 20 μm, mode 36 μm). The dust deposited at Yeroham contained approximately 40% fine sand, while that at Jerusalem contained less than 2% (Fig. 6.3).

Local desert dust deposited by summer haboobs at Tempe in Arizona was reported by Péwé *et al.* (1981) to contain about 8% fine sand, 78% silt and 14%

clay, with a median size of 21 μm. The size distribution of this material is broadly similar to that deposited by dust storms in Turkmenstan (Petrov, 1968) but slightly coarser than the dust from the Arabian Gulf and Red Sea analysed by Cailleux (1961, 1963; Table 6.2).

Table 6.1
Percentage of particles larger than 16 μm and smaller than 4 μm in atmospheric dust samples.

Area	$\% > 16\mu$m	$\% < 4\mu$m	Reference
Oceanic atmosphere			
NE Equatorial Atlantic	3·5	75	Chester *et al.* (1972)
E Equatorial Atlantic	0·6	88	Chester *et al.* (1972)
E Atlantic	0·5	88	Chester *et al.* (1972)
SE Atlantic	0·4	87	Chester *et al.* (1972)
E Atlantic Grp I	3·0	73	Chester and Johnson (1971*a*)
Grp II	0·5	84	Chester and Johnson (1971*a*)
Grp III	0·5	84	Chester and Johnson (1971*a*)
Barbados	0·35	77	Delany *et al.* (1967)
Continental atmosphere			
Beijing, China	25	22	Liu Tung-sheng *et al.* (1981)
Yeroham, Negev	83	9	Yaalon and Ginzbourg (1966)
Jerusalem, Israel	53	28	Yaalon and Ginzbourg (1966)
Arizona	62	16	Péwé (1981*b*)
Kansas	77	13	Swineford and Frye (1945)
Kano, Nigeria	50–70	12–30	McTainsh and Walker (1982)
SE Australia	14	40	Walker and Costin (1971)
Yorkshire	18	5	Wheeler (1985)
Southern France 1972	45	37	Bucher and Lucas (1984)
1975	19	38	Bucher and Lucas (1984)
1980	4	50	Bucher and Lucas (1984)

Table 6.2 Comparison of grain size of some dust samples from different parts of the world.

	Ashkabad, Turkmenstan (average of 5; Petrov, 1968)	Arabian Gulf (Cailleux, 1961)	Red Sea (Cailleux, 1963)	Beijing, China (Liu Tung-sheng et al., 1981)	Tempe, Arizona (Péwé et al., 1981)
>50 μm	18·39	8·6	—	3·0	12·0
10–50 μm	49·44	19·1	6·0	37·0	61·0
5–10 μm	7·54	12·0	14·1	27·0	11·0
1–5 μm	10·34	50·0	67·0	33·0	8·0
<1 μm	9·36	10·3	12·9	—	8·0

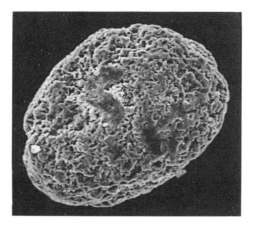

Figure 6.2. SEM micrograph of a windblown carbonate grain collected from a fissure near the summit of a granite inselberg, Sinai, in February 1982. The large diameter of the grain (approximately 150 μm) indicates that it was probably transported in saltation or modified saltation. (Photo by M. Coudé-Gaussen.)

The size distribution of local dust is strongly controlled by that of the source materials. In central and western Texas, for example, Rabenhorst *et al.* (1984) found that dust derived from wind erosion of local shallow skeletal soils developed on limestone, marl, calcareous shale and sandstone contained on average 65% particles <5 μm and only 9% particles >20 μm. Walker and Costin (1971) also found that the median diameter of dust derived from

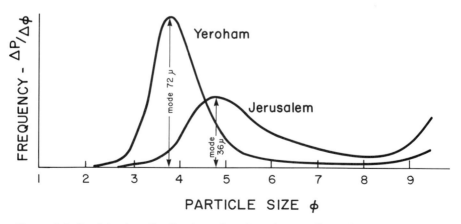

Figure 6.3. Particle size distribution of aeolian dusts collected at Jerusalem and Yeroham (northern Negev) during the easterly dust storm of November 1958. (After Yaalon and Ginzbourg (1966), Sedimentology 6, 315–332; © Elsevier Publishing Co.)

drought-affected fields in southern Australia and deposited in the Canberra–Mount Kosciusko area had a median diameter of only 4 μm.

In areas close to the source, the mean size of dust typically decreases with height (Chepil and Woodruff, 1957; Nickling, 1983). Gillette *et al.* (1978) measured the size of dust at ground-level and at heights of 2–5 km above the surface of wind-eroded soils in Texas, New Mexico, Colorado and Oklahoma. They found that suspended dust 1 m above the ground was bimodal, with a coarse mode in the 40–80 μm range and a fine mode composed of 1–15 μm particles. Dust collected 2–5 km above the ground was unimodal and mostly in the 1–30 μm size range.

6.3 GRAIN SIZE OF FAR-TRAVELLED DUST

Dust transported over large distances is composed mostly of particles <16 μm in size because only such small particles can remain in suspension for long periods. When dust plumes pass over land areas they may pick up local coarser components, but this is not the case with transport over the oceans.

Liu Tung-sheng *et al.* (1981) reported that dust which had been transported up to 3000 km from the deserts of northwest China to Beijing contained more than 90% particles <30 μm. The dust contained approximately 22% clay and had a median size of 9 μm, which is slightly finer than the late Pleistocene Malan loess at Luochuan on the Loess Plateau (Fig. 6.4). Yang Shao-jin *et al.*

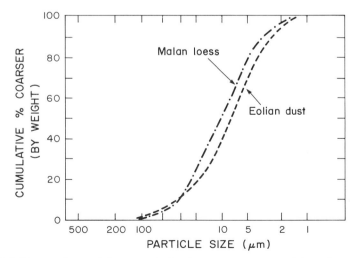

Figure 6.4. Comparison of the cumulative frequency curves of dust deposited at Beijing on 18 April 1980, and Malan loess from Luochuan, Shaanxi Province. (© After Liu Tung-sheng et al. (1981), Geol. Soc. Am. Spec. Pap. 186, pp. 149–158.)

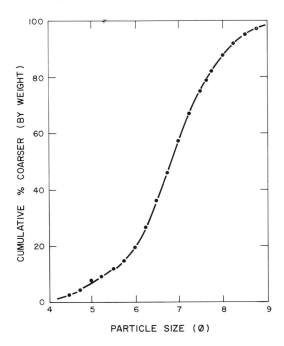

Figure 6.5. Particle-size distribution of Saharan dust which fell in England in 1968. (After Pitty (1968), Nature 220, 364–365; © Macmillan Journals Ltd.)

(1981) reported that dust haze in the same storm around Beijing was mostly finer than 1 μm.

Pitty (1968) reported that the median size of Saharan dust which fell on Britain in 1968 (a similar transport distance of about 3000 km) was also 9 μm, with about 15% of the dust composed of >20 μm particles (Fig. 6.5). Chester *et al.* (1971, 1972) found that the mean grain size of Saharan dust decreased rapidly over a short distance from the coast of northwest Africa. Dust collected inshore off the coast of Morocco contained only 3·7% particles >16 μm and 71% clay (<4 μm). At a distance of 400 km from the coast the clay content increased to 92·5%. These measurements were not, however, taken following a major dust storm. Delany *et al.* (1967) reported that dust at Barbados contained 0.35% particles >16 μm and 76.7% particles <4 μm. Prospero *et al.* (1970) also reported that 80–95% of Barbados dust is usually <5 μm, but coarser dust falls occur occasionally (Fig. 6.6). Dust collected in June 1967 contained an anomalously large amount of 5–20 μm material and was related by Prospero *et al.* to particularly vigorous wind-storm conditions in the Saharan source region. Prospero and Bonatti (1969) reported that dust col-

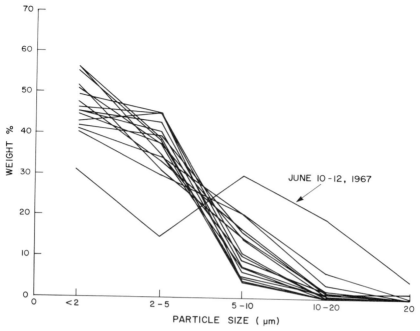

Figure 6.6. Particle-size distribution of 18 representative airborne dusts collected at Barbados, West Indies. Most of the dusts contain few particles larger than 10 μm, but a sample collected on 10–12 June 1967 contained an unusually large amount of 5–20 μm-size material, and was related to a particularly vigorous dust storm in Africa with subsequent rapid transport across the Atlantic in the Saharan Air Layer. (After Prospero et al. (1970), Earth Planet. Sci. Lett. 9, pp. 287–293; © Elsevier Publishing Co.)

lected over the eastern equatorial Pacific and derived from arid areas of South and Central America contained few or no particles >20 μm and had a modal size of 2–5 μm. Dust in areas very remote from the main continental sources, such as the Arctic (Darby *et al.*, 1974) and the Hawaiian Islands (Shaw, 1980), is predominantly <5 μm in size but may contain up to 15% larger material. In general, fine tropospheric dust which has been transported more than a few hundred kilometres from the source tends to show a uniformity of grain size and composition (Junge, 1969).

Johnson (1976) pointed out that some discrepancies in the size of oceanic dust reported by different authors may result from the different methods of collection and measurement employed. In general, the grain size of dust collected by airborne mesh samplers has been found to be finer than that of dust deposited on the surfaces of ships. This may be because deposition on ships occurs mainly by gravitational settling (dry fallout), which tends to under-represent the finer particles, whereas airborne mesh samplers theoretically sample the total suspended sediment in the atmosphere and should be more representative of wet dust deposition. In reality the collection efficiency of the meshes decreases with decreasing particle size, and the collected dust does not exactly reflect the actual particle size of dust in the atmosphere (Parkin *et al.*, 1970).

6.4 MIXTURES OF LOCAL AND FAR-TRAVELLED DUST

A number of authors have reported dust samples with markedly bimodal and sometimes polymodal grain-size distribution (e.g. McTainsh and Walker, 1982; Fig. 6.7). There are two possible ways in which this can arise. First,

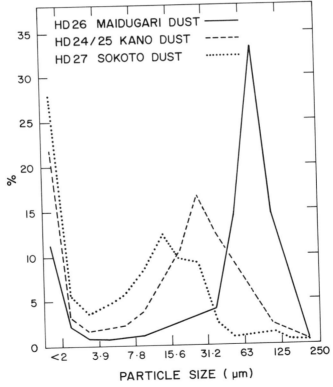

Figure 6.7. Relationship of Harmattan dust particle size to distance downwind from the source area. (After McTainsh and Walker (1982), Zeit. Geomorph. NF 26, 417–436; © Gebruder Borntraeger.)

clay-size particles which are transported mainly as silt-size aggregates and coatings on larger grains may become disaggregated during the grain-size analysis procedure. Second, dust samples at particular locations may be mixtures of local coarse and far-travelled fine components. In the case of dusts collected in arid and semi-arid continental areas both explanations are likely to apply to some degree. In the case of Harmattan dust, for example, Whalley and Smith (1981) observed coarse silt-sized aggregates of fine silt and clay in dust collected at Kano in northern Nigeria. At this and other sites in northern Nigeria, McTainsh and Walker (1982) reported considerable amounts of sand in Harmattan dust, with some grains up to 200 μm in size, which clearly must be a local component. Incorporation of local coarse material can occur wherever a local source is available, but is particularly likely in and around urban areas where bare, disturbed surfaces are exposed and where wind turbulence is enhanced around buildings.

6.5 MINERAL COMPOSITION OF DUST

The main constituents found in continental dusts derived from soils and sediments are quartz, feldspars, calcite, dolomite, micas, chlorite, kaolinite, illite, smectite, mixed-layer clays, palygorskite, heavy oxide and silicate minerals, gypsum, halite, opal, amorphous inorganic material and organic matter (including bacteria, fungal spores, pollen grains, seeds, stem tissue and ash). The precise composition of any given dust is, of course, dependent on the nature of the source material, but coarse-grained dusts are usually rich in quartz, feldspar and carbonate minerals, while far-travelled dusts are typically enriched in fine-grained micas and clays. Table 6.3 summarizes the main minerals which have been reported in dusts from a number of different parts of the world. The appearance of a typical dust collected by an air filter sampler in the Canary Islands is shown in Fig. 6.8. A much finer dust deposited by rain in eastern England is illustrated for comparison in Fig. 6.9.

Quartz, feldspars and micas are virtually ubiquitous; they tend to be least abundant in local dusts derived mainly from outcrops of carbonate rocks and sabkhas and in very fine-grained, far-travelled dusts rich in clay minerals. For example, Glaccum and Prospero (1980) found that illitic clays comprised 50% of fine Saharan dust samples collected at Sal Island (Cape Verde Islands) and 64% at Barbados and Miami, while quartz comprised 19% of the dust at Sal Island and 14% at Barbados and Miami (Fig. 6.10). Glaccum and Prospero attributed these changes to the more rapid fallout of quartz (relative to the clays) because of its larger mass median diameter and, hence, greater settling velocity.

Prospero and Bonatti (1969) also demonstrated that quartz and feldspar are present mainly in the >5 μm- and 2–5 μm-size fractions of dust collected over

Table 6.3
Reported mineral composition of some windblown dusts.
X = dominant constituent, x = important constituent.

Quartz	Feldspars	Mica/illite	Chlorite	Kaolinite	Smectite/mixed layer	Palygorskite	Calcite	Dolomite	Gypsum	Halite	Other	
X	X	x					X	x	X	x		Kuwait (Khalaf *et al.*, 1985)
X	x	x		x								N. Nigeria (McTainsh and Walker, 1982)
X	x	x	x	x	x		x					Arctic (Darby *et al.*, 1974)
X	x	x	x	x	x		x	x	x			S. Israel (Yaalon and Ginzbourg, 1966)
X				X	X							N. Nigeria (Adetungi and Ong, 1980)
X	x	x	x	x	x						A	N. Nigeria (Wilke *et al.*, 1984)
X	x	x		x	x							N. Nigeria (Whalley and Smith, 1981)
X	x	X		x		x	X	x				NW. England (Pringle and Bain, 1981)
X		x					x			x	H,C	Ireland (Vernon and Reville, 1983)
X	x	X	x			x	x					Scotland (Bain and Tait, 1977)
X	x	X		x	x		x	x				N. England (Wheeler, 1985)
X	x	x	x	x		x	x	x			H	Netherlands (Schoorl, 1973)
X	x	X	x	x	x		x				H	Equatorial Atlantic (Folger, 1970)
X	x	X	x	x			x					Barbados (Glaccum and Prospero, 1980)
X	x	X	x	x	x							E. Equatorial Pacific (Prospero and Bonatti, 1969
X	x	X	x	x								N. Atlantic (Windom and Chamberlain, 1978)
X	x	X	x	x	x		x	x			H,G	E. Equatorial Atlantic (Chester *et al.* 1972)
X	x	X	x								A,T	E. South Atlantic (Chester *et al.*, 1972)
X	x	X	x	x			x	x			H,T	E. Atlantic (Chester and Johnson, 1971*b*)
X	X	x					x				Gl	Ashkabad, Turkmenstan (Petrov, 1968)
X		x	x				x			x		New Mexico (Hoidale and Smith, 1968)
X	x	X	x				x				A	Beijing, China (Liu Tung-sheng *et al.* 1981)

Other: A = amphibole, H = haematite, G = goethite; Gl = glauconite; T = talc; C = cristobalite.

the eastern equatorial Pacific, and that the $<2\,\mu$m fraction is composed mainly of smectite, fine-grained micas, kaolinite, chlorite and mixed layer clays (Fig. 6.11).

The clay mineral composition of aeolian dusts shows only a weak general relationship with the climate of the source area since a high proportion of the clays are derived from older geological formations rather than formed authigenically in soils. Palygorskite is probably the most diagnostic clay mineral in

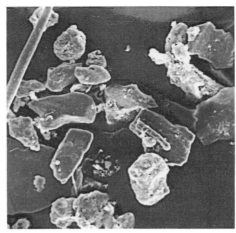

Figure 6.8. SEM micrograph showing dust collected on a filter at Fuerteventura, Canary Islands, on 14 April 1984. The grains are mostly quartz, feldspars and carbonates. Picture width = 300 μm. (Photo by M. Coudé-Gaussen.)

terms of source area, since it forms mainly in alkaline arid conditions in the presence of salts and free silica, and is unstable in other diagenetic environments (Singer and Galan, 1985; Shadfan *et al.*, 1985). The presence of palygorskite in dustfalls over Europe has been cited as partial evidence of a North African origin (Schoorl, 1973; Bain and Tait, 1977; Pringle and Bain, 1981). In local North African dusts, palygorskite commonly occurs as silt and fine sand-size pellets (Coudé-Gaussen and Blanc, 1985).

Figure 6.9. SEM micrograph of North African dust deposited in eastern England in 1984. (Sample collected by K. S. Richards.)

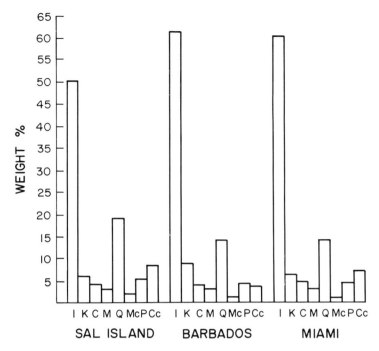

Figure 6.10. Mineralogy of Saharan dust collected at Sal Island (Cape Verde Islands), Barbados and Miami. I = mica/illite; K = kaolinite; C = chlorite; M = montmorillonite (upper limit 5%); Q = quartz; Mc = microcline; P = plagioclase; Cc = calcite. (After Glaccum and Prospero (1980), Mar. Geol. 37, 295–321; © Elsevier Publishing Co.)

Chester *et al.* (1972) found that dusts in the eastern Atlantic derived from the more humid tropical parts of Africa contained more kaolinite and smectite than dusts derived from the sub-tropical desert regions of the Sahara and the Namib. Dusts from these arid regions were found to be richer in illite, while the relative abundance of chlorite in the dusts showed little variation with climate of the source area (Fig. 6.12). Not all desert-derived dusts are dominated by illite, however. North African dusts transported over the Mediterranean have a dominant kaolinite component in the clay fraction (Chester *et al.*, 1977), while those transported over the eastern Mediterranean and Israel are rich in smectite (Yaalon and Ganor, 1973).

The amount of carbonate present in aeolian dusts varies substantially, reflecting the varying areal extent of carbonate rocks and sediments in different source areas. Both dolomite and calcite are significant constituents of Saharan

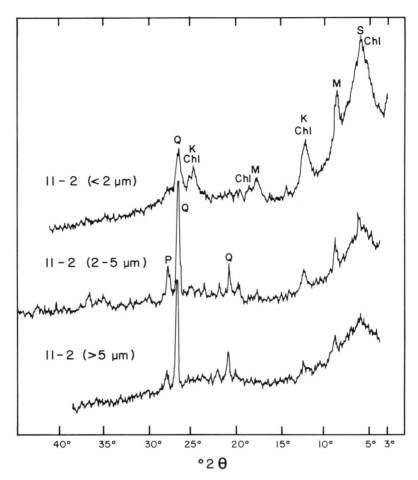

Figure 6.11. X-ray diffraction spectra for different-size fractions of dust collected in the eastern equatorial Pacific atmosphere. (Modified from Prospero and Bonatti (1969), J. Geophys. Res. 74, 3362–3371; © American Geophysical Union.)

dust transported over the northeast and equatorial Atlantic but they are virtually absent in dust over the eastern South Atlantic (Chester *et al.*, 1972; Johnson, 1979). Dolomite, together with calcite and gypsum, is a major constituent of dust in many parts of the Middle East (Fig. 6.13).

Gypsum occurs in some dusts as primary grains deflated from sabkhas or ancient evaporites (Hoidale and Smith, 1968), but in other cases it is a secondary product formed by reaction between calcite and sulphate dissolved

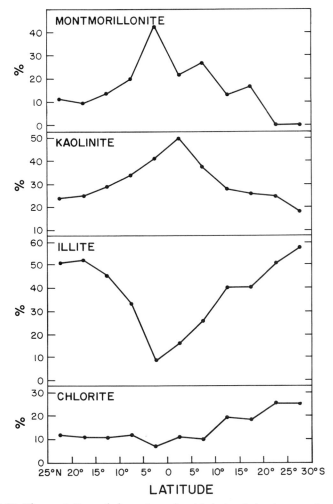

Figure 6.12. The variation of clay minerals with latitude in dusts collected off the west coast of Africa. The distribution of clay minerals (as a percentage of the total <2 μm fraction clays) is plotted against latitude; all values are averaged for each 5° of latitude. (After Chester et al (1972), Marine Geol. **13**, 91–106; © Elsevier Publishing Co.)

in atmospheric moisture (Moharram and Sowelim, 1980; Glaccum and Prospero, 1980). Halite in dust also may be primary or secondary. Schroeder *et al.* (1984–85) and Schroeder (1985) described halite cementing fine grains in dust eroded from saline soil crusts near the Red Sea, and Yaalon and Ginzbourg

Figure 6.13. SEM micrographs showing quartz (a) and dolomite (b, c, d) silt grains in shamal dust from Qatar. (Samples collected by P. Davison.)

(1966) reported halite in Jerusalem dust which was probably derived from salt flats in the Dead Sea area. On the other hand, halite reported in dustfalls over Ireland (Vernon and Reville, 1983) and many other areas is probably formed by evaporation of seawater droplets.

Talc has been reported as a minor constituent of both local and far-travelled dust (Windom *et al.*, 1967; Delany *et al.*, 1967; Parkin *et al.*, 1970; Riseborough *et al*, 1968). Talc is widely used as a carrier in pesticides and its origin in dust is probably mainly agricultural.

Dusts typically contain about 2–5% amorphous iron and organic phases, plus 3–8% amorphous aluminium. The lower figures are characteristic of dusts which have not been carried great distances, while far-travelled fine dusts are richer in amorphous material (Johnson, 1979).

Even dusts which have been transported very long distances frequently contain substantial numbers of freshwater diatoms, phytoliths, fungal spores and opaque organic spherules (Ehrenberg, 1851; Game, 1964; Delany *et al.*, 1967; Folger *et al.*, 1967; Darby *et al.*, 1974). Because such particles are frequently hollow or composed of low atomic weight compounds they form some of the largest grains in far-travelled dust. Folger (1970) reported phytoliths up to 41 μm long and diatoms up to 58 μm long in air samples from the equatorial Atlantic. The great majority of phytoliths are composed of opal, although quartz and calcite forms are also known (Baker, 1960). Most freshwater diatoms are also composed of opaline silica (Kolbe, 1957).

Pollen grains up to 20 μm in diameter are dispersed over very wide areas and are a common minor constituent of dusts (Melia, 1984; Hooghiemstra and Agwu, 1986). Fungi and bacteria can also be transported over large distances, either independently or adhering to the surfaces of mineral particles (Padu and Kelly, 1954). Heavy aerial falls of the 'manna lichen' *Lecanora esculenta* have been reported from a number of places following severe wind storms, particularly in western and west-central Asia (Donkin, 1981). The thalli of granulose *Lecanora*, which are insecurely attached to the calcareous soil or bedrock on which they grow, are easily eroded by the wind at the end of a dry season. Deposition normally occurs some tens or hundreds of kilometres away from the source. In rare instances deposits of windborne *Lecanora* have been reported to cover an area 8–16 km in diameter to a depth of 15 cm. The deposited material consists of a mixture of organic hyphae and small calcium oxalate crystals (formed by the action of oxalic acid released by the lichens on the calcium carbonate substrate).

In parts of the world affected by recent volcanic activity, such as the North Island of New Zealand and the Andes, reworking of tephra deposits means that surface-derived dusts contain significant amounts of glass shards, plagioclase, feldspar, pyroxenes, amphiboles, magnetite. The quartz content of such dusts is low, but tridymite may be present.

6.6 CHEMICAL COMPOSITION OF DUST

Surprisingly little reliable information is available regarding the major and trace element composition of dust derived from soils and sediments. The majority of analyses reported in the literature are partial or refer to samples which have been extensively pretreated to remove carbonates, dissolved salts and organic matter. The major element compositions of dust samples from a number of different continental areas of the world are shown in Table 6.4.

As would be expected, the major element composition is closely related to the mineral composition of dust. Dust in Israel has high CaO and CO_2 contents compared with other samples, reflecting an abundance of detrital calcite and lesser amounts of dolomite. Harmattan dust in northern Nigeria is characterized by a high SiO_2/Al_2O_3 ratio and a high alkali content. This is related to the presence of both abundant quartz (some of which is fine sand of local origin) and a high proportion of smectite and mixed-layer illite – smectite in the clay fraction (Adetungi and Ong, 1980). The Harmattan dust is also notable for its high organic nitrogen and carbon contents, although the origin of these high values is uncertain (Lepple and Brine, 1976; Wilke *et al.*, 1984).

The Si/Al ratios of the dust samples listed in Table 6.4 range from 2·89 (Ireland) to 5·71 (Zaria, Nigeria). This compares with estimates of Si/Al in the average crustal rock ranging from 3·41 (Mason, 1966) to 4·02 (Turekian, 1977) and Si/Al in the average soil of 4·65 (Bowen, 1966). Thus the Turkmenian, Israeli and northern Nigerian dusts show no aluminium enrichment relative to the average crust or soil, whereas the much finer dusts from the Red Sea (cf. Table 6.2) and Ireland show aluminium enrichment of 18–60%. The aluminium enrichment in the finer dusts is related to the higher clay mineral content, since the Si/Al ratio of clay minerals ranges from 1·04 (kaolinite) to 2·07 (smectite). Rahn (1976) reported that the Si/Al ratio of fine-grained mineral aerosols on average is 35% lower than that of the average crust and 50% lower than that of the average soil.

Under most conditions particulate aluminium in the atmosphere is almost exclusively associated with aluminosilicate minerals and is therefore often used as an indicator of the amount of crustal material in a mixed aerosol population. Soil aerosol concentrations in air have been computed from aluminium concentrations on the assumption that soil aerosols have an aluminium content equal to that of average soils, i.e. 6–8% (Duce *et al.*, 1980; Uematsu *et al.*, 1983; Chester, 1985). Aluminium has also been used by a number of authors as a crustal reference element against which the relative enrichment of other elements in aerosols can be compared (iron is also sometimes used as a reference element – e.g. Kushelevsky *et al.*, 1983). Chester *et al.* (1984a) calculated enrichment factors for various elements in suspended dust collected over the Tyrrhenian Sea using the following equation: $EF_x = (C_{xp}/C_{Alp})/(C_{xc}/C_{Alc})$, in which C_{xp} and C_{Alp} are the concentrations of an element x and aluminium, respectively, in the aerosol, and C_{xc} and C_{Alc} are their concentrations in average crustal material. Enrichment factors close to unity are taken to indicate that an element has a mainly crustal origin, and those greater than about 10 are considered to indicate that a substantial proportion of an element has a non-crustal source.

Chester *et al.* (1984a) found significant differences in elemental concentrations and enrichment factors in aerosols over the Tyrrhenian Sea. In aerosols derived from the northern Mediterranean land areas nickel, copper, zinc, lead and cadmium showed enrichment factors >10, indicating an urban pollution source. Iron, chromium and manganese showed enrichment factors of between 1 and 10. In aerosols derived from the Sahara only manganese, lead and cadmium showed enrichment factors >10, indicating that non-crustal contributions are a minor component of the total aerosol (Table 6.5).

6.7 OXYGEN ISOTOPE RATIOS OF QUARTZ IN DUST

The oxygen isotopic ratios of 1–10 μm-size quartz have been used to identify the source of such particles in long-range dusts and aeolian additions to soils and ocean sediments (Rex *et al.*, 1969; Syers *et al.*, 1969; Jackson *et al.*, 1971, 1972, 1973, 1982; Clayton *et al.*, 1972; Mockma *et al.*, 1972; Gillette *et al.*,

Table 6.4

Major element composition of windblown dusts. Methods of pre-treatment and analysis vary.

	Turkmenstan (Petrov, 1968)	Arabian Gulf (Cailleux, 1963)	Red Sea (Cailleux, 1961)	Israel (Ganor, 1975)	Nigeria (McTainsh & Walker, 1982)	Arizona (Péwé et al., 1981)
SiO_2	56·91	41·70	56·05	33·73	66·03	57·92
Al_2O_3	10·90	11·00	15·50	6·72	11·08	12·71
Fe_2O_3	4·49	7·45	8·45	3·74	4·45	4·72
FeO	—	1·26	—	—	—	—
MgO	0·78	3·90	3·30	3·52	0·82	3·01
CaO	4·80	12·65	6·35	23·35	0·13	2·01
Na_2O	1·60	0·70	2·15	0·73	0·91	1·93
K_2O	0·91	0·95	1·70	0·82	2·04	2·63
TiO_2	1·06	0·60	1·40	0·63	0·73	0·74
P_2O_5	1·30	0·22	0·26	0·28	0·17	—
MnO	0·54	0·14	0·15	0·10	0·10	0·07
SO_3	2·84	0·38	—	—	—	—
CO_2	—	—	—	21·48	—	—
H_2O	—	—	—	4·06	—	2.14
LOI	10·84	18·50	—	—	12·79	11·64
Total	96·97	99·45	95·31	99·16	99·25	99·52
N	—	—	—	—	—	—
C	—	—	—	—	—	—
Cl	—	—	—	—	—	—
Si/Al	4·56	3·32	2·94	4·39	5·21	3·98
SiO_2/Al_2O_3	5·22	3·79	3·62	5·02	5·96	4·56

Table 6.4 continued

	Ireland (Vernon and Reville, 1983)	Kano, Nigeria (Nov–Dec) (Wilke et al., 1984)	Kano, Nigeria (Oct–May) (Wilke et al., 1984)	Zaria, Nigeria (Nov–Dec) (Wilke et al., 1984)	Zaria, Nigeria (Oct–May) (Wilke et al., 1984)
SiO_2	58·50	57·19	59·05	57·45	65·04
Al_2O_3	17·73	12·11	11·32	10·64	9·97
Fe_2O_3	3·46	5·30	4·63	4·34	3·78
FeO	–	–	–	–	–
MgO	0·24	0·81	0·75	0·81	0·62
CaO	0·87	3·61	3·01	2·88	1·90
Na_2O	2·77	1·46	1·30	2·14	1·12
K_2O	0·75	2·95	2·87	3·26	2·95
TiO_2	–	0·83	0·81	0·82	0·92
P_2O_5	–	0·25	0·22	0·18	0·18
MnO	–	0·08	0·08	0·09	0·08
SO_3	1·22	–	–	–	–
CO_2	9·36	4·99	5·47	6·38	4·18
H_2O	3·90	9·74	8·94	9·00	7·30
LOI	–	–	–	–	–
Total	98·80	99·32	98·45	97·99	98·04
N	0·24	0·37	0·34	0·47	0·33
C	–	2·22	1·69	1·52	2·22
Cl	0·94	–	–	–	–
Si/Al	2·89	4·13	4·56	4·72	5·71
SiO_2/Al_2O_3	3·30	4·72	5·22	5·40	6·52

Table 6.5

Elemental concentrations and enrichment factors in the total aerosol of the eastern Mediterranean atmosphere (filter-collected samples). Population A aerosols are derived from the northern fringe of the Mediterranean. Population B aerosols are derived from North Africa. Concentrations in ng m^{-3} of air. (After Chester et al. (1984a), Atmos. Env. **18**, 929–935; © Pergamon Press.)

Population	Sample No.	Al Conc.	Al Ef.	Fe Conc.	Fe Ef.	Mn Conc.	Mn Ef.	Cr Conc.	Cr Ef.	Ni Conc.	Ni Ef.	Cu Conc.	Cu Ef.	Zn Conc.	Zn Ef.	Pb Conc.	Pb Ef.	Cd Conc.	Cd Ef.
A	F1	80	1·0	99	1·8	3·4	3·7	1·5	16	2·1	29	1·4	25	9·0	133	16	1333	0·17	886
	F2	97	1·0	127	1·9	3·4	3·0	0·71	6·1	0·85	9·7	1·7	25	8·4	102	16	1100	0·11	473
	F3	75	1·0	81	1·6	2·4	2·8	0·42	4·7	0·84	12	0·86	17	5·0	79	6·4	569	0·08	445
	F4	90	1·0	149	2·4	5·2	5·0	1·2	11	—	—	3·2	51	13	170	7·5	556	0·26	1203
	F9	216	1·0	128	0·87	7·2	2·9	1·4	5·4	1·4	7·2	2·6	18	21	115	8·9	275	1·6	3087
Average		112	1·0	117	1·7	4·3	3·5	1·0	8·6	1·3	15	2·0	28	12	120	11	767	0·44	1219
B	F5	249	1·0	175	1·0	2·8	1·0	1·1	3·7	1·1	4·9	3·0	18	1·5	7·1	3·2	86	0·07	117
	F6	1396	1·0	865	0·91	15	0·93	5·9	3·5	3·9	3·1	1·8	1·9	9·0	7·6	9·0	43	0·39	116
	F7	1283	1·0	709	0·81	10	0·67	4·4	2·9	2·3	2·0	0·90	1·0	3·1	2·9	7·7	40	0·35	114
	F8	5072	1·0	3247	0·94	46	0·78	8·2	1·3	5·1	1·1	4·9	1·4	20	4·7	9·6	13	0·41	34
Average		2000	1·0	1249	0·92	18·5	0·84	4·9	2·9	3·1	2·8	2·7	5·6	8·4	5·6	7·4	46	0·31	96

1978). The oxygen isotope ratio of quartz in any given rock reflects the temperature of quartz crystallization. Igneous quartz typically has $\delta^{18}O$ values of 8–10‰, metamorphic quartz $\delta^{18}O$ values of 10–16, and low-temperature (sedimentary) quartz $\delta^{18}O$ values of 14–33‰ (Jackson, 1981). Once established, the $\delta^{18}O$ value is unlikely to change at temperatures below 350° C. Most soils and sediments contain a mixture of quartz particles derived from different source rocks. The oxygen isotopic signatures of quartz in dusts should therefore reflect the nature of the parent mixtures. According to Jackson (1981), 1–10 μm quartz in mixed detrital sediments in the southern hemisphere has a $\delta^{18}O$ value of about $12 + 2‰$, whereas that in the northern hemisphere has a $\delta^{18}O$ value of about $19 \pm 2‰$. This difference is due to the presence of a large proportion of cherty, rock-derived quartz of low-temperature origin and high $\delta^{18}O$ values in the northern hemisphere. The $\delta^{18}O$ values of 1–10 μm quartz particles in dusts reflect these source differences. Reported $\delta^{18}O$ values in North Pacific aeolian-derived pelagic sediments and aeolian components in Hawaiian and Southeast Asian soils range from 17 to 21‰. By contrast, $\delta^{18}O$ values for aeolian components in South Pacific pelagic sediments, Australasian soils and Antarctic ice range from 13 to 16‰.

6.8 MAGNETIC DIFFERENTIATION OF ATMOSPHERIC DUSTS

Most atmospheric dusts contain small amounts of magnetic components derived either from soils or from industrial sources. Recently, mineral magnetic measurements have been used as a means of identifying the source of aerosols (Thompson and Oldfield, 1986; Hunt, 1986). Chester *et al.* (1984*b*) showed that the ferrimagnetic mineral component of >5 μm-size particles from the Mediterranean atmosphere is derived from two main source regions, the urban and industrial centres of southern Europe, and the deserts of North Africa. Oldfield *et al.* (1985) were able to demonstrate that 'summer' dusts of Saharan origin deposited at Barbados can be distinguished on the basis of magnetic characteristics from 'winter' dusts derived from South America. The red Saharan dusts showed high coercivities of remanence, 'harder' isothermal remanent magnetization (IRM)/saturation IRM (SIRM) quotients and higher SIRM/χ (low field susceptibility) values consistent with a high haematite contribution regarded to be typical of surface materials in arid and semi-arid regions. The grey South American dusts were found, on average, to have a higher frequency-dependent component ($\chi_{fd}\%$) in their total low-field susceptibility and to be almost entirely reverse-saturated at fields <400 *mT*. These characteristics were considered by Oldfield *et al.* (1985) to be consistent with derivation largely from fine secondary ferrimagnetic grains at the surface of magnetically 'enhanced', possibly burnt, soils.

Chapter Seven

IMPLICATIONS OF DUST DEFLATION, TRANSPORT AND DEPOSITION

The implications of dust blowing can be considered in terms of three basic aeolian processes: (1) deflation; (2) transport; and (3) deposition. Although this book is primarily concerned with fine dust particles which are transported in suspension, in many situations the effects of sand and dust movement must be considered together.

7.1 DEFLATION PROBLEMS

7.1.1 Undermining of roads and structures

Deflation, the removal of sand and dust from bare surfaces by the wind, is a problem of major economic significance in arid and semi-arid areas. Scouring and undermining of buildings, pipelines, roads, railway lines and telegraph poles is not uncommon. The Arizona Department of Transport (1974–75) estimated that deflation loss of fines from unpaved roads in Arizona was of the order of 8–80 kg km^{-1} yr^{-1}. Without frequent road maintenance, deflation causes pothole formation and development of coarse lag deposits on unpaved surfaces. The effects are increased tyre wear and damage to vehicle suspensions, paintwork and windscreens.

7.1.2 Wind erosion of soils

Soil erosion by wind is a serious problem. Carter (1977) estimated that more than \$15 billion had been spent on soil conservation in the United States up to 1976. Although erosion by water is the most significant mechanism of soil loss

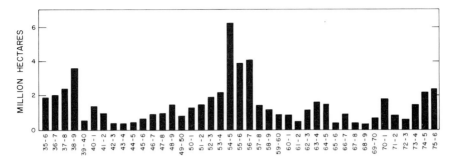

Figure 7.1. Amounts of land damaged annually by wind erosion in the Great Plains: seasons 1935/36–1975/76 inclusive. Data for the period 1943/44–1952/53 were obtained from reports of the Great Plains Council. All other data were obtained from SCS reports. (The number of counties reporting may vary from year to year). (After Kimberlin et al. *(1977), Trans. Am. Soc. Agric. Eng.* **20**, *873–879;* © *American Society of Agricultural Engineers.)*

in the United States, several hundred thousand square kilometres of land have suffered serious wind damage since the 1930s (Kimberlin *et al.*, 1977). The amount of land damaged annually ranged from 4050 km^{-2} in 1968–69 to over 60 000 km^{-2} in 1954–55 (Fig. 7.1). According to Kimberlin *et al.*, wind damage to fields becomes visible to the eye when the rate of soil loss reaches 3360 t km^{-2} yr^{-1}.

Selective removal of fines by the wind causes loss of essential plant nutrients, a reduction in moisture-holding capacity, and leads to the formation of infertile stony lag deposits. These factors combine to reduce soil fertility and crop yields (Fryrear, 1981). Yields of sorghum at Dalhart, Texas were calculated to have been reduced by 63% between 1908 and 1938 due to wind erosion (Lyles, 1975). Restoration of fertility in wind-eroded soils requires expensive fertilizer applications (Eck *et al.*, 1965).

During a severe wind storm which affected the San Joaquin Valley of California in December 1977, soil was deflated in places to a depth of 0·6 m, exposing the underlying bedrock (Wilshire *et al.*, 1981). Comparable soil losses occurred in northwest Texas and New Mexico during February 1977, when winds gusting to 100 km h^{-1} eroded soil to a depth of up to 1 m in 12 h (McCauley *et al.*, 1981).

7.1.3 The 'dust bowl' years of the 1930s

The soils of the Great Plains suffered very severe wind erosion during the 'dust bowl' years of the 1930s. During a single disastrous storm on 11 May 1934, an estimated 272 × 10^6 t of soil were transported by the wind (Bennett, cited by

Kimberlin *et al.*, 1977). In 1937 the US Soil Conservation Service estimated that severe wind erosion had affected 43% of a 65 000 km^{-2} area in the Southern Plains states. During the mid 1930s there were, on average, nine storms per month during the first four months of each year (the main dust-storm season), although they were not all intense or long-lasting. Lockertz (1978, p. 560) described conditions typical of the more severe dust storms:

> *During a bad dust storm any semblance of normal activity was out of the question. Homes, barns, tractors and fields were buried under drifts up to 25 feet high. The sky could turn completely black in a matter of minutes, and at times dust obscured the sun for several days. Some people actually thought they were seeing the end of the world. Even wet towels stuffed in the cracks of windows could not keep the dust out, and across the room an electric light might look no brighter than the tip of a cigarette. Everything in the house – even food in the refrigerator – was covered with dust. To be able to breathe, people covered their faces with wet cloths, but continuously breathing the damp air only aggravated the effects of the dust. Each storm was followed by many cases of serious lung damage, and some proved fatal.*
>
> *(1978, p. 560)*

Although it is difficult to separate the contributions to the 'dust bowl' catastrophe made by the economic recession and droughts of the 1930s, the combined result was that in some southern Plains counties more than half of all farm families were receiving social relief in 1935 (Lockertz, 1978; Worster, 1979).

7.2 PREDICTION OF SOIL EROSION BY WIND

7.2.1 Wind erosion equation

A wind erosion equation designed to predict soil loss from fields under different conditions was developed by Chepil and his associates (Chepil and Woodruff, 1963; Woodruff and Siddoway, 1965). The equation takes the general form:

$$E = f(I, C, K, L, V) \qquad [7.1]$$

where

I = a soil erodibility index
C = a local wind erosion climatic factor
K = a measure of local surface roughness
L = equivalent width of field (the maximum unsheltered distance across the field along the prevailing wind erosion direction)
V = equivalent quantity of vegetation cover

The five main variables in the equation were obtained by grouping and converting eleven primary variables which govern soil erodibility (see Woodruff and Siddoway, 1965). The relationships between the different variables and their effect on soil loss are complex and were initially displayed in a series of graphs and tables. A computer solution was subsequently published by Skidmore *et al.* (1970).

7.2.2 Climatic index of wind erosion

Chepil *et al.* (1962, 1963) also devised a climatic index of wind erosion:

$$C = 100\ U^3/(P - E)^2 \qquad [7.2]$$

where

U = the average annual wind velocity at a standard height of 10 m
$P - E$ = the effective precipitation index of Thornthwaite (1948)

The index is based on the following premises: (1) the intensity of wind erosion varies with the cube of the velocity (Bagnold, 1941); (2) inversely as the square of moisture content at the soil surface (Chepil, 1956); and (3) inversely as the square of the effective precipitation (Chepil *et al.*, 1962). The equation is standardized using the average annual value of $C = U^3/(P - E)^2$ for Garden City, Kansas ($=2 \cdot 9$):

$$C_1 = 100U^3/(P - E)^2/2 \cdot 9 \qquad [7.3]$$

The distribution of values of C_1 in the United States is shown in Fig. 7.2.

Yaalon and Ganor (1966) applied the index to Israel and suggested that areas of potential dust deflation and deposition can be identified by comparing local values of the index with standard values (Table 7.1).

Table 7.1
Wind erosion risk values of the
Climatic Index (C_1) in Israel.

Wind erosion risk	C_1
very low	0–17
low	18–35
intermediate	36–71
high	72–150
very high	>150

(After Yaalon and Ganor, 1966.)

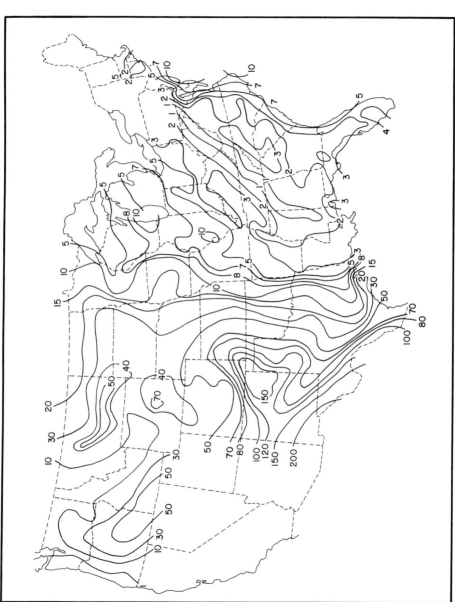

Figure 7.2. Annual climatic factor (C_1) for certain areas of the United States. (After Kimberlin et al. (1977), Trans. Am. Soc. Agric. Eng. **20**, 873, 879. © American Society of Agricultural Engineers.)

A predictive equation for the number of dust storms was obtained by Chepil *et al.* (1963) by computing a regression between 3-year average values of the climatic index, C_3, and the annual number of dust storm days:

$$N = a + bC_3 \qquad [7.4]$$

where

$N =$ the total annual number of dust storms
a and b are constants equal to $-4 \cdot 1$ and $0 \cdot 24$ respectively
C_3 is the 3-year running average value of C_1

7.2.3 Assessment of blowing dust hazard

Assessment of potential blowing dust hazard is important at the planning stage of many civil engineering and agricultural development projects (Cooke *et al.*, 1982; Jones *et al.*, 1986). Assessment techniques fall into four main groups: (1) analysis of meteorological data; (2) analysis of remote sensing imagery; (3) surface mapping; and (4) monitoring of processes. The applicability of each of these techniques depends on data availability, the terrain characteristics, the proportions of sand and dust in the mobile material, and the time and funds available (Jones *et al.*, 1986).

The assessment of blowing dust hazard is more difficult than the assessment of blowing sand due to: (1) uncertainty regarding the choice of threshold velocity to be used in construction of 'dust roses' from meteorological data; and (2) the difficulty in identifying dust-yielding surfaces from air photographs or on the ground. In a study of blowing sand and dust around 'Alpha', Jones *et al.* (1986) attempted to overcome the first of these difficulties by determining the wind speeds which were responsible for reducing visibilities to <3000 m at the local meteorological station (20 knots). They found that the 'dust rose' showed close similarity to the 'sand rose' constructed for winds >11 knots (Fig. 7.3). Analysis of air photographs and ground surveys allowed identification of 18 surface types which were classified as high, medium, or low in terms of their potential to generate blowing sand and dust. Extensive areas of fine-grained wadi sediments and disturbed desert pavements were identified as the main potential sources of blowing dust (Fig. 7.4).

A general procedure was suggested by Jones *et al.* (1986) for assessment of dust hazard around urban areas. The first step involves relating the 16 segments of the 'dust rose' to the outline of the urban area so as to produce overlapping rays of differing width, with each ray drawn at the mean orientation for each particular segment. This procedure makes the assumption that wind conditions around the perimeter of the town are uniform and similar to those at the meteorological station. Ideally, data from several different obser-

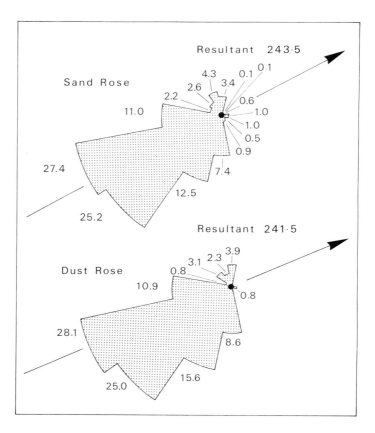

Figure 7.3. Sand and dust rose for 'Alpha' airport using 1980 and 1981 data, showing the percentage of mean annual drift potential of winds from 16 sectors. (After Jones et al. (1986), Quart. J. Eng. Geol. 19, 251–270; © The Geological Society.)

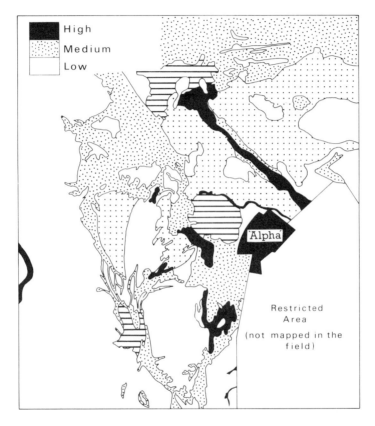

Figure 7.4. Map of estimated dust-drift potential for area to the north and west of 'Alpha'. (After Jones et al. (1986), Quart. J. Eng. Geol. 19, 251–270; © The Geological Society.)

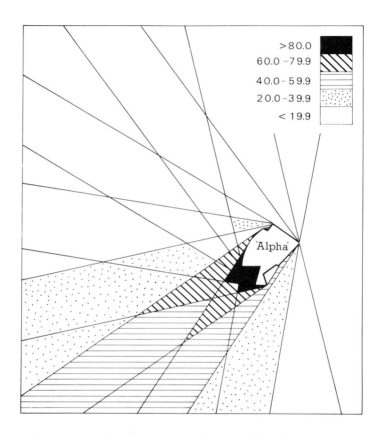

Figure 7.5. Dust source significance map developed by relating each segment of the dust rose to the outline of 'Alpha' and adding values where rays overlap. The higher the value, the greater the probability that dust generated will blow on to the town – the maximum possible value is 100 because the dust-rose values were expressed as percentages of total dust-drift potential. It is logical, therefore, that the highest values should be close to the edge of the western side of the town. (After Jones et al. (1986), Quart. J. Eng. Geol. **19**, 251–270; © The Geological Society.)

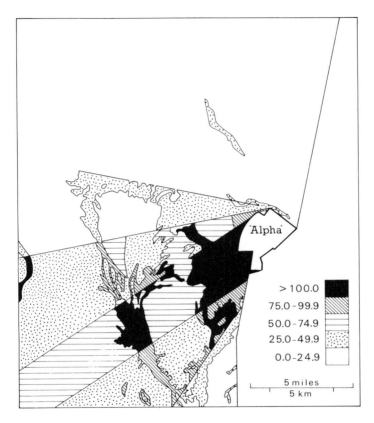

> 100.0
75.0 - 99.9
50.0 - 74.9
25.0 - 49.9
0.0 - 24.9

5 miles
5 km

Figure 7.6. Dust management map for 'Alpha', prepared by multiplying the values on the dust source significance map (Fig. 7.5) by 1, 2 or 3 depending on whether the areas were estimated to be of 'low', 'medium' or 'high' drift potential (Fig. 7.4). No significance could be attached to the absolute values (the maximum possible score is 300) but their relative magnitude indicates which areas could prove particularly problematic, in terms of dust generation for the town, and therefore require careful management. (After Jones et al. (1986), Quart. J. Eng. Geol. 19, 251–270; © The Geological Society.)

vation sites are required. The values for each segment of the 'dust rose' are then applied to the respective rays and the values summed for areas where the rays overlap (Fig. 7.5). Large values indicate a high likelihood that dust generated by surface disturbance will be blown on to the town. The dust-generating potential of different surfaces is then taken into account by laying the dust drift map over the dust potential map and multiplying the summed segment values by 1, 2 or 3 depending on whether areas have been classified as having low, medium or high dust-generating potential (Fig. 7.6).

7.3 PROBLEMS ASSOCIATED WITH THE TRANSPORT OF SAND AND DUST

7.3.1 Abrasion damage to crops

Abrasion by blowing sand and dust is responsible for major damage to crops and natural vegetation. Woodruff (1956) reported abrasion damage to winter wheat in the Great Plains, and Armbrust (1968) found that even low rates of soil movement and short periods of exposure could cause damage to cotton seedlings. Wilshire et al. (1981) described extensive damage to root crops, vineyards and citrus groves during the 1977 San Joaquin wind storm. Damage included defoliation due to sandblasting, root exposure and uprooting of trees. The mechanics and effects of dust movement through vegetation are considered in detail by Chamberlain (1975).

7.3.2 Abrasion damage to vehicles, buildings and structures

Sandblasting also causes major damage to vehicles and buildings; this includes stripping of paint and chromework and frosting of glass doors and windows. In the 1977 San Joaquin windstorm, concrete highway survey markers were severely eroded, wooden fence posts were sculptured into bizarre shapes or worn through by wind-driven sand, and asphalt curbing along highways was deeply scoured leaving asphalt 'yardangs' up to 3·5 cm high (Wilshire et al., 1981). A number of cattle trapped in an exposed location were asphyxiated; hair and skin was sandblasted from their hindquarters.

7.3.3 Damage to engines

In dusty areas engines of all types require more frequent maintenance and have shorter life-spans than in less dusty areas. Clements et al. (1963) reported that in the Second World War the tanks and vehicles used by General Patton's troops in their Arizona desert training grounds had a higher than average rate of engine replacement due to abnormal cylinder wear. Aircraft are also affected, particularly small planes and helicopters which use unpaved airstrips in desert terrain. Dust clogs air filters, contaminates fuel, scores turbine blades and cylinders, scratches commutators and causes electrical short-circuits, often even when protective measures are employed. Helicopter engine failure due to dust was reported to have contributed to the failure in 1979 of the attempt to rescue American hostages in Iran (Carter, 1979).

7.3.4 Accidents resulting from reduced visibility

Road accidents are commonly caused in arid areas when visibility is suddenly reduced by blowing dust. The frequency of dust-related accidents was so great on sections of Highways 8 and 10, northwest of Tucson, Arizona, that a Dust Storm Alert System was introduced by the Arizona Department of Transportation (Buritt and Hyers, 1981). Investigations revealed that accidents were most frequent on stretches of road bordered by abandoned farmland where the surface had been disturbed by off-the-road vehicles, or where disturbance was caused by grazing cattle (Hyers and Marcus, 1981; Buritt and Hyers, 1981).

Poor visibility due to dust has also caused a number of serious aircraft accidents. In January 1973, for example, a Royal Jordanian Airlines Boeing 707 crashed in thick dust haze at Kano airport in northern Nigeria, killing 176 passengers and crew.

From simultaneous measurements of daytime visibility and dust concentration, Chepil and Woodruff (1957) showed that for visibilities ranging from 10 km to only a few metres the visibility, V, in km, is related to the dust load, C, in $\mu g\ m^{-3}$ by the expression:

$$V = 7080c^{-0.8} \qquad\qquad [7.5]$$

While the above relationship appears generally valid for visibilities of < 10 km in areas close to a dust source, for greater visibilities with lower dust loadings and finer particles the relationship must eventually become asymptotic to the Rayleigh limit caused by molecular scattering ($V = 340$ km for pure air at sea-level). The experimental data reported by Patterson (1977) span the gap between the pure air and dusty atmosphere extremes (Fig. 7.7). Hall (1981) estimated that the relationship shown in Fig. 7.7 adequately predicts visibility, related to dust load, within a factor of 2 in three out of four cases.

7.3.5 Electrification during sand and dust storms

The impact of windblown particles produces large electrostatic charges (Stow, 1969). During the dust bowl conditions of the 1930s in the Great Plains of the United States, ignition systems of some vehicles would not operate during dust storms unless the chassis were grounded by wires or chains (Choun, 1936; Sidwell, 1938). Clements et al. (1963) reported that in Saudi Arabia, electro-

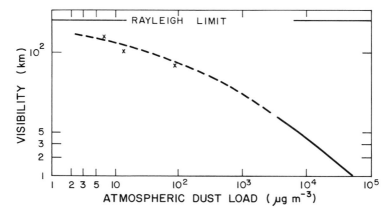

*Figure 7.7. Visibility versus atmospheric dust load, combining the empirical expression by Chepil and Woodruff (1957), V = 7080C$^{-0.8}$ (solid line) for low visibilities with the data from Patterson (1977), (x's and dashed line). (After Hall (1981), Atmos. Env. **15**, 1929–1933; © Pergamon Press.)*

static charges of as much as 150 kV had made telephone and railway telegraphic communications inoperable during sand storms. Although not usually dangerous, electrostatic charges are annoying to personnel working in desert areas affected by blowing sand and dust. Local intense electrical fields generated by dust storms can also seriously interfere with radio communications, adding to the hazard for aircraft on take-off and landing in dusty conditions.

7.3.6 Effects of dust on human health

Inhaled dust can be responsible for a number of short-term and long-term respiratory diseases. If dusts are inhaled, particles larger than 10–15 μm are trapped by very fine hairlike projections (cilia) and mucus which lines the upper respiratory tract. Some of the mucus and trapped dust is moved into the oral cavity and expectorated, but most of the material is passed into the oesophagus and through the gastrointestinal tract (Wagner, 1980). The capture efficiency of the upper respiratory tract decreases with decreasing particle size, and spherical particles < 10 μm or fibres of < 3 μm in diameter can pass down into the lower respiratory tract where some may become trapped.

The reaction of the lungs to mineral dust depends on the dosage and the nature of the dust inhaled. The natural mechanisms of the lungs are capable of rejecting small amounts of dust, but if the dosages are high these mechanisms may become overloaded and lung damage results. The damage is of two main types: (1) scarring of the alveolar walls (fibrosis); and (2) distension of the alveoles, leading to the formation of large spaces in which the walls of the alveoli are disrupted and the capillary beds are destroyed (emphysema). Dust particles composed of, for example, tin, iron or pure carbon are relatively non-cytotoxic and the degree of fibrosis produced is slight, but much more serious damage is caused by more cytotoxic materials. Quartz dust causes the formation of nodules in the lymph glands and severe scarring of the lungs (silicosis), but rarely causes death by itself. Crocidolite and amosite asbestos and other respirable, short, straight, fibrous minerals, including palygorskite and the zeolite erionite, which cannot easily be engulfed and removed by the macrophages, cause extensive tissue scarring and may result in cancer (Elmes, 1980). Mineral fibres which are naturally curly, such as chrysotile asbestos, do not pose such a serious threat because they tend to get caught and immobilized in the airways and alveoli.

Reported amounts of dust responsible for severe cases of fibrosis vary considerably. Nagelschmidt (1960, 1965) reported that the lungs of South Wales coal-workers suffering from pneumoconiosis contained up to 50 g of dust, and the lungs of a Witwatersrand gold-miner suffering from massive silicotic fibrosis contained 4–10 g of dust, of which 1–3 g was quartz. The lungs of a British factory worker suffering from severe asbestosis were reported to contain 4 g of dust, of which asbestos fibres comprised < 1 g.

Not all cases of dust-induced lung damage are industry-related. Baris et al. (1978) reported a case from Turkey where the inhabitants of two villages showed an unusually high incidence of mesothelioma due to inhaled fibrous erionite derived from local tuffaceous rock used for building construction. High incidences of silicosis and pneumoconiosis have also been reported amongst inhabitants of desert regions (e.g. Bar-Ziv and Goldberg, 1974). Green et al. (1981), suggested that some constituents of volcanic ash, notably glass shards < 10 μm in size, might constitute a significant pneumoconiosis risk to inhabitants in regions of active volcanism (see also Akematsu et al., 1982; Martin et al., 1983, and Nicol et al., 1985).

Organisms responsible for many infectious diseases are commonly present in dust. Such organisms may settle on the skin and infect it, be breathed into the respiratory passages, or be swallowed. Many germs can survive dormant for long periods in dust; these include anthrax, tetanus, brucellosis and psittacosis. In Arizona *Coccidioides immitis* is a common airborne fungus in dust which causes valley fever in man and certain animals. There are over 27 deaths each year in Arizona due to this disease alone, and the estimated loss of personal income and medical care costs due to valley fever in 1980 were conservatively estimated at $320 million (Leathers, 1981). Other organic constituents of dust, such as pollen and fungal spores, are widespread causes of allergies and less-serious diseases.

7.3.7 Possible weather and climate modification

Airborne dust causes scattering of incoming short-wave solar radiation (Fig. 7.8). However, because dust is also responsible for absorbing some solar radiation as well as outgoing long-wave terrestrial radiation, whether the net effect is a warming or cooling of the atmosphere is still a matter of debate.

7.3.7.1 Atmospheric dust and solar radiation

Humphreys (1913) expressed the view that fine volcanic dust particles 'shut out solar radiation manyfold more effectively than they hold back terrestrial radiation'. Lamb (1970) also concluded that volcanic dust-veils in the lower stratosphere could lead to a cooling of global climate, although the effects would probably be relatively short term. Ash emissions from the Krakatoa (1883) and Katmai (1912) eruptions resulted in a 10–20% global decrease in

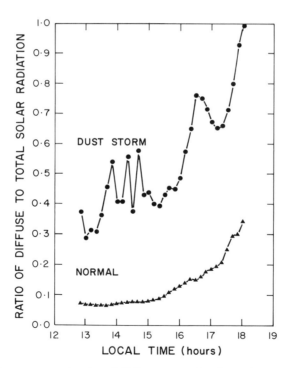

Figure 7.8. The measured ratio of diffuse to total solar radiation received at a horizontal surface (Phoenix, Arizona) on 3 February 1971, the day of a dust storm, and on 10 February 1971, a normal or average cloudless day. (After Idso (1972), Weather **27**, 204–208; © Royal Meteorological Society.)

solar radiation lasting 1–2 years. Lamb (1971) also showed that many of the coldest and wettest summers in Britain, such as 1695, 1725, the 1760s, 1816, the 1840s, 1879, 1903 and 1912, coincided with times of high volcanic dust inputs to the atmosphere, while the warm temperatures of the 1920s, 1930s and 1940s corresponded with a period when there were few volcanic eruptions.

A number of authors have suggested that an increase in atmospheric turbidity due to industrial pollution is responsible for an observed cooling trend in the northern hemisphere in the 1950s and 1960s (Bryson, 1967; McCormick and Ludwig, 1967; Schneider, 1971). However, Charlson and Pilat (1969) pointed out that these authors had considered only the effects of scattering of solar radiation by atmospheric dust, and that additional consideration of absorption indicated that either heating or cooling could occur, depending on the relative magnitudes of the extinction coefficients due to absorption (a) and backscatter (b). Atwater (1970) suggested that the critical ratio (p_c) of absorption to backscatter (a/b) was dependent on the albedo of the underlying surface, and concluded that warming would probably occur above an urban area while cooling is more likely in desert and prairie environments. Reck (1974a, b) concluded that aerosol-induced heating would always occur over surfaces of albedo $>0·6$ and that cooling would result over surfaces with albedo of $<0·35$.

7.3.7.2 Atmospheric dust and thermal radiation

It was widely believed until the early 1970s that sub-micron particles in the atmosphere have a negligible effect on long-wave terrestrial radiation (e.g. Mitchell, 1971). However, observations made in Arizona by Idso (1972, 1973b, 1975) provided evidence that thermal radiation can be increased by dust in the lower atmosphere. Idso (1972) concluded that, although solar radiation was significantly reduced by suspended dust, there was a net 2·6% increase in the total solar and thermal radiation received at the surface. During a haboob in May 1973, the dust was not dense enough to reduce the incoming solar radiation significantly, but incoming thermal radiation was found to be 10% higher (Idso, 1975). Partly on the basis of these observations, and partly based on comparisons of population growth and temperature trends at Phoenix since 1946, Idso (1974b) proposed a general model of 'thermal blanketing' to explain climatic warming due to atmospheric aerosols. The model was subsequently modified and extended by Idso and Brazel (1977) who concluded that as the atmospheric dust concentration rises, the net radiation balance initially produces warming of the Earth's surface, but under conditions of frequent and severe dust storms, with very high atmospheric dust concentrations, the warming influence is reversed. In a later study, Idso and Brazel (1978) pointed out that these results apply only to low-level tropospheric aerosols, and that in the case of stratospheric aerosols, as produced by volcanic activity, the backscatter of solar radiation prevails over the thermal blanketing effect, producing net surface cooling.

7.3.7.3 Dust and processes of cloud and rain formation

Fine dust particles in a moist atmosphere can act as condensation nucleii for cloud and rain formation (Murty and Murty, 1973; Maley, 1982). Close to the sea, aerosols of marine origin provide most of the condensation nuclei, but in continental interiors, where the concentration of marine aerosols is low, terrestrial dust particles play a relatively more important role (Mason, 1971; Twomey, 1977). Continental aerosols contain numerous Aitken nucleii (<0.3 μm in diameter) which favour the formation of large numbers of moisture droplets. Few droplets in continental-type clouds are $>20\,\mu$m in size, whereas maritime cloud droplets display a wider range of sizes, with the maximum frequency in the 20–25 μm-size range (Squires and Twomey, 1960). Small continental-type cloud droplets have a lower coagulation efficiency, such that droplet growth to the size of raindrops (up to 100 μm) is less likely than in maritime-type clouds. The concentration of dust in the continental atmosphere is therefore important in inhibiting or retarding the rainmaking process and in favouring the persistence of clouds, which tend to reduce evaporation from the Earth's surface. Additionally, by lowering the radiative exchange and increasing the atmospheric stability, atmospheric dust may have the further effect of impeding the vertical air motions necessary for cloud formation (Maley, 1982; though see el Fandy, 1949). Under such conditions, cloud and then rain formation require lifting of air by strong dynamic motions of the atmosphere (e.g. associated with jet streams). Maley (1982) has suggested that changes in atmospheric dust concentration may have affected the nature of rainfall in tropical North Africa during the late Pleistocene and Holocene, and that this in turn affected runoff characteristics and the nature of fluvial erosion/sedimentation.

7.3.8 Dispersion and deposition of radionuclides

It has been suggested that aerosolic dusts may act as scavengers for radioactive ions released into the atmosphere by nuclear tests or accidents at nuclear power plants (Syers et al., 1969, 1972). The dispersion behaviour of radionuclides may therefore be partly controlled by that of larger dust particles.

Fission products resulting from an atomic test in Nevada in 1955 were carried by aerosolic dust and deposited in rainfall in Japan (Miyake et al., 1956). [210]Po and its precursor ^{210}Pb have also been identified in dust particles and in rainfall (Francis et al., 1968; Windom, 1969), although not all is from anthropogenic sources. [210]Pb is also produced by decay of gaseous [222]Rn emanating from continental soils, and from volcanic sources (Turekian and

Cochran, 1981). Syers *et al.* (1972) pointed out that the extent of uptake of radioactive ions by dusts and the strength with which they are held should be strongly influenced by the mineralogical composition of dust, since it is well established that cation exchange selectivity is determined largely by mineralogical composition. These authors examined the mineralogical characteristics. ^{137}Cs content and ^{137}Cs retention characteristics of a range of dusts collected from the atmosphere, mud rains, snow and ice. Five of the seven aerosolic dusts and a dust-bearing Hawaiian soil were found to contain ^{137}Cs. The highest activity of ^{137}Cs was found in a dust of Australian origin collected from a glacier in New Zealand. This dust was also found to have a high content of micaceous vermiculite. Tests showed the ^{137}Cs to be tightly held by the vermiculite, and it was concluded by Syers *et al.* that little of the ^{137}Cs entering soils and water supplies in such dusts is likely to be taken up by plants and aquatic life unless the vermiculite suffers partial or complete decomposition. Syers *et al.* (1969) also suggested that scavenging by atmospheric dust particles is likely to reduce the potential health hazard associated with fallout since such particles can be readily washed off the skin or other surfaces on which they are deposited.

7.3.9 Dispersion of toxic metals and other pollutants

Dust released into the atmosphere from industrial plants and mines frequently contains large amounts of heavy metals which may pose serious local toxicity problems. For example, Goldberg (1971) reported that about one-quarter of the world's mercury production of $9 \cdot 2 \times 10^9$ g yr^{-1} escapes to the atmosphere in stack gases emanating from electrolytic cells wheres saline solutions are decomposed to produce chlorine and sodium carbonate. Mining and smelting operations are responsible for large, though poorly quantified, inputs of a wide range of metals into the atmosphere. Significant amounts of atmospheric lead also result from car exhaust emissions. Where local topographic and meteorological circumstances combine to limit dispersion of these inputs, severe local pollution problems result, but since most of the pollutants are of small size, they are normally dispersed over a wide area and precipitated mainly in rainfall. Lead concentrations in the mixed layer of the world's oceans were reported by Goldberg (1971) to have increased from $0 \cdot 01$–$0 \cdot 02$ to $0 \cdot 7$ μg km^{-1} of seawater since the introduction of lead tetraethyl as an anti-knock chemical in petrol. Long-range transport of toxic organic species, including chlorinated hydrocarbons, has also been documented (Risebrough *et al.*, 1968). In some instances where relatively large particulates of toxic material are exposed to wind erosion, as on industrial stock piles and mine tailings, the coarser particles are transported only relatively short distances and severe local pollution problems can result close to the source.

7.4 IMPLICATIONS OF DUST DEPOSITION

Dust deposition has a number of important geomorphological and geological implications, some of which, in turn, have economic and human significance.

7.4.1 Desert varnish formation

Desert varnish is a dark, typically brown or black, manganese and iron-rich coating up to 100 μm thick, which occurs on the upper exposed rock surfaces in arid regions. An orange coating frequently covers buried portions of boulders and cobbles resting on the desert surface. Several different theories of desert varnish formation have been proposed (see Whalley, 1983), but there is evidence that in at least some cases, deposition of windborne dust plays an important role. Potter and Rossman (1977) analysed desert varnish scraped from localities in the Mojave Desert, California, using infrared spectroscopy, X-ray diffraction and electron microscopy. They found that the coats were rich in clay minerals, particularly illite, montmorillonite, or mixed-layer illite–montmorillonite, with small amounts of kaolinite and chlorite. They noted that in some instances, e.g. where varnish was developed on coarse-vein quartz cobbles and boulders, none of the varnish material could be derived from alteration of the rock itself, and suggested that wind transport of clays may be significant in varnishing outcrops where large distances or topography make water transport unlikely. Allen (1978) also concluded that wind-transported dust and clay are major constituents of varnished surfaces in the Sonoran Desert. The same conclusion was reached by Perry and Adams (1978), who noted a lack of correlation between the composition of varnish and underlying substrate in Arizona, Utah and Idaho. Potter and Rossman (1977) and Allen (1978) suggested that dust is captured by rock surfaces wetted by dew or sporadic rainfall. Subsequent wetting and drying, involving partial dissolution and reprecipitation are responsible for the formation of a hard varnish coat.

7.4.2 Case-hardening of rock outcrops

Hardened surface crusts commonly develop during weathering of sedimentary rocks in arid and semi-arid areas. The case-hardened layers, which range from a few millimetres to a few centimetres in thickness, are erosion-resistant relative to the underlying host rock due to secondary cementation and infilling of pore spaces by calcite, clay or other minerals. Some case-hardening results from alternate wetting and drying, which causes soluble constituents to be drawn to the surface by evaporation. In other cases a large proportion of material appears to be derived from sources external to the rock. For example, Conca and Rossman (1982) concluded that case-hardened crusts developed on the Aztec Sandstone in Nevada were cemented by calcite, colemanite and kaolinite of aeolian origin.

7.4.3 Duricrust development

A number of authors have suggested that airborne dust can provide a major source of ions for the development of duricrusts. Dust is often rich in silica, much of it present as highly soluble disordered quartz or biogenic opal forms, and could be important for silcrete formation in arid and semi-arid environments (Summerfield, 1983). Waugh (1970) suggested that fine aeolian abrasion products in dust provided the source of silica for early diagenetic cementation of the Penrith Sandstone (Lower Permian) of northwest England. Whalley *et al.* (1987) demonstrated that high concentrations of dissolved silica (up to 1600 p.p.m.) can be generated when experimentally abraded quartz sand grains are allowed to stand in distilled water. The enhanced solubility of the abraded quartz compared with that of most natural quartz (6 p.p.m.) may have been due to the formation of a disrupted surface lattice layer during the abrasion experiments (cf. Lidstrom, 1968; Iler, 1979).

Calcium carbonate in dust has also been suggested to be responsible for the formation of some calcretes (see Goudie, 1983b). Blumel (1982) described calcretes of aeolian origin in eastern Namibia and reported the presence of aeolian components in calcretes of southeast Spain. Evidence of an aeolian origin was provided by the fact that carbonate crusts occur on siliceous substrates and on hillcrests where groundwater and/or surface-water flow could not have been responsible for introducing the carbonate. Aeolian transport was also regarded by Lattman (1973) to be the main source of calcium carbonate in cemented non-carbonate alluvial fan sediments in Nevada.

Coque (1955) and Mensching (1964) suggested that gypsum crusts developed on fans in southern Tunisia owe their origin to wind transport of gypsiferous dust from nearby lake basins. A similar explanation was suggested by Jessup (1960) to account for the development of gypsum crusts in parts of South Australia. In Namibia, extensive gypsum crusts have formed by evaporative concentration of marine sulphate dissolved in fog (Watson, 1985).

7.4.4 Effects on weathering

Deposition of salts dissolved in precipitation or present as solid particulates in dust can significantly affect weathering rates of both natural outcrops and building stones. As discussed in Chapter 2, salt weathering is the major weathering process on many alluvial fans and rock surfaces in arid regions. Goudie (1977) reported severe salt damage to the ancient city of Mohenjo Daro, Pakistan, partly due to windborne salts deflated from salinized ground.

7.4.5 Sediment diagenesis

7.4.5.1 Alluvial fan sediments

Walker *et al.* (1978) emphasized the significance of infiltrated clay in the early diagenesis of late Cenozoic alluvial fan sediments in Baja California. According to Walker (1976), mechanically infiltrated clay is an almost ubiquitous constituent of coarse alluvium of Quaternary and late Tertiary age throughout the arid regions of North America. While some clay is formed by *in-situ* weathering of silicate minerals, and some detrital fine material is introduced by periodic floods, a significant proportion is of aeolian origin. The clay is infiltrated into the sediment by percolating rain and flood waters, and commonly accumulates on the under surfaces of buried pebbles by gravity settling, producing a geopedal fabric. In older sediments virtually all of the grains are coated by stained clay which also partly fills the pores.

The depth of accumulation of the fines depends on the depth of wetting and age of the sediment, but is generally of the order of 5–50 cm. The infiltrated clay is frequently grey or pale yellowish-brown when deposited, but becomes progressively redder with time due to the formation of authigenic haematite. Some of the oxidized iron is derived from *in-situ* weathering of unstable silicate minerals (Walker *et al.*, 1978), but some appears to be released by weathering of the deposited airborne dust. Bowman (1982) observed ferric enrichment with increasing age in alluvial fan sediments along the western side of the Dead Sea, and attributed it to aeolian deposition. In the same fan sequence, Amit and Gerson (1986) showed that the texture of the A_v and B horizons becomes finer with time, and that in the older fan terraces the clay content reaches 20–25%. These authors also attributed most of the silt and clay to airborne dust deposition.

7.4.5.2 Dune sediments

In sand dune terrain, deposition of dust and airborne salts can form a surface crust which tends to stabilize the sand by raising the threshold velocity required for particle entrainment. Temporary stabilization by crusting may, in turn, allow colonization of the sand surface by lichens, fungi and vascular plants, ultimately leading to complete stabilization of the dunes.

Some of the deposited dust and salts are infiltrated further into the dunes by rain, and may have a significant effect on the early diagenesis of the sands (Pye and Tsoar, 1987). Some stabilized linear dunes in the northern Negev contain >10% silt and clay near the surface, though the amount of fines decreases to 5–6% at 40 cm depth. Vegetated linear dunes in Rice Valley, California, reportedly contain 4–6·6% fines, and in central Australia up to 5% fines (Breed and Breed, 1979).

The depth of dust infiltration is dependent on the amount and frequency of rain, on the grain size of the sand, and on the size of the dust. In Israel, Orev (1984) found that 1 mm of rain penetrated to a depth of 50 mm in dune sand. Dincer *et al.* (1974) reported that 1 mm of rain penetrated to a depth of 7 mm in well-graded fine dune sand (mean size 150 μm) and to 20 mm in medium dune sand (mean size 300 μm). In the northern Negev, where the average annual rainfall is about 100 mm (an additional 30 mm of moisture is provided by condensation in the upper sand layers), rain penetrates the sand dunes to a depth of 1 m, compared with 30 cm in neighbouring loess (Tsoar and Zohar, 1985).

Experiments using sand columns with different particle-size characteristics demonstrated that fine silt and clay particles are more readily translocated through sand columns than medium and coarse silt particles, and that the rate of translocation of fines decreases with the size of sand (Wright and Foss, 1968).

Since different grain-size fractions in dust vary in terms of their mineral composition, mineralogical fractionation is to be expected during infiltration of fines through dune sands. Pye and Tsoar (1987) showed that the coarse silt fraction of infiltrated dust collected at a depth of 10 cm in a stabilised Negev dune was composed largely of quartz, feldspars and calcite with some dolomite, while the finer fractions were progressively enriched in calcite and clay minerals (Fig. 7.9). The mineral composition of the <63 μm fraction as a whole was similar to that of unweathered loess in the northern Negev. Scanning electron microscopy examination confirmed that the calcite grains are on average smaller than quartz and feldspar grains (Fig. 7.10). After deposition on the dune surface, carbonates and clay minerals are translocated further into the sands than the quartz and feldspar.

Deposited airborne salts have different relative solubilities, with the result that chloride is translocated to a greater depth than sulphate, which in turn is translocated further than bicarbonate (Yaalon, 1964b). However, after periods of wetting, soluble salts may be drawn to the surface by evaporation, forming a thin crust of cemented quartz grains (Fig. 7.11).

During drying the salinity and alkalinity of moisture films in the uppermost metre or so of sand can become very high. These conditions favour dissolution and reprecipitation of silica, with the result that the framework sand grains become coated with mixtures of salts, infiltrated clay and amorphous silicate gel (Fig. 7.12). The sands more than 30 cm below the surface generally never dry out entirely, and moisture films containing concentrated salt solutions are in continuous contact with grain surfaces. Under these conditions silica on the surfaces of quartz grains is dissolved and locally reprecipitated when the solutions become supersaturated, first with respect to amorphous silica. In this way 'waxy' or 'scaly' coats, referred to by Folk (1978) as 'turtle-skin silica coats', are formed. Quartz grains in stabilized dune sands, while often rounded at edges and corners, generally do not show fresh surface textural features such

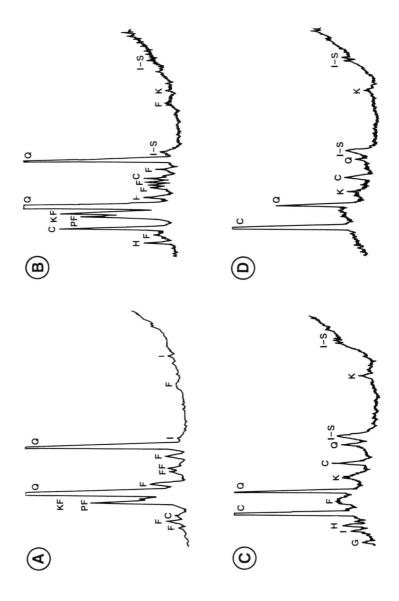

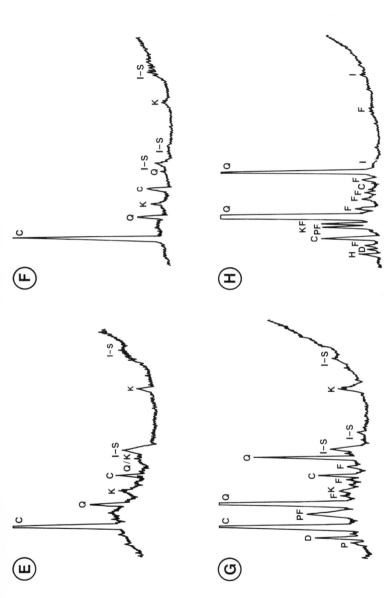

Figure 7.9. X-ray powder diffraction traces (CuK$_\alpha$ radiation) of different-size fractions of dust in the upper 10 cm of a stable linear dune, northern Negev, Israel ((a)–(f)), of a sample of unweathered 'Upper Loess' at Netivot, northern Negev (g), and a bulk sample of cemented dune sand crust, Negev (h): Q = quartz; PF = plagioclase feldspar; KF = K-feldspar; C = calcite; D = dolomite; H = halite; K = kaolinite; I–S = mixed-layer illite smectite plus mica (M), illite (I) and smectite (S). All samples are air-dry. (After Pye and Tsoar (1987).), in I. Reid and L. Frostick (eds.) Desert Sediments Ancient and Modern; © The Geological Society.)

Figure 7.10. SEM micrograph showing the 20–63 μm fraction from a Negev linear dune. Carbonate grains (C) are on average smaller than the quartz and feldspar grains (Q). (Scale bar = 10 μm.)

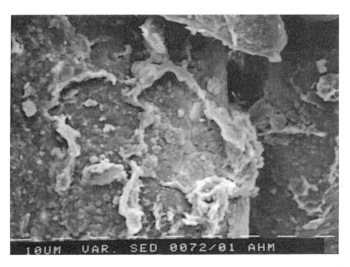

Figure 7.11. SEM micrograph showing quartz sand grains cemented by halite and amorphous aluminosilicate material in the surface crust of a linear dune, northern Negev. (Scale bar = 10 μm.)

Figure 7.12. Coating composed mainly of amorphous silica on the surface of a quartz sand grain from a dune surface crust, northern Negev. (Scale bar = 10 μm.)

as 'frosting' (Kuenen and Perdok, 1962) or 'upturned plates' indicative of aeolian abrasion (cf. Krinsley and McCoy, 1978).

Energy-dispersive X-ray microanalysis (EDXRA) showed that the siliceous coatings on Negev dune quartz grains also contain varying amounts of aluminium, iron, potassium and calcium (Pye and Tsoar, 1987). Generally, the redder the grains the higher the content of iron and aluminium in the coatings. A similar relationship was observed by Walker (1979) in Libyan dune sands. X-ray diffraction, transmission electron microscopy and Mossbauer studies have shown that the iron is present as very finely divided haematite, or mixtures of haematite and a poorly crystalline hydrated ferric oxide precursor. Although the reddening of dune sands has been suggested by some authors to be time-dependent, the rate of haematite formation is affected by many mineralogical and environmental factors, such that degree of reddening is often a poor guide to dune age (Gardner and Pye, 1981; Pye, 1983c).

7.4.6 Ice ablation due to windblown dust

Deposited dust reduces the albedo of snow and ice, resulting in warming and accelerated melting of the surface during summer (Davitaya, 1969). Hattersley-Smith (1961) reported that the combined effect of a warm spring and a dust layer deposited in May 1960 caused ablation of ice to begin two weeks earlier than usual on the Gilman Glacier, Ellesmere Island. Some authors have suggested that extensive dust blankets resulting from major volcanic eruptions might alter the albedo of the global snow cover sufficiently to set in train a sequence of events leading to large-scale melting of the polar ice caps, but there is no real evidence to support the hypothesis.

7.4.7 Aeolian contributions to sabkhas, lakes and fluvial sediments

Ali and West (1983) reported that windborne dust forms a significant proportion of the fine grained sediment found in saline depressions between lithified aeolianite ridges in northern Egypt. The material was found to be mineralogically similar to recent dust deposits in this area and in southern Israel, consisting mainly of quartz, feldspars, calcite, dolomite, and clay minerals. The presence of weak lamination and carbonate stringers derived from the aeolianite ridges indicated some syn- or post-depositional reworking of the deposited dust by running water.

Dust eroded from interfluves by surface wash accumulates in closed depressions, valley bottoms and lakes where it may become mixed with other siliclastic or biogenic sediment. Yaalon and Dan (1974) pointed out that in extremely arid parts of the Negev an absence of protective vegetation cover prevents accumulation of dust on interfluves. The dust is reworked by surface wash during rainstorms and deposited in small alluvial fans, wadi beds and playas (Issar *et al.*, 1984; Bowman *et al.*, 1986). Localized 'pocket' deposits are preserved on some jointed or karstified bedrock interfluves. In slightly less arid parts of the Negev where the vegetation cover is thicker, deposited dust is retained as a continuous cover on the interfluves. Loess blankets the whole landscape but is often thicker on north-facing slopes which experience cooler temperatures and retain more moisture, thereby supporting denser vegetation growth. In areas of dissected terrain the loess thickness generally decreases downslope, indicating that reworking by slopewash is important, but in plateau areas dust has accumulated without much reworking (Dan *et al.*, 1973). Large amounts of fluvially-reworked loess are found in the lower reaches of many wadis in the northern Negev (Gardner, 1977; Sneh, 1983).

In the Belarabon area of New South Wales, valley fills developed on siliceous bedrock contain carbonate and allochthonous clay minerals which are thought to represent reworked aeolian dust (Wasson, 1982*b*). Similar *in-situ* fine-grained calcareous material, ('parna'), occurs widely on interfluves in southeastern Australia (Butler, 1956, 1982; Butler and Hutton, 1956; Blackburn, 1981; Dare-Edwards, 1984).

According to Macleod (1980), the alluvial Red Beds which fill valleys in Epirus, Greece, have a particle-size distribution and mineralogy which indicate they have been derived by erosion of 'terra rossa' soils which, as discussed below, contain much aeolian material. Many of the valley fills in the Mediterranean region described by Vita-Finzi (1969) may have such an origin.

The possible importance of fluvially-reworked windblown dust in contributing to ancient continental sediments has recently been recognized. For example, Holm *et al.* (1986) suggested that the dolomitic mudstones and siltstones of the Mercia Mudstone Group (Upper Triassic) in southwest England contain substantial amounts of locally reworked aeolian material.

Aeolian components are important in the sediments of lakes located close to major dust sources, e.g. ice-marginal lakes downwind of braided outwash plains. Water bodies have 100% capture efficiency for particles settling on their surfaces. Dust deposited on the surrounding hillsides may also accumulate in lake basins after being reworked by fluvial action. Allen (1986) reported that some late Quaternary lake sediments in Greece contained large amounts of siliclastic fine silt and clay of probable aeolian origin. The exact source of the dust is uncertain, although the mineral composition is consistent with a North African origin (cf. Chester *et al.*, 1977; Macleod, 1980).

Bowler (1978) reported the occurrence of red, clay-coated fine sand and silt-size quartz grains (wüstenquartz) in late Pleistocene lake sediments downwind of formerly active dunefields in southeastern Australia, and concluded that the finer grains were winnowed from the dunes and deposited in the lakes during a period of increased late-glacial aridity.

7.4.7 Dust additions to soils

There is now strong evidence that many soils contain significant amounts of airborne dust, and that some soils are composed almost entirely of such material. Singer (1967) reported that basalt-derived soils in Israel contain 10–50% quartz in the silt fraction, with a modal size of 17–51 μm. Rex *et al.* (1969) and Jackson *et al.* (1973) reported up to 45% quartz in soils overlying quartz-free volcanic rocks in Hawaii, with 70% quartz in the 2–10 μm fraction. Large amounts of silt-size quartz and mica also occur in basalt depression fills in northeast Sardinia (Sevink and Kummer, 1984). Although some quartz may form authigenically in soils developed on mafic rocks (Eswaran, 1972; Robinson, 1980), chemical mass balance calculations indicate that not all of the quartz can have such an origin. The presence of large amounts of potassic mica in basaltic and andesitic soils also strongly suggests an allochthonous origin, since such micas are absent in the underlying rocks and true pedogenic mica is very rare. Dymond *et al.* (1974) demonstrated that micas in Hawaiian soils yield K/Ar ages which are approximately 100 million years greater than the age of volcanism on the island. The $^{87}Sr/^{86}Sr$ ratios of the mica-bearing soils were also found to be much greater than would be expected if they were derived only by weathering of Hawaiian lavas.

Dan and Yaalon (1966) concluded that the red, clay-rich 'hamra' soils developed on sandy parent materials in the coastal plain of Israel owe their characteristics primarily to addition of fine-grained aeolian dust. The parent sands were shown to contain few feldspars or other weatherable minerals which could alter to produce sufficient clay. The main clay species in these soils, montmorillonite and kaolinite, are also the dominant clay mineral constituents of modern dust in Israel. It was concluded that the carbonate originally present in the deposited dust had been leached during formation of the red hamra soils.

Syers *et al.* (1969) and Bricker and Mackenzie (1971) suggested that red soils overlying coastal carbonates in Bermuda and the Bahamas represent weathered accumulations of aeolian dust. A similar origin has been proposed for terra rossa soils developed on carbonate rocks in the Mediterranean region and elsewhere. Yaalon and Ganor (1973) found that in Israel >90% of the acid-insoluble residue of various limestones and dolomites is composed of clay-size material, whereas the overlying soils contain 40–50% silt. Macleod (1980) also reported a high silt/clay ratio in terra rossa soils developed on pure limestones in Epirus, Greece. Since the average acid-insoluble residue content of the limestones was found to be 0.15%, it would require dissolution of 130 m of limestone to produce 40 cm of soil (the average thickness on plateau areas in Epirus). Macleod also demonstrated that the insoluble residue of the limestones contains coarse and fine sand in addition to silt and clay, but the terra rossa soils contain only 2% material $>63 \mu$m. The $<2 \mu$m fractions of the soils were found to be composed mainly of kaolinite with small amounts of mica, vermiculite, quartz and iron oxide. North African dust transported northwards across the Mediterranean by sirocco winds is also rich in kaolinite and mica (Chester *et al.*, 1977). The vermiculite in the Epirus soils may be a post-depositional alteration product of illite, chlorite and montmorillonite, while some of the kaolinite may also have formed *in-situ* at the expense of 2:1 layer silicate minerals (Macleod, 1980).

Rapp (1984) concluded that terra rossa in Spain, Italy and other parts of the Mediterranean is composed mainly of weathered aeolian dust derived from North Africa. This conclusion was affirmed by Rapp and Nihlén (1986) and Nihlén and Solyom (1986), who also reported African dust trapped in high-altitude snow patches on Crete and the Peloponnese.

Danin *et al.* (1983) showed that dolomitic rocks near Jerusalem contain 1% insoluble residue, and calculated that solution of 20 m of dolomite would be required to form a terra rossa soil 40 cm deep. They found the imprints of endolithic lichens on slightly weathered dolomite rock surfaces beneath up to 65 cm of soil, supporting the hypothesis that terra rossa in this area has formed largely by aeolian accumulation. Extrapolating from the present dust deposition rate at Jerusalem (36μm yr^{-1}; Yaalon and Ganor, 1975), they calculated that the soils could have formed within about 10 000 years.

Chapter Eight

DUST DEPOSITION IN THE OCEANS

Ocean sediments contain three main components: (1) detrital terrigenous sediment grains; (2) marine biogenic material; and (3) authigenic minerals formed at or below the sea-bed. Detrital terrigenous material is supplied to the oceans by ice-rafting, fluvial input, coastal erosion, and wind transport. Direct riverine supply rarely extends more than 100–200 km offshore, and terrigenous sediment is supplied to the deep ocean basins mainly by turbidity currents, ice-rafting and wind transport.

The mineralogy and grain-size distributions of late Cenozoic ocean sediments have been studied extensively in the past two decades, and this work has shown conclusively that in some areas aeolian dust comprises up to 80% of the total accumulated sediment. Aeolian dust input has been most important in the eastern Atlantic and northwest Pacific, but is also significant in the northwest Indian Ocean, southwest Pacific and eastern equatorial Pacific.

Dust accumulations in ocean sediments have also provided much useful information about changes in global wind systems and environmental changes on the continents during the Quaternary. The oceanic sedimentary record has the advantage of being relatively complete compared with the continental record, and relatively good stratigraphic and age control is provided by the biogenic components in ocean sediments, so it is often possible to correlate fluctuations in terrigenous sediment input over quite large areas. It has also proved possible to relate variations in continental dust input to changes in ice volume as indicated by oxygen isotope data obtained from foraminifera in deep-sea cores (e.g. Parkin and Shackleton, 1973). However, the story of late Cenozoic dust supply to the oceans is far from complete. The environmental changes which influenced the dust flux from West Africa to the eastern Atlantic during the late Quaternary have been extensively investigated, but even in this area details of the picture remain unclear. Less is known about late Quaternary environmental changes in Asia and the associated effects on dust supply to the

171

northwest Pacific, while knowledge about the Quaternary history of dust deposition in other areas is still fragmentary.

8.1 THE EFFECT OF DUST DEPOSITION ON THE COMPOSITION OF MODERN OCEAN SEDIMENTS

Several authors have demonstrated mineralogical and grain-size similarities between Saharan dust and North Atlantic sediments (Delany *et al.*, 1967; Beltagy *et al.*, 1972; Windom, 1975; Johnson, 1979). Maps of ocean bottom sediment composition prepared by Biscaye (1965), Griffin *et al.* (1968) and Windom (1975) all show a zone of kaolinite and quartz-rich sediments in the eastern equatorial Atlantic (Figs 8.1 and 8.2). Much of the kaolinite in the eastern equatorial Atlantic is supplied by rivers from the wet tropical parts of Africa, but the 'tongue' rich in kaolinite and iron-stained quartz which extends southwest from the coast of Sierra Leone, Liberia and the Ivory Coast coincides with the trajectory of 'winter' dust plume originating in the southern

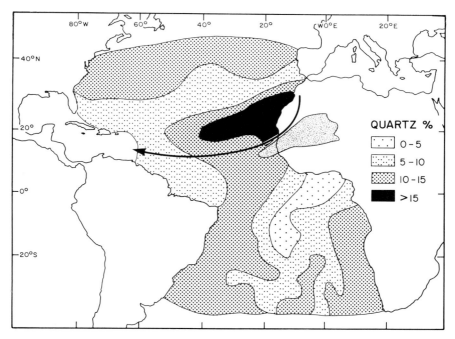

Figure 8.1. Quartz distribution of equatorial Atlantic sediments. Arrow and shaded area indicate the trajectory and dust veil of the dust storm studied by Prospero et al. (1970). (After Windom (1975), J.Sedim. Petrol. **45**, *520–529;* © *Society of Economic Paleontologists and Mineralogists.)*

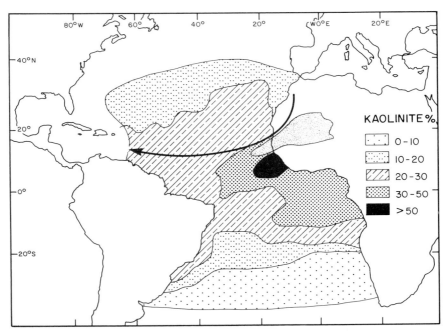

Figure 8.2. Kaolinite distribution in the <2 μm-size fraction of equatorial Atlantic sediments. Arrow and shaded area indicate the trajectory and the dust veil studied by Prospero et al. (1970). (After Windom (1975), J. Sedim. Petrol 45, 520–529; © Society of Economic Paleontologists and Mineralogists.)

Sahara and Sahel regions. A second 'tongue' of quartz and illite-rich sediments extends west from the arid coasts of Senegal, Mauritania and Spanish Sahara. The sediments of this second 'tongue' are also relatively rich in plagioclase feldspar and dolomite (Fig. 8.3). Modern dust outbreaks passing over the Spanish Sahara in summer also contain significant amounts of dolomite (Johnson, 1979). Eastern Atlantic sediments north of 20° N are particularly rich in unstained quartz, plagioclase, illite and chlorite derived mainly from basement rocks in the southern Atlas Mountains (Diester-Haas, 1976; Lange, 1982). Pollen, fungal spores, opal phytoliths and freshwater diatoms are important constituents of ocean bottom sediments all along the West African continental margin, with the maximum frequencies observed between 8 and 20° N (Fig. 8.4).

Ruddiman and McIntyre (1976) evaluated the relative importance of different potential sources of terrigenous detritus in the North Atlantic north of latitude 45° and concluded that, although ice-rafting has been most important throughout the Quaternary, there is probably a subsidiary aeolian component, possibly derived from the loess areas of North America. This is consistent with

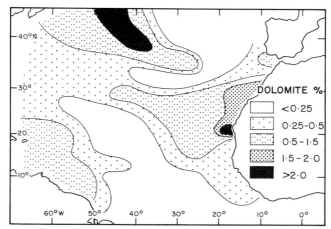

Figure 8.3. The distribution of dolomite in surface sediments of the equatorial North Atlantic. (After Johnson (1979), Mar. Geol. **21**, M17–M21; © Elsevier Publishing Co.)

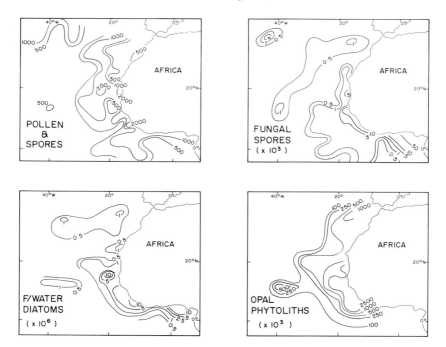

Figure 8.4. Distribution of pollen and spores, fungal spores, freshwater diatoms and opal phytoliths in surface sediments of the eastern equatorial Atlantic. (After Melia (1984), Mar. Geol. **58**, 345–371; © Elsevier Publishing Co.)

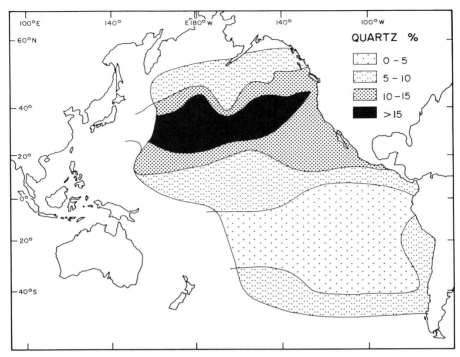

Figure 8.5. Quartz distribution in Pacific Ocean sediments. (After Windom (1975), J. Sedim. Petrol. **45**, 520–529; © Society of Economic Paleontologists and Mineralogists.)

observations that dust plumes from North America periodically reach the mid North Atlantic Ocean (Folger, 1970; Windom and Chamberlain, 1978). Robinson (1986) presented evidence that African dust also sometimes reaches the central North Atlantic.

In the North Pacific, Rex and Golberg (1958) were the first to recognize a zone of relatively quartz-rich bottom sediments extending from southern Japan, across the Hawaiian Islands, and almost reaching the west coast of North America (Fig. 8.5). Illite distribution shows a similar pattern (Griffin *et al.*, 1968). This zone is coincident with the trajectories of Asian dust plumes recorded by satellite tracking (Chung, 1986), surface water sampling (Uematsu *et al.*, 1985) and shipboard or airborne dust monitoring (Ferguson *et al.*, 1970; Duce *et al.*, 1980; Uematsu *et al.*, 1983; Parrington *et al.*, 1983). The source areas of dust in North Pacific sediments are the deserts of Asia and the loess regions of northern China. Blank *et al.* (1985) compared the mineralogy of modern dusts with that of surface marine sediments in the western North Pacific and found a very close similarity with the exception of one mineral,

Table 8.1
Comparison of mean compositions of < 2 μm fraction in dust and ocean bottom
sediments from the North Pacific Ocean.

		Quartz	Plagioclase	Smectite	Illite	Kaolinite	Chlorite	Illite/quartz	Kaolinite/chlorite	Illite/kaolinite
Aerosols (N = 6)	$\bar{x}$	10·5	11·2	1·1	39·5	15·5	2·7	3·81	5·81	2·66
	Sx	0·49	0·86	0·26	1·92	1·16	0·15	0·29	0·38	0·29
Sediments (N = 12)	$\bar{x}$	9·9	9·7	3·1	38·7	16·4	3·0	3·90	5·72	2·61
	Sx	0·27	0·41	0·44	0·89	1·18	0·15	0·15	0·49	0·24

(After Blank et al. (1985), Nature 314; 84–86; © Macmillan Journals Ltd.)

smectite, which was found to be more abundant in the ocean sediments than in the sampled dust (Table 8.1). These authors noted that smectite in the dust may have been underestimated by the mesh sampling technique used, but it is possible that the additional smectite observed in the ocean bottom sediments owes its origin to alteration of volcanic rocks within the Pacific Basin. According to Windom (1969) and Leinen and Heath (1981), as much as 75–95% of the surface sediment in this part of the Pacific is probably derived from atmospheric dust fallout.

The South Pacific Ocean receives little runoff from the surrounding continents, and consequently the sediments are dominated by volcanic material derived from within the ocean basin (Griffin et al., 1968). However, a 'tongue' of bottom sediments rich in quartz, illite and kaolinite extends from southeast Australia towards New Zealand (Fig. 8.6; Griffin et al., 1968; Glasby, 1971; Thiede, 1979). This zone also corresponds with the trajectories of modern dust plumes which have been observed to travel from Australia towards New Zealand and beyond (Kidson, 1930; Gabites, 1954; Healy, 1970). Glasby (1971) suggested that topographic features like the Dampier Ridge serve as barriers to the transport of terrigenous material by turbidity currents from the Australian continental shelf, and that most of the non-biogenic pelagic sediment on the Lord Howe Rise, which has accumulated at a rate of $1-2\,\mu\text{m yr}^{-1}$, is probably of aeolian origin. The main source of dust is the arid interior of Australia.

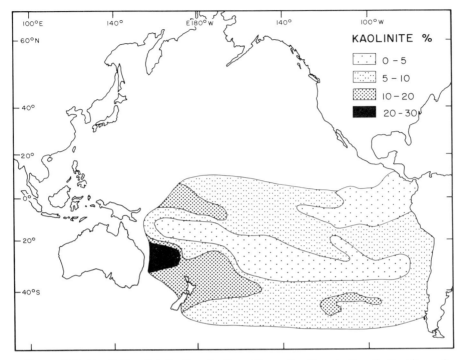

*Figure 8.6. Kaolinite distribution in the <2 μm fraction in South Pacific sediments. (After Windom (1975), J. Sedim. Petrol. **45**, 520–529; © Society of Economic Paleontologists and Mineralogists.)*

In the equatorial eastern Pacific, a 'tongue' of quartz-rich sediments extends northwestwards from the coast of Peru and Colombia (Fig. 8.7; Molina-Cruz, 1977; Scheidegger and Krissek, 1982). While not all of the quartz is necessarily aeolian, the quartz concentrations and other mineralogical characteristics of the bottom sediments are similar to those of airborne dusts collected in this area which were probably derived partly from the deserts of Peru and northern Chile (Prospero and Bonatti, 1969).

In the Indian Ocean, Griffin *et al.* (1968) and Kolla and Biscaye (1973) identified a 'tongue' of kaolinite-rich sediments extending northwestwards from the coast of Western Australia into the southeast Indian Ocean (Fig. 8.8). This zone coincides with the trajectory of easterly winds which blow offshore from the arid regions of Western Australia (Brookfield, 1970).

In the northern Indian Ocean, Goldberg and Griffin (1970) concluded that aeolian dust derived from the southern Asian landmasses forms a significant component in bottom sediments of the Bay of Bengal, but in this area the

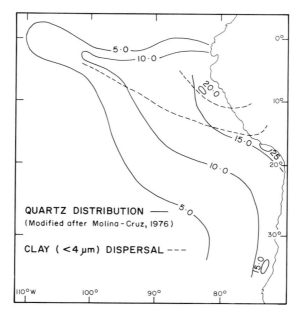

Figure 8.7. Occurrence of quartz-rich sediment (calculated on an opal- and carbonate-free basis), in the eastern equatorial Pacific. The quartz-rich lobe is associated with quartz-rich sediment sources in central and northern Peru. (© After Scheidegger and Krissek (1982), Geol. Soc. Am. Bull. **93**, 150–162.)

aeolian contribution appears to be masked by the high sediment load supplied by the Ganges and other rivers (Kolla and Biscaye, 1973, 1977).

Kukal and Saadallah (1973) concluded that aeolian admixtures are important in the sediments of the Persian Gulf on the basis of three lines of evidence cited as: (1) a high rate of deposition in recent dust storms; (2) a similar grain-size/carbonate content relationship between the marine sediments and modern dust storm sediments; and (3) 'pseudo sand' grains composed of aggregated material are present both in the marine sediments and modern dust.

8.2 OCEANIC DUST DEPOSITION DURING THE LATE CENOZOIC

8.2.1 The eastern Atlantic

A considerable number of workers have examined variations in the abundance of aeolian components in eastern Atlantic sediments with a view to obtaining

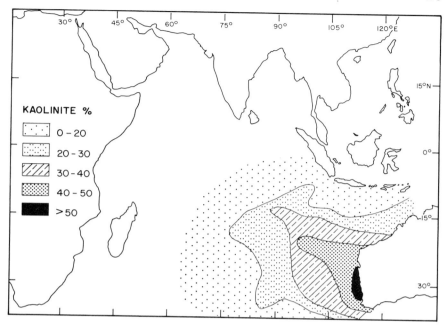

*Figure 8.8. Kaolinite distribution in the <2 μm-size fraction of sediments in the southeast Indian ocean. (After Windom (1975), J. Sedim. Petrol. **45**, 420–429; © Society of Economic Paleontologists and Mineralogists.)*

evidence of Quaternary climate and vegetation changes in neighbouring parts of Africa. There is general agreement that the dust flux to the eastern Atlantic increased during cold periods and decreased during warm periods, but interpretations differ regarding the nature of the environmental changes in northwest Africa which were responsible for these fluctuations. In part this is due to the difficulty in separating the effects of changes in dust supply, which reflect changes in rainfall, vegetation cover, runoff processes and lake levels, from the effects of changes in the position and strength of the prevailing wind systems.

Hays and Peruzza (1972), working on sediments from two cores near the Saharan coast west of Dakar, noted an increase in terrigenous material relative to biogenic calcium carbonate content during glacial periods and concluded that trade wind velocities may have been higher at such times. Parkin and Shackleton (1973) examined the dust content and oxygen isotope signatures of marine foraminifera in a core taken just north of the Cape Verde Islands (21° 18′ N, 22° 41′ W) which contained an essentially complete record extending back to the Brunhes–Matuyama magnetic reversal (0·7 million years ago). Using a winnowing model described more fully by Parkin (1974), it was concluded that variations in the size of aeolian quartz found at different depths in the core provide an indication of changes in the average velocity of the dust-transporting winds. Parkin and Shackleton found that the size of dust increased during periods of increased ice volume (i.e. cold stages) indicated by the oxygen isotope data (Fig. 8.9). On this basis they concluded that stronger trade winds were a feature of glacial stages, probably due to a steeper temperature gradient between the equator and the poles. Subsequent grain-size and mineralogical analysis of a core from south of the Cape Verde Islands (8° 18' N, 22° 45' W) led Parkin and Padgham (1975) to conclude that, although during glacial stages the trade winds north of Cape Verde were more vigorous than present, the wintertime Harmattan was weaker than present in the region south of Cape Verde. They also concluded that the land north of Dakar remained desert, and was especially arid during glacials, while to the south of Dakar conditions oscillated between desert during interglacials and savanna during glacials.

Different conclusions were reached by Parmenter and Folger (1974) who examined changes in the abundance of terrigenous biogenic material (opal phytoliths and freshwater diatoms) over the last 1·8 million years in two cores taken to the north and one to the south of the Cape Verde Islands (located respectively at 19° 41' N, 24° 35' W, 21° 4' N, 28° 2' W, and 8° 38' N, 22° 2' W). They found that, in the last 200 000–300 000 years, fewer land-derived biogenic particles reached the northern area than reached the southern area. In the southern core, maximum concentrations of biogenic dust components coincided with, or slightly preceded, intervals with a higher percentage of cold marine foraminifera species. Minimum frequencies of biogenic dust particles coincided with high percentages of excess warm marine foraminifera species. The mineral grain content ($>10\,\mu$m) was found to be approximately twice as high during cold periods as during warm periods. These authors concluded that soils in the dust source area for the southern core were more susceptible to deflation during glacials than during interglacials. They noted that, whereas mineral grain abundance increased by a factor of two between interglacial and glacial stages, the phytolith and diatom concentrations increased by a factor of 50, suggesting that increased aridity played a major role in accelerating the rate of dust supply from the source area.

Bowles (1975) observed from cores located at 8–9° N that during the last 600 000 years a relatively constant background of clay mineral deposition has existed on which are superimposed large oscillations in quartz input relative to illite. He concluded that the quartz/illite variations reflect past changes in the dust-transporting capacity of the trade winds, and that high inputs of quartz corresponded with cold periods. However, Bowles also recognized that the relationship between quartz/illite variations and wind intensity is complicated

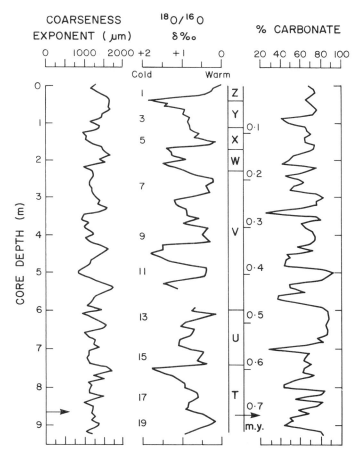

Figure 8.9. The coarseness exponent compared with the $^{18}O/^{16}O$ deviations from present-day values and the total carbonate percentage, for core V23–100. The Z, Y, X . . . column is the Globorotalia menardii zonation. The 1, 3, 5 . . . column numbers only the warm periods in the climatic stages. The ages (in million years) are linearly interpolated between the top of the core and the magnetic reversal at the bottom. (After Parkin and Shackleton (1973), Nature **245**, *455–457;* © *Macmillan Journals Ltd.)*

by the fact that dust emission is also influenced by climatic factors other than wind, notably continental moisture budget.

Diester-Haas (1976) examined the carbonate content, quartz content, and quartz/illite ratios in sediments from a number of cores in the eastern Atlantic and found a significant difference between areas north and south of latitude 20° N. In the northern region, high frequencies of unstained desert quartz grains derived from northwest Africa were found in Holocene sediments, low frequencies in sediments of last glacial age, and high frequencies in sediments which accumulated during the last interglacial. In the southern region frequencies of red desert quartz grains derived from the southern Sahara and Sahel generally showed an opposite relationship with the highest frequencies being present in sediments of last glacial age. Diester-Haas interpreted this as indicating a humid last glacial climate in northwest Africa north of latitude 20° N, possibly due to a southward displacement of the zone of Mediterranean climate, while the southern Sahara must have been drier at this time due to a southwards shift of the sub-tropical anticyclone and Northeast Trade Wind belt.

Kolla *et al.* (1979) confirmed that the zone of quartz-rich sediments in the equatorial eastern Atlantic was more extensive at the time of the last glacial maximum (18 000 yr BP) than at present. Within this zone, which extended to 5° N and 45° W, two quartz maxima were identified adjacent to the African coast at 18 000 BP, one centred at about 23° N and the second centred at 15° N. This second maximum was the more intense of the two at 18 000 BP, and represents a southward shift of 8° compared with the Holocene maximum. Kolla *et al.* suggested that the southward shift of the zone of maximum quartz dust sedimentation at the glacial maximum corresponded with a southward shift in the southern margin of the active desert dune belt (Fig. 8.10). Both dune activation and dust deflation may have been favoured by increased vigour of the trade winds.

Sarthein and Koopman (1980) demonstrated that, whereas at present Saharan dust is deposited mainly in a 'hook'-shaped zone projecting from the coast at latitude 18° N and extending north to 23° N (Fig. 8.11a), during the last glacial maximum large amounts of silt-size dust were blown into the sea all along the coast between 10° and 25° N (Fig. 8.11b). However, they concluded that the centre of dust deposition remained located between 15–20° N at the glacial maximum, when extensive active dunefields occupied adjacent parts of Mauritania, northern Senegal and Mali (Sarnthein, 1978; Sarnthein and Diester-Haas, 1977). During the middle Holocene (6000 yr BP), which is known from oxygen isotope and other evidence to have been warmer and more humid than present in tropical Africa and the equatorial Atlantic (Sarnthein, 1978; Nicholson and Flohn, 1980), dust sedimentation in the eastern Atlantic was much reduced at all latitudes compared with the present, but still remained centred at about 18° N (Sarnthein *et al.*, 1981). Sarnthein and

---

Figure 8.10. Distribution of weight per cent quartz (carbonate-free) in Atlantic sediments of the last glacial (18 000 yr BP) times. (After Kolla et al. (1979), Quat. Res. **47**, *642–649; © Academic Press.) (Desert sand dune areas (black) on the African continent are after Sarnthein and Diester-Haas (1977).)*

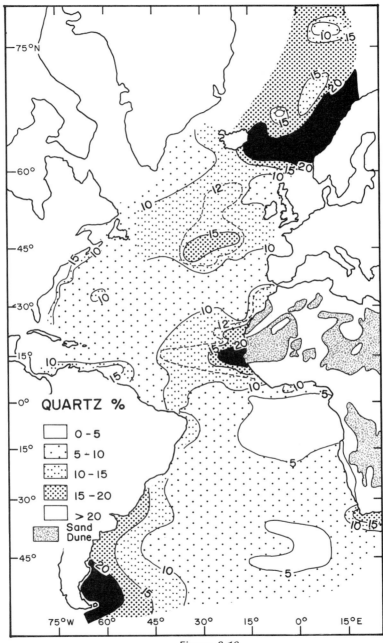

Figure 8.10.

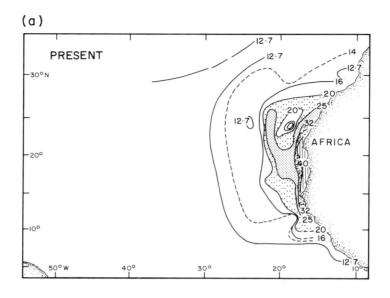

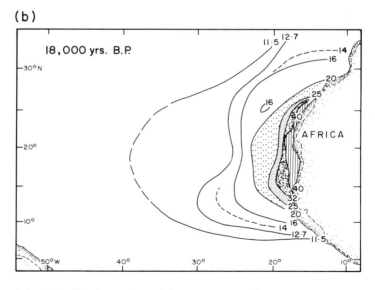

Figure 8.11. Distribution of modal grain sizes of terrigenous silt (>6 μm, carbonate and opal-free) in: (a) surface sediments of the southeast North Atlantic; and (b) sediments of the Last Glacial (18 000 yr BP). (Modified after Sarnthein and Koopman (1980), Paleoecol. Africa and the Surrounding Islands *12*, 239–253; © A. A. Balkema.)

Koopman (1980) therefore concluded that : 'the Glacial sediment distribution patterns clearly reveal that the Saharan dust outbreaks to the Atlantic Ocean did not undergo noteworthy latitudinal shifts in position from Interglacial to Glacial conditions, despite varying climatic zonations on land' (1980, pp. 246–247). Fossil dune evidence suggests that during the last glacial maximum the Sahara desert extended its margins both north and south (Sarnthein, 1978). Increased aeolian deposition of pale quartz, illite and chlorite in the sea north of 20° N indicated that the Northeast Trades were more vigorous at this time, but the oceanic dust records indicate that zonal dust-transporting wind speeds over the eastern equatorial Atlantic were reduced by more than a third (Sarnthein *et al.*, 1981).

Pokras and Mix (1985) examined phytolith and freshwater diatom (*Melosira*) abundances in cores from the eastern and equatorial Atlantic. They found that abundances of *Melosira* north of 10° N at 18 000 BP were essentially the same as at present, implying that transport in the summer dust plume was similar to modern levels, but in the Gulf of Guinea diatom abundances were much lower, suggesting that the winter plume was less extensive. The highest abundance of phytoliths in both modern and last glacial sediments was found centred at approximately 10° N. This corresponds with the present-day belt of tall-grass savanna in northwest Africa. However, in the Gulf of Guinea, phytolith abundances at 18 000 BP were greater than at present, suggesting a southward expansion of the grass belt during glacial time.

Examining the record for the last 150 000 years, Pokras and Mix showed that in the northern area freshwater diatoms were most abundant in deposited dust during glacials and stadials (oxygen isotope stages and sub-stages 2, 4, 5b, 5d and 6) and the upper Holocene (stage 1). The upper Holocene is anomalous in that it was found to contain the only interglacial or interstadial diatom maximum during the last 150 000 years (Fig. 8.12). Phytoliths were found to be most abundant at times of relatively low ice volume in the middle Holocene, (stage 3, and sub-stages 5a and 5c).

In the southern area, maximum diatom abundances occurred just after ice volume minima (sub-stages 5e, 5c, and 5a) and in the late Holocene (Fig. 8.13). Phytolith abundances were high during stage 2, lower stage 3 and/or upper stage 4, and sub-stage 5b, but low in the mid Holocene and sub-stage 5e interglacial periods.

Pokras and Mix concluded from the inverse correlation of phytolith abundances in the northern and southern cores that both margins of the savanna grass belt probably moved southward in relatively arid glacial times and northward in relatively humid interglacial times. They accepted that the changing pattern of phytolith abundance might also be partly explained by a change in relative wind strengths; winds may have been more northerly (intensified meridional flow) during glacials, and more easterly (stronger zonal

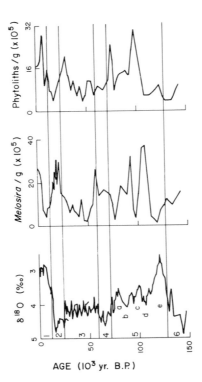

Figure 8.12. Variation in: (a) $\delta^{18}O$, based on Cibicides wuellerstorfii; (b) the abundance of freshwater diatoms (Melosira); and (c) opal phytoliths, in core V30–49, off the coast of northwest Africa (the northern area investigated by Pokras and Mix (1985)). Melosira maxima indicate arid times at glacial maxima, when lakes dried-up and were deflated. Abundant Melosira in the late Holocene are unique in the last 150 000 years, reflecting anomalous aridity for this area in full interglacial conditions. Phytolith maxima indicate northward movement of the grass belt during interglacial and interstadial maxima, with the possible exception of sub-stage 5e. (After Pokras and Mix (1985), Quat. Res **24**, 137–149; © Academic Press.)

Figure 8.13. Variation in: (a) $\delta^{18}O$, based on Globigerinoides sacculifer (with sac, 415–500 μm), core V30–40; (b) Melosira, core V30–40; (c) Melosira, core RC13–205; (d) Melosira, core V27–248; and (e) opal phytoliths, core V27–248, in the eastern equatorial Atlantic (the southern area investigated by Pokras and Mix (1985)). Melosira maxima in this area occurred during ice-growth phases at the stage 5e/5d, 5c/5b and 5a/4 boundaries, and in the late Holocene. Phytolith maxima in this area indicate glacial aridity and movement of the grass belt southward, replacing woodlands in tropical West Africa.

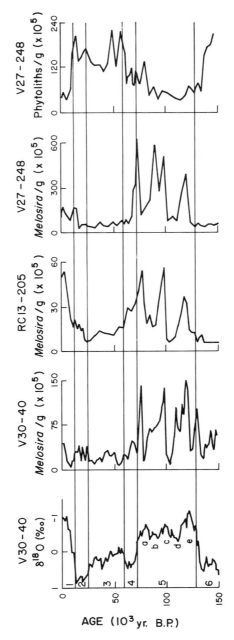

Figure 8.13.

flow) during interglacial times. The high abundances of freshwater diatoms in the northern area during glacial times is consistent with increased aridity indicated by desiccation of lakes bordering the western Sahara between 24 000 and 12 000 yr BP (Maley, 1977; Street and Grove, 1979; Servant and Servant-Vildary, 1980; Nicholson and Flohn, 1980; Street-Perrott and Harrison, 1984) and with evidence of increased input of mineral dust into the eastern Atlantic (Kolla *et al.*, 1979; Sarnthein *et al.*, 1981).

The mechanisms responsible for late glacial aridity in the southern Sahara and Sahel are still disputed. The model of stable latitudinal wind belts proposed by Sarnthein *et al.* (1981) implies a long-term average constant summer position of the Inter Tropical Convergence Zone (ITCZ) and the adjacent upper air high-pressure systems over the western Sahara. However, other authors have concluded that the palaeoenvironmental evidence provided by lake sediments, fluvial sediments and palaeosols on land in northwest Africa is best explained in terms of a 5–8° shift in the position of the ITCZ in response to expansion of the polar ice caps and differential temperature changes between the northern and southern hemispheres (Rognon and Williams, 1977; Nicholson and Flohn, 1980; Nicholson, 1982; Newell and Kidson, 1979; Pokras and Mix, 1985). Yet other authors have suggested that glacial aridity could be a relatively local response to colder ocean temperatures off West Africa, related to greater coastal upwelling (Thiede, 1977).

In the southern cores studied by Pokras and Mix (1985), higher abundances of freshwater diatoms indicate greater aridity in the source area of the winter dust plume during ice-growth phases after interglacial/interstadial maxima, while low diatom abundances indicate relatively humid conditions during periods of deglaciation. The suggestion of relatively wet conditions during the last glacial/early Holocene transition is consistent with high outflow indicated by sediments of the Niger River (Pastouret *et al.*, 1978) and with high lake levels near the equator between 12 500 and 7500 yr BP (lakes further north only attained high levels after 10 000 yr BP), (Talbot and Hall, 1981; Street-Perrott and Roberts, 1983).

Pokras and Mix (1985) noted that the abundance of diatoms in their southern cores showed a close inverse relationship with summer insolation maxima and minima in response to precessional amplitude variations (Berger, 1978), and suggested that there might be a causal connection. Pokras and Mix further pointed out that the late Holocene is the only time in the last 150 000 years when arid conditions have prevailed simultaneously in both tropical northwest Africa and central equatorial Africa during interglacial times. Sarnthein and Koopman (1980) also noted a general increase in late Holocene dust deposition in east Atlantic sediments. The reasons for this are not certain. Decreasing summer insolation over the last 6000 years may have decreased the strength of the African monsoon, reducing rainfall in equatorial Africa (Kutzbach, 1981), but this cannot explain greater aridity in northwest tropical Africa.

A major uncertainty concerns the magnitude of human impact on the environment and climate in West Africa (Charney *et al.*, 1975), which may have become significant as much as 5000–8000 years ago (Nicholson and Flohn, 1980).

Aeolian dust has also been identified as a significant component in pre Quaternary sediments of the North Atlantic Ocean. Lever and McCave (1983) analysed the clay mineral composition and quartz silt-size distribution in over 100 pelagic sediment samples taken from eight time planes between the early Cretaceous and late Miocene. In the time planes studied, these authors found a slight decrease in the size of quartz silt towards the centre of the ocean from both sides. The clay mineralogy was also found to differ from one side of the Atlantic to the other. These trends were interpreted to indicate input of aeolian material from both sides of the Atlantic, although the input from Africa was about twice that from North America. Lever and McCave concluded that the zone of maximum aeolian input, as indicated by the coarsest, least sorted, most positively skewed aeolian quartz, has remained in palaeolatitudes 20–30° N since early Cretaceous times.

Some northward shift in the zone of maximum African dust deposition should in fact be expected in pre Quaternary times. Results obtained during the Deep Sea Drilling Project (DSDP) showed that the Antarctic ice cap had almost reached its present volume by 12–14 million yr BP and had a volume larger than the present during the Messinian (5–6 million yr BP) (Kennett, 1977), whereas large-scale continental glaciation in the northern hemisphere did not begin until about 3 million yr BP, followed by formation of the Arctic sea-ice (Shackleton and Opdyke, 1977). Flohn (1981) argued that this temperature asymmetry between the hemispheres would have displaced the 'meteorological equator' (ITCZ) from about latitude 6° N today to about 10° N in the late Tertiary. The sub-tropical anticyclones and trade wind belts would also have shifted northward. During the Messinian, arid conditions prevailed in the Mediterranean and southern Europe, and evaporites formed as far north as Austria. The northern limit of the continental arid belt lay at approximately 47° N in the early and mid Tertiary, but moved south to 42° N in the Miocene–Pliocene and to about 38° N in the Pleistocene. During the Miocene much of the now arid southern Sahara experienced a humid or sub-humid tropical climate (Maley, 1980). This area is therefore unlikely to have provided a major dust source until the Pleistocene.

8.2.2 The North Pacific

Only very limited information is available regarding fluctuations in dust input to the North Pacific during the Quaternary. No detailed core records have been obtained close to the Asian coast, where dust deposition rates are highest and where the most detailed picture of depositional fluctuations is likely to be found.

Bonatti and Arrhenius (1965) reported a sound-reflecting under-bottom layer composed of silty clay in the eastern North Pacific off the coast of Baja California. The quartz-rich silt was interpreted as aeolian dust deposited during an earlier period when easterly dust transport occurred on a larger scale than at present. The dust source was not identified with certainty, but was inferred to be the deserts of northwest Mexico and/or the southwest United States. A radiocarbon date from organic-rich clayey sediment above the silty layer gave an age of $>40\,000$ ^{14}C yr BP.

Janecek and Rea (1985) investigated the evidence for Quaternary dust deposition in a core from the Hess Rise in the central North Pacific. They found that aeolian mass accumulation rates showed relative maxima centred at about 45 000, 210 000, 310 000, 420 000, 600 000 and 720 000 years ago (Fig. 8.14a). During two periods, one centred at 120 000 and the other at 540 000 yr BP, aeolian mass accumulation rates were severely reduced. Smaller reductions in aeolian mass accumulation rates occurred in periods centred at 270 000 and 400 000 yr BP. However, aeolian sedimentation rates were calculated for only a limited number of stratigraphic horizons, and this, combined with the effects of bioturbative mixing, may mean that glacial/interglacial fluctuations are to some extent averaged out. Down-core variations in the median size of the aeolian material showed no systematic relationship with fluctuations in aeolian mass accumulation rate, indicating that the aeolian load reaches an equilibrium (i.e. no further change in size with increasing distance of transport) at approximately 2000 km from the source (Rea *et al.*, 1985).

In a second core, located in the eastern equatorial Pacific north of the Galapagos Islands (within the trade wind belt), Janecek and Rea (1985) found that aeolian mass accumulation rates were less than half those on the Hess Rise. Periods of increased aeolian sedimentation, with one exception, were associated with low carbonate accumulation and equated with interglacial times (Fig. 8.14b). A weak inverse relationship was evident between the amount of dust deposited and median grain size.

Janecek and Rea (1985) concluded that, since aeolian accumulation rates in the North Pacific and eastern equatorial Pacific were generally higher during interglacial times, this suggests that the dust source areas in Asia and Central America were wetter and better vegetated during glacial times. However, as discussed in Chapter 9, other evidence indicates this interpretation is not entirely correct. Janecek and Rea (1985) found that the median size of dust deposited in the eastern equatorial Pacific was slightly coarser during glacial periods, and concluded that trade winds were more vigorous at such times. Similar conclusions were reached by Molina-Cruz (1977) on the basis of sedimentary parameters from several other eastern equatorial Pacific cores.

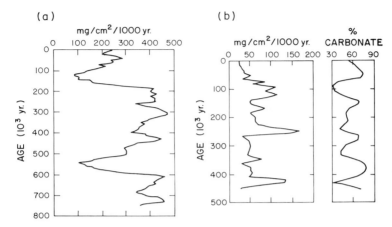

Figure 8.14. (a) *Aeolian mass accumulation rates in the later Quaternary at site KK75–02 in the North Pacific;* (b) *Aeolian mass accumulation rate and percentage of calcium carbonate at DSDP Site 503. Low carbonate intervals represent interglacial time and high carbonate intervals represent glacial times. (After Janecek and Rea (1985), Quat. Res. 24, 150–163; © Academic Press.)*

The strength of the North Pacific westerlies, on the other hand, appears to have fluctuated to a much lesser degree during glacial/interglacial cycles.

A longer-term record of aeolian input to the North Pacific was provided by two apparently continuous cores which span the entire Cenozoic (Leinen and Heath, 1981; Janecek and Rea, 1983; Rea *et al.*, 1985). The first of these cores, from DSDP site 576, located about 1500 km southeast of Japan (32·4° N, 164·3° E) indicated an aeolian mass accumulation rate approximately twice as high as that recorded in the second core (GPC3) located about 500 km north of Hawaii (30·3° N, 157·8° W). Being located closer to the Asian dust source, the median size of dust in core 576 was found to be slightly coarser than in core GPC3. The temporal pattern of mineralogy, mass flux, and grain size of both the total aeolian component and the extracted quartz fraction in the two cores showed a high level of agreement, indicating, according to Rea *et al.* (1985), a broad-scale uniformity of aeolian processes in the North Pacific over the last 70 million years (Fig. 8.15). Core GPC3 was sampled at shorter intervals than core 576, and therefore shows considerably more detailed fluctuations. Both cores indicate a substantial increase in aeolian deposition beginning in the late Tertiary, following a period of low aeolian accumulation rates during the Eocene and Oligocene. The grain size of the deposited dust in both cores also reached a minimum in the Eocene. Superimposed on this pattern in core GPC3 are two sharp peaks of increased aeolian deposition at 15–16 million years ago and 21–23 million years ago.

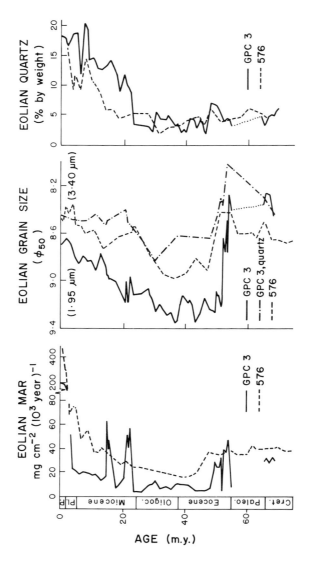

Figure 8.15. Aeolian mass accumulation rate (MAR), median grain size of both the total aeolian component and of quartz grains extracted from core CPC3 by J. P. Dauphin, and percentage of quartz in aeolian sediments from the North Pacific Ocean. (After Rea et al. (1985), Science **227**, 721–725; © American Association for the Advancement of Science.)

These results, together with data from other DSDP cores in the North Pacific (Rea and Janecek, 1982; Janecek and Rea, 1983), indicate major changes in the strength of the westerlies during the last 70 million years. The low aeolian accumulation rates and fine grain size of dust deposited in the early Tertiary are consistent with independent evidence that the global climate at this time was relatively warm and humid. Wind intensity began a long-term overall increase beginning about 35 million years ago and reached a maximum during the Pleistocene. A post Eocene increase in wind strength supports the idea of thermal asymmetry between the hemispheres (Flohn, 1981) associated with polar cooling and development of the Antarctic ice cap about 38 million years ago.

However, data from cores located east of Tahiti (Rea and Bloomstine, 1986) provide only weak evidence of an increase in the vigour of the eastern equatorial Pacific trade winds during the late Cenozoic. The mass accumulation rate of dust has been uniformly low in this area since the late Oligocene, but the median grain size of the aeolian material shows an increase from $3 \cdot 4\,\mu$m in older sediments to $7 \cdot 2\,\mu$m in younger material. Rea and Bloomstine (1986) considered this indicative of a significant increase in the intensity of atmospheric circulation about $10 \cdot 5$ million years ago, but they observed no obvious response to the onset of northern hemisphere glaciation $2 \cdot 5$–3 million years ago.

Rea *et al.* (1985) pointed out that data from North Pacific cores 576 and GPC3 indicate deposition of relatively coarse aeolian material during the Cretaceous and Palaeocene, which is contrary to the widely held view of the Cretaceous as a period characterized by warm climate and sluggish ocean circulation. At DSDP site 463, located west of Hawaii, Rea and Janecek (1981) found higher aeolian mass accumulation rates in the early Cretaceous (Aptian and Albian) than in the mid and late Cretaceous. Although pole-to-equator temperature gradients are estimated to have been about half the present value (Douglas and Savin, 1975), and atmospheric circulation should therefore have been less intense, the climate of the early to mid Cretaceous was probably the hottest and driest experienced during the whole of Phanerozoic time (Frakes, 1979). The high dust deposition rates of the early Cretaceous may therefore have been a response to increased aridity rather than increased windiness.

8.2.3 The southwest Pacific

In marine sediments of the southwest Pacific, Thiede (1979) observed a higher input of quartz-rich dust from Australia at 18 000 yr BP compared with the present, and attributed this to increased aridity during the last glacial period. This interpretation is consistent with palaeoenvironmental evidence from the Australian mainland. Lake levels in southeast Australia fell significantly between 25 000 and 12 000 yr BP, clay dunes formed on their downwind (eastern) margins, and desert dune ridges were reactivated (Bowler, 1975; Bowler *et al.*, 1976). Bowler (1978) suggested that development of these aeolian features was related to more frequent summer outbreaks of hot, dry continental air travelling southwards from the desert interior of Australia. Such outbreaks were probably related to the passage of low-pressure centres across the Great Australian Bight. These depressions are likely to have taken a more northerly track than at present due to northward displacement of the Antarctic Polar Front and belt of westerlies during Antarctic cold stages. Steep pressure gradients would have formed between these depressions and sub-tropical high-pressure cells been located temporarily over northeast Australia and the Coral Sea, creating a synoptic situation similar to that associated with the severe Australian dust storm of October 1928 (Kidson, 1929). Late glacial northwesterly winds were responsible for the formation of longitudinal dunes on King Island (Bass Strait) and in northeast Tasmania, and were probably also responsible for increased dust transport to the Tasman Sea. Some dust was also deposited on land, forming the 'sheet parna' deposits of western New South Wales and Victoria (Butler, 1974; Dare-Edwards, 1984).

After 12 000 yr BP lake levels in southeast Australia began to rise again, dunes became stabilized, and the rate of dust deposition in the southwest Pacific declined. These changes were probably related to a reduced frequency of dry, northwesterly air outbreaks due to southward movement of the Antarctic Polar Front and westerly depression tracks associated with retreat of the Antarctic ice cap (Bowler, 1978).

8.2.4 Summary of the ocean core evidence

The evidence from ocean cores generally suggests that dust input during the Quaternary has been influenced both by changes in aridity in the source areas and by changes in the strength and position of the major global wind systems. The wider body of palaeoenvironmental evidence accumulated in the past two decades indicates that there was no spatially uniform pattern of environmental change in the tropics and sub-tropics associated with advances and retreats of the polar ice caps. Even within individual regions, the pattern of environmental change did not necessarily follow the same pattern from one glacial/interglacial cycle to the next. Consequently, inputs of dust to the oceans during the Quaternary were not rhythmic, but varied in a more irregular manner according to the interaction of global and regional climatic trends. Spectral analyses suggest there is only a very weak correspondence between the timing of dust fluctuations and that of changes in the Earth's orbital parameters (e.g. Janecek and Rea, 1985).

Considering the Cenozoic as a whole, there is evidence that a general increase in intensity of the atmospheric circulation led to increased oceanic dust deposition in the late Tertiary and Quaternary compared with the early Tertiary. The increase in circulation intensity may have been related to polar cooling and build-up of the Antarctic ice sheet after about 38 million yr BP. However, data are still sparse, particularly for the southern oceans, and conclusions regarding the nature and causes of atmospheric circulation changes remain tentative.

8.3 RECORD OF QUATERNARY DUST DEPOSITION PRESERVED IN ICE CORES

Windom (1969) examined late Holocene dust accumulations in the Antarctic and Greenland ice sheets and five temperate glaciers in New Zealand, Mexico and North America. Measured accumulation rates were found to range from 0.0001 mm yr^{-1} (Antarctica) to 0.001 mm yr^{-1} (Mexico). Dust accumulation rates in the Antarctic and Greenland were interpreted to reflect global dust fallout rates, since these areas are remote from the major dust source, whereas accumulation rates at the other sites probably reflect fallout of dust partly from local sources. Hogan (1975) also concluded that Antarctic aerosols are 'globally representative'.

Studies of variations in microparticle concentrations with depth in ice cores have shown that global fallout rates during the Holocene have generally been much lower than during the later part of the last glacial period. Thompson and Mosley-Thompson (1981) documented a marked increase in the abundance of particles >0.6 μm in diameter at the Holocene–last-glacial boundary in cores from Dome C (East Antarctica), Byrd Station (West Antarctica) and Camp Century (Greenland) (Fig. 8.16). The ratio of the average microparticle concentration in the late glacial ice strata to that in the Holocene strata was found to be $6:1$ for the core from Dome C, $3:1$ for the core from Byrd Station, and $12:1$ for the core from Camp Century. Geochemical analyses of microparticles and ice in the Dome C core reported by Petit *et al.* (1981) indicated that continental and marine inputs were respectively 20 and 5 times higher during the late glacial than during the Holocene. Subsequent investigations by Briant *et al.* (1982), using scanning electron microscopy and X-ray microanalysis to characterize individual microparticles, confirmed that the majority are of continental rather than volcanic origin, in both the Holocene and last glacial parts of the core. A large proportion of the particles in both Holocene and last glacial ice were found to consist of illite-type micas ($>25\%$), but a higher proportion of the last glacial microparticles consisted of quartz (15.4% compared with 5.4% in the Holocene ice). The increase in continental dust deposition in East Antarctica was attributed by these authors to the widespread

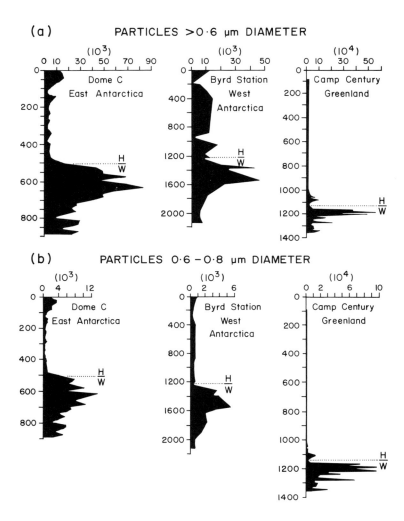

Figure 8.16. Changes in the concentration of terrestrial dust particles >0.6 μm with depth in ice cores from Antarctica and Greenland: (a) the average concentration of particles with diameters greater than or equal to 0·6 μm for all samples in each section; and (b) the concentration of particles with diameters between 0.6 and 0.8 μm in the cleanest 10% of the samples in each section of three deep ice-cores encompassing the end of the last major glaciation. The transition into the Holocene (designated H/W) is based upon the $\delta^{18}O$ record for each of the cores. (After Thompson and Mosley-Thompson (1981), Science **212**, 812–815; © American Association for the Advancement of Science.)

development of aridity and possibly windier conditions during late glacial times.

In the Camp Century ice core, Hammer *et al.* (1978) found evidence that deposition of late glacial dust was markedly seasonal, with apparent peaks in the 'spring'.

Within the Holocene, a period of increased microparticle deposition during the Little Ice Age (between approximately AD 1450 and 1850) was identified in the South Pole ice core by Mosley-Thompson and Thompson (1982). Particularly high concentrations of microparticles were recorded around AD 1830, 1780, 1460 and 1150. These authors did not, however, determine the proportions of volcanic and continental dust particles present. Some of the dust, at least, is volcanic, since a prominent peak corresponding to the 1883 Krakatoa eruption was found.

The high-resolution nature of the microparticle record in ice sheets with high ice-accretion rates has allowed 'dating' of ice layers using a technique similar to varve counting in glacial lake sediments (Hammer *et al.*, 1978; Mosley-Thompson and Thompson, 1982).

Chapter Nine

LOESS

The term 'loess' derives from the German *löss* which was first used in the early 1820s by the German tax inspector and amateur geologist Karl Caesar von Leonard (1823–24) to describe loose, friable, silty deposits along the Rhine Valley near Heidelberg (Kirchenheimer, 1969). Lyell (1834) brought the word loess into widespread English usage and was responsible for stimulating international interest in its properties and genesis after he visited the Rhine Valley in 1833. During his second visit to America in 1845–46, Lyell observed the loess along the Mississippi Valley and noted its similarity to the Rhine Valley deposits (Lyell, 1847). Lyell did not recognize the aeolian origin of loess, and considered that it was most likely to be a fluvial deposit. Virlet-d'Aoust (1857) may have been the first to propose the aeolian hypothesis (Russell, 1944a), but von Richthofen (1877–85, 1882) was responsible for popularizing it in his writings on China. In China, the term 'huangtu' has long been used to describe loose, earthy deposits, and the relationship between huangtu and windborne dust was apparently recognized at least 2000 years ago (Liu Tung-sheng *et al.*, 1985a).

9.1 DEFINITION OF LOESS

There have been many attempts to define loess on the basis of its physical and mineralogical properties, but none has proved entirely satisfactory. Many earlier workers described loess as a 'uniform, yellow, calcareous silt', or 'unstratified, homogeneous, calcareous silt, light yellow or buff in colour'. However, more recent work has shown that loess is not always calcareous, it may be clayey or sandy, and the colour varies considerably; dark grey, yellow, brown, white and red loess deposits are known. A great deal of effort has been expended in attempts to distinguish true aeolian loess from loess-like sediments. The latter have been variously designated by such terms as river loess, lake loess, marine loess, loess loam, alluvial loess, proluvial loess, colluvial loess and deluvial loess. The German literature is particularly rich in terms used to describe loess and loess-like deposits in different topographic settings and with slightly different physical appearance. In some cases the loess-like deposits represent wind-deposited dust which has been redeposited by water or slope processes, while in other cases aeolian transport appears to have played little or no part in the formation of the deposits. Partly due to the way in which terminology has been employed, confusion has arisen regarding the origin of fine-grained deposits in different areas. In an attempt to clarify the situation, Pye (1984*a*) proposed a series of simplified definitions which are adopted here as follows.

Loess is defined simply as a terrestrial windblown silt deposit consisting chiefly of quartz, feldspar, mica, clay minerals and carbonate grains in varying proportions. Heavy minerals, phytoliths, salts and volcanic ash shards are also sometimes important constituents. In a fresh (i.e. unweathered) state, loess is typically homogeneous, non- or weakly stratified and highly porous. Most commonly it is buff in colour, but may be grey, red, yellow or brown. When dry, loess has the ability to stand in vertical sections and sometimes shows a tendency to fracture along systems of vertical joints, but when saturated with water the shear strength is greatly reduced and the material is subject to subsidence, flowage and sliding. The grain-size distribution of 'typical' loess shows a pronounced mode in the range 20–40 μm (5·7–4·65 ϕ), and is positively skewed (i.e. towards the finer sizes). Typical loess often contains up to 10% fine sand (>63 μm), but in cases where the sand content exceeds 20% the term 'sandy loess' is appropriate. Up to 20% clay (<4 μm) is not unusual in typical loess; if the sediment contains more than 20% clay it can be described as clayey loess.

Wind-deposited primary loess not infrequently shows some evidence of syn-depositional reworking by raindrop impact, surface wash, soil creep or other soil processes. A faint downslope stratification may be apparent, and the loess thickness may increase in a downslope direction. However, the distance of reworking is generally only a few metres or tens of metres. Reworked loess, on the other hand, consists of primary wind-deposited material which has been eroded and redeposited by running water and/or slope processes at some considerable distance from its original site of deposition. Reworked loess usually accumulates in valley bottoms, lakes and on river terraces.

Weathered loess is primary loess whose sedimentary characteristics have been markedly modified by weathering, soil formation and diagenesis. Weathered loess is usually decalcified and contains more clay (up to 60%) than unweathered loess. Thick loess profiles often contain alternating layers of weathered and unweathered loess. The weathered layers may represent a single soil profile formed during a hiatus in loess sedimentation, or consist of several superimposed palaeosols (known as a *pedocomplex*) formed during slow or intermittent dust deposition.

Sediments in some parts of the world, e.g. northern Nigeria (Bennett, 1980; Smith and Whalley, 1981; McTainsh, 1984), are composed of mixtures of deposited dust and other sediment (dune sand, soil, or alluvial fill). The term *loessoid* can be used to describe such sediments.

Loess-like deposits are sediments which possess many of the sedimentological properties of aeolian loess but which have not been transported by the wind at any stage in their history. They include overbank silts, lacustrine silts and some colluvial deposits. Differentiation of such deposits from reworked loess is often difficult, but may be possible on the basis of geochemical, mineralogical and textural criteria.

Loessite is a term originally proposed by J. B. Woodworth to describe lithified loess rock in the sedimentary record. However, only a very few ancient loessites have been recognized (e.g. Edwards, 1979). It is uncertain whether this is due to the fact that loess was never deposited on a large scale in pre Quaternary times, whether it was deposited but not preserved, or simply whether adequate criteria are not available to distinguish it from other siltstones.

9.2 LOESS DISTRIBUTION

According to Pecsi (1968a), loess and loess-like deposits cover about 10% of the world's land surface area. However, the area covered by primary aeolian loess is considerably less, probably nearer 5%. The thickest and most extensive loess blankets occur in China, Soviet Central Asia, the Ukraine, Central and Western Europe, the Great Plains of North America and Argentina (Fig. 9.1). In China, thick loess covers more than 273 000 km^2 in the Loess Plateau of Shanxi, Shaanxi and Gansu Provinces (Barbour, 1927, 1935; Liu Tung-sheng and Chang Tsung-yu, 1964; Derbyshire, 1978, 1983a; Liu Tung-sheng et al., 1985a; CHIQUA, 1985; Fig. 9.2). Thick, extensive loess also occurs further west around Lanzhou (Derbyshire, 1983b) and on the foothills surrounding the Tarim and Jungarr Basins (Derbyshire, 1984b). In total, primary and reworked loess covers more than 1 million km^{-2} in China (Liu Tung-sheng et al., 1985a).

In Soviet Central Asia, loess is found in a broad belt extending from Kirghiz SSR and Kazakhstan, through Uzbekistan and Tajikistan to Turkmen SSR (Penck, 1930; Mavlyanov, 1958; Lukashev et al., 1968; Dodonov et al., 1977; Dodonov, 1979, 1984; Lazarenko, 1984). Loess occurs on terraces and slopes in river valleys draining the Altai, Ghissars and Pamirs, and blankets many of the alluvial fans, foothills and plains on their northwestern margins (Fig. 9.3).

Extensive areas of loess occur in the Ukraine and South Russian Plain (Ivanova and Velichko, 1968; Gerasimov, 1973; Zolotun, 1974; Kraev, 1975; Veklich, 1979). The Ukraine loess belt extends westwards into the Baltic

Figure 9.1. The global distribution of major loess occurrences. (After Pye (1984a), Prog. Phys. Geog. *8*, 176–217; © Edward Arnold.)

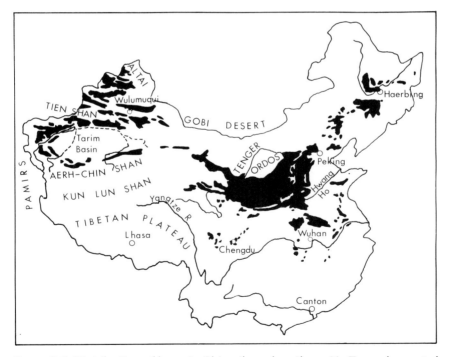

Figure 9.2. Distribution of loess in China (based partly on Liu Tung-sheng et al. (1981)). (After Pye (1984a), Prog. Phys. Geog. *8*, 176–217; © Edward Arnold.)

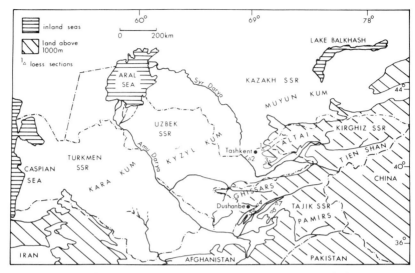

Figure 9.3. Map showing major mountain ranges, rivers and desert areas of Soviet Central Asia. Loess occurs mainly along the valleys of the Syr Darya, Amu Darya and their tributaries, and along the foothills and steppes on the western side of the Altai, Ghissars and Pamirs.

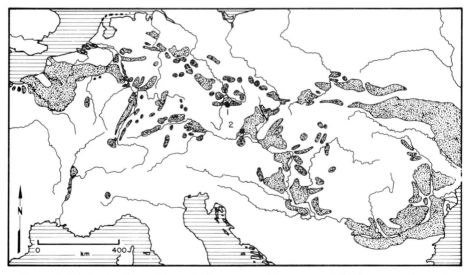

Figure 9.4. Distribution of loess in Europe and location of the loess sections at Cerveny Kopeč (1) and Krems (2). (After Haesaerts (1985), Bull. Ass. Franç. Étude Quat 2–3, 105–115; © French Quaternary Association.)

socialist republics, Poland and East Germany (Maruszczak, 1965, 1985; Cegla, 1972). Important areas of loess are found in the Danube Basin (Fink, 1968; Mojski, 1968a, b; Markovic-Marjanovic, 1968; Lozek, 1968a, b; Pecsi, 1968b, 1985; Smalley and Leach, 1978), in Bavaria, along the Rhine (Brunnacker, 1968; Buraczyński, 1978, 1982), and in Belgium and northern France (Audric, 1973; Jamagne *et al.*, 1981; Lautridou *et al.*, 1984; Haesaerts, 1985; Balescu *et al.*, 1986; Fig. 9.4). Localized continuations of the northwest European loess belt are found in the Channel Islands and southern England (Pitcher *et al.*, 1954; Catt, 1977, 1978, 1979, 1985; Lill and Smalley, 1978; Eden, 1980; Burrin, 1981).

In North America the largest area of loess forms a belt which extends from the Rocky Mountains to western Pennsylvania, with a southern extension along the Mississippi Valley into Louisiana (Thorp and Smith, 1952; Frye *et al.*, 1962; Ruhe, 1976; Lugn, 1960, 1962, 1968; Fig. 9.5). Loess also occurs locally in Washington and Oregon (Treasher, 1925; Thiesen and Knox, 1959), New Mexico, Texas (Gile, 1979), Canada (Dumanski *et al.*, 1980; Catto, 1980;

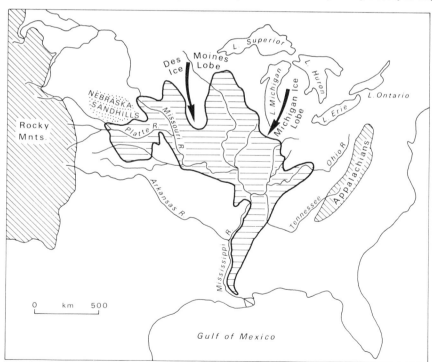

Figure 9.5. Distribution of loess in the Great Plains and Mississippi Valley of North America in relation to major late Pleistocene ice lobes, river valleys, and the Nebraska Sandhills. Based partly on Thorp and Smith (1952). (After Pye (1984a), Prog Phys. Geog. **8**, 176–217; © Edward Arnold.)

Lim and Jackson, 1984), and Alaska (Tuck, 1938; Péwé, 1951, 1955, 1968, 1975; Trainer, 1961).

Loess is extensive in the Pampas region of Argentina (Terruggi, 1957; Riggi, 1968), although little is known about its composition and history. Smaller but still significant deposits are found in several parts of Siberia (Péwé *et al.*, 1977; Péwé and Journaux, 1983; Volkov and Zykina, 1984; Tomirdiaro, 1984), New Zealand (Young, 1967; Bruce, 1973*a*, *b*; Selby, 1976; Smalley and Davin, 1980; Eden, 1982), the northern part of the Indian sub-continent (Pias, 1971; Bal and Buursink, 1976; Kusumgar *et al.*, 1980; Agrawal *et al.*, 1980; Pant *et al.*, 1983; Williams and Clark, 1984; Rendell *et al.*, 1983; Rendell, 1985), and Israel (Ginzbourg and Yaalon, 1963; Ginzbourg, 1971; Yaalon and Dan, 1974). Localized pockets of loess or sandy loess are found in other parts of the Middle East (Ferrar, 1914; Fookes and Knill, 1969; Kukal and Saadallah, 1970, 1973; Doornkamp *et al.*, 1980), Tunisia (Rathjens, 1928; Brunnacker, 1973; Coudé-Gaussen *et al.*, 1982; Rapp and Nihlén, 1986), southern Europe (Brunnacker and Lozek, 1969; Orombelli, 1970; Brunnacker, 1980), northern Nigeria (Bennett, 1980), Greenland (Hobbs, 1931), and Spitzbergen (Bryant, 1982). Typical silty quartz-rich loess is virtually absent in Australia, although localized pockets of carbonate-rich loess have been reported and thin deposits of clayey loess (parna) are common in the southeast (Butler, 1956, 1974; Gill, 1973; Gill and Reeckman, 1980; Gill and Segnitt, 1982; Dare-Edwards, 1984).

9.3 THICKNESS AND MORPHOLOGY OF LOESS DEPOSITS

Loess typically forms a blanket of varying thickness, covering a variety of relief features including steep valley slopes, plateaux, terraces, alluvial fans and pediments (Figs. 9.6 and 9.7). The surface morphology of loess sheets is strongly controlled by the underlying topography. In Mississippi, for example, where loess was deposited on undulating terrain formed by fluvial dissection of late Tertiary clays and Pleistocene sandy gravels, the surface of the loess is also undulating (Fig. 9.8), giving rise to structures termed 'pseudo-anticlines' (Priddy *et al.*, 1964; Fig. 9.9). Loess which has been eroded from the upper slopes often forms weakly stratified accumulations on the lower slopes.

Loess is found over a very wide altitudinal range, occurring near sea-level in northwest Europe to more than 2000 m in parts of Central Asia and China. Loess thicknesses exceeding 300 m have been recorded near Lanzhou in China (Derbyshire, 1983*b*), but 80–120 m is more common on the Loess Plateau. In parts of Soviet Central Asia loess exceeds 200 m in thickness (Dodonov *et al.*, 1977; Davis *et al.*, 1980). In the Danube Basin, Argentina and parts of North America, loess thicknesses occasionally exceed 60 m, but 20–30 m is more

Figure 9.6. Steep loess-covered slopes in South Tajikistan.

Figure 9.7. Loess bluffs capping fluvial terrace sediments of the Chirchik River, Uzbekistan.

Figure 9.8. Undulating loess topography developed on dissected Tertiary and Quaternary fluvial sediments, north of Vicksburg, Mississippi.

Figure 9.9. 'Pseudoanticline' structure in loess overlying undulating pre-loess topography, Vicksburg, Mississippi: (1) loess of late Last Glacial age: and (2) loess of earlier Last Glacial age mixed with sands from underlying Pleistocene fluvial layer (PFL).

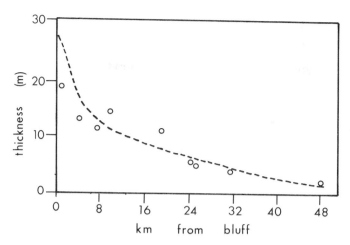

Figure 9.10. Thinning of loess with distance from the bluffs (dust source) in southern Mississippi. (After Snowden and Priddy (1968), Missis. Geol. Surv. Bull. 111, 13–203; © Mississippi Department of Natural Resources.)

typical. Elsewhere, loess is generally less than 30 m thick. Large thicknesses occur in areas where the dust supply has been unusually large and/or sustained over a very long period, and locally in areas where the blowing dust was trapped within a limited geographical area by vegetation or some other surface roughness element.

As pointed out by Penck (1930) and Smalley (1972), many loess deposits occur downwind of major river valleys which provided a source of silt during earlier periods of the Quaternary. During cold periods these valleys received large amounts of poorly-sorted glacial outwash sediment. Dust blown from exposed channel bars during times of low flow, mainly in summer and autumn, was trapped close to the source by vegetation and topographic obstacles. In such cases, e.g. the Mississippi Valley (Simonson and Hutton, 1954; Waggoner and Bingham, 1961; Krinitzsky and Turnbull, 1967; Frazee et al., 1970; Handy, 1976), the thickness of loess is greatest immediately adjacent to the riverine source and decreases regularly in the downwind direction, creating a wedge-shaped deposit in cross-section (Fig. 9.10).

9.4 PHYSICAL AND CHEMICAL PROPERTIES OF LOESS

9.4.1 Grain size

'Typical' unweathered loess is composed mainly of particles between 10 and 50 μm in size (Browzin, 1985; Tsoar and Pye, 1987), and usually has a

Table 9.1
*Median grain size of loess in different parts of the world. Methods of
sample pre-treatment and analysis vary so results are not exactly
comparable.*

Area	Range in median size (ϕ)	Average (ϕ)	Sample size	Reference
Nebraska	4·42–4·56	4·47	4	Swineford and Frye (1951)
Kansas	4·29–5·59	5·00	43	Swineford and Frye (1951)
France	4·85–6·55	5·59	4	Swineford and Frye (1951)
Germany	4·95–5·70	5·29	8	Swineford and Frye (1951)
Argentina	4·00–5·10	4·38	12	Teruggi (1957)
Siberia	4·18–6·88	5·10	26	Péwé and Journaux (1983)
New Zealand	4·40–6·35	5·51	33	Young (1964)
Mississippi	4·99–5·93	5·76	42	Snowden and Priddy (1968)
Tajikistan	5·70–6·93	6·31	11	Goudie *et al.* (1984)
China	5·10–7·17	6·36	12	Derbyshire (1983*b*)

prominent mode between 20 and 30 μm. The median size of loess from several different parts of the world is shown in Table 9.1, and a number of cumulative grain-size curves are shown in Fig. 9.11. In most cases the reported median size is between 20 and 30 μm, although the Nebraska loess is coarser while the Chinese and Tajikistan loess is finer. It is important to realize that a continuum of aeolian sediments exists between dune sands, 'typical loess' and clayey loess. A number of authors have maintained that saltation and suspension are mutually exclusive sediment transport processes which lead to a spatial separation of deposited aeolian sand, silt and clay (Bagnold, 1941; Franzmeier, 1970). However, this is not always the case. As discussed in Chapter 3, there is no sharp distinction between saltation and suspension, and many fine sand and coarse silt grains are transported in modified saltation or short-term suspension. A transition from silty sand sheets to sandy loess has been observed in several parts of the world including the northwest Negev (Ravikovitch, 1953; Fig. 9.12), the northern fringe of the Chinese Central Loess Plateau (Kes, 1984), and the area southeast of the Nebraska Sandhills (Lugn, 1962, 1968). Silty sand and sandy silt deposits, known as *flöttsand* and *sandloess*, are quite common downwind of Pleistocene inland dunefields and coversands in Europe (Dowgiallo, 1965; Thorez *et al.*, 1970). In Nebraska, the sand content of the Peorian loess decreases from 60% on the southern margin of the Sandhills to less than 5% on the southeastern margin of the Loess Hills (Lugn, 1968). The thickness of loess shows a parallel decrease from about 28 to less than 6 m, although there is a secondary increase in thickness on the south side of the

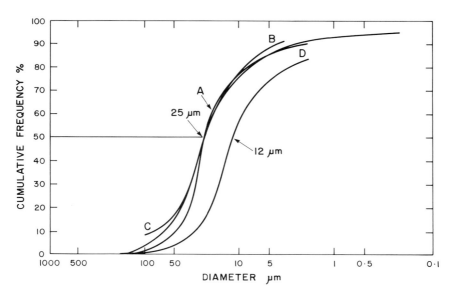

Figure 9.11. Cumulative grain-size curves of loess from: A – Siberia (Péwé, 1981b); B – Kansas (Swineford and Frye, 1945); C – Alaska (Péwé, 1951); D – Tajikistan (Goudie et al., 1984).

Figure 9.12. Sandy loess overlain by Holocene coastal dune sands (Gaza Strip).

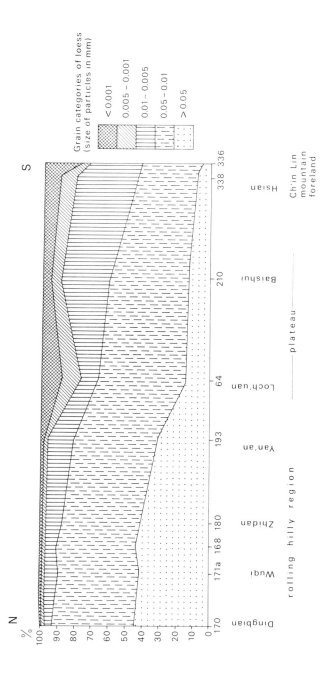

Figure 9.13. Changes in grain size of loess across the Loess Plateau (Dingbian – Hsian, China). (After Kes (1984), in M. Pecsi (ed.) Lithology and Stratigraphy of Loess and Palaeosols, pp. 104–111; © Geographical Research Institute, Hungarian Academy of Sciences.)

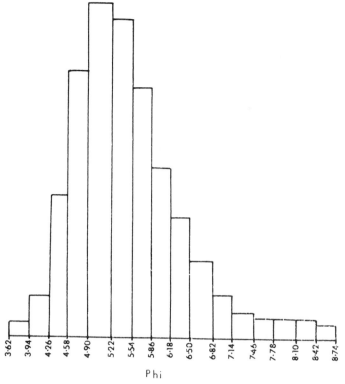

Figure 9.14. Grain size histogram of a sample of unweathered late Pleistocene loess from Vicksburg, Mississippi (analysed by Coulter Counter). (After Pye (1984a), Prog. Phys. Geog. 8, 176–217; © Edward Arnold.)

Platte River which provided a subsidiary dust source (Kollmorgen, 1963). On the Chinese Loess Plateau, sandy late Pleistocene Malan loess shows a transition to 'typical' silty loess and then to clayey loess in a downwind (southeasterly) direction (Liu Tung-sheng and Chang Tsung-yu, 1964; Kes, 1984; Fig. 9.13). The Malan loess also thins to the southeast, although the converse is true of the older Lishih and Wucheng loess sheets, which become thicker (though still finer) towards the southeast. Liu Tung-sheng and Chang Tsung-yu (1964) suggested this was due to erosion of the older loess sheets in the northwest, though the evidence is not conclusive.

Unweathered loess is often regarded as a poorly sorted sediment, with ϕ sorting values (Folk and Ward, 1957), between 1 and 3. Skewness, which is almost always positive (indicating a tail of finer grains), generally lies between $+0.3$ and $+0.7$. A grain-size histogram typical of unweathered loess is shown in Fig. 9.14. It has not been established what proportion of the fine silt and clay

Figure 9.15. Pedogenetically-altered loess section at Netivot, in semi-arid region of southern Israel.

grains are transported as individual particles and what proportion as aggregates which become disaggregated during the size analysis procedure. However, SEM examination of modern dusts suggests that the proportion of aggregated fine material is significant (Gillette and Walker, 1977; Whalley and Smith, 1981; Rabenhorst et al., 1984).

Loess formed by dust deposition at great distance from the source is finer than 'typical' loess. However, the higher clay content of loess at great distance from the source can be partly due to post-depositional weathering, since more slowly deposited dust is affected to a greater extent by weathering. Silt-size feldspars and other unstable silicates undergo alteration to clays, carbonate grains are dissolved, and even quartz may suffer some size reduction. On the other hand, loess deposited rapidly close to the source is exposed to surface weathering for a shorter period, and there is less post-depositional modification of the primary grain size characteristics. Where the distal parts of loess deposition belts lie in areas of humid or sub-humid climate, the loess is likely to be particularly clay-rich and will show pronounced pedogenetic horizonation. In the semi-arid fringe of southern Israel, for example, much of the late Pleistocene loess contains over 60% clay and displays numerous pedogenetic carbonate horizons (Bruins, 1976; Bruins and Yaalon, 1979; Fig. 9.15). Pye and Tsoar (1987) proposed a schematic model which attempts to explain the changes in loess character which occur along a hypothetical environmental gradient between a warm desert and a semi-arid area (Fig. 9.16). The mean size of deposited dust decreases with distance from the desert source, while the proportion of clay minerals and carbonate increases with distance since these

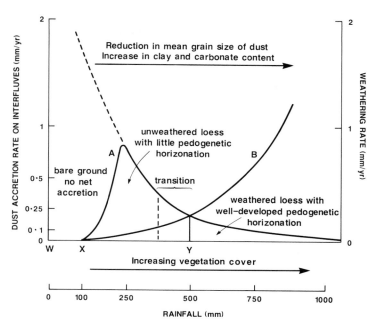

Figure 9.16. Schematic model showing expected changes in the character of loess deposited along an environmental gradient between warm desert and sub-humid climatic conditions. Curve A indicates the dust accretion rate on interfluve sites. No significant thickness of dust accumulates in the hyper-arid zone between the desert dust source W and point X, where the rainfall is sufficiently high to support dust-trapping vegetation. The dashed extension of curve A represents the expected dust accretion rate if effective dust-trapping vegetation was present adjacent to the dust source. Under these circumstances the rate of dust accretion would decrease regularly with distance from the source. The mean grain size of dust also decreases with distance from the source, while the clay mineral and carbonate contents increase. Curve B shows the increase in rate of weathering of the deposited dust (notionally expressed as depth of leaching in mm yr^{-1}) as the rainfall increases away from the desert source. At distances from the source greater than point Y the rate of weathering (decalcification) exceeds the rate of dust accretion, and the deposited loess will show evidence of pronounced weathering and pedogenetic horizonation. Between point Y and the desert margin of the loess belt the loess will show much weaker weathering and pedogenetic horizonation. Changes in either the rate of dust accretion or rainfall could cause shifts in the position of Y; if repeated, this will produce alternating loess and soil horizons. (After Pye and Tsoar (1987), in I. Reid and L. Frostick (eds.) Desert Sediments Ancient and Modern; © The Geological Society.)

minerals are enriched relative to quartz and feldspars in the finer-size fractions. As the rainfall increases towards the semi-arid area, the rate of syn-depositional weathering and leaching increases. At some distance from the source, desig-nated Y in Fig. 9.16, the weathering rate (indicated by depth of decalcification) becomes equal to the dust accretion rate. Beyond this point the entire loess section will display significant evidence of weathering and soil development (increased clay content, leaching of primary carbonate, development of secon-dary nodular carbonate horizons). Conversely, upwind of point Y the loess will appear relatively fresh, with limited carbonate leaching and little horizonation. Changes in either the dust deposition rate or rainfall would cause a shift in the position of Y and, if repeated several times, would produce a stratigraphic sequence composed of alternating weathered and relatively unweathered loess. Dust deposition need not cease altogether in order to produce a weathered horizon, but the rate of accretion must fall below the weathering rate. During a depositional hiatus weathering has an opportunity to progress to a greater degree (and to a greater depth), and the clay content of the near-surface horizons will increase substantially.

9.4.2 Grain shape

The shape of grains in loess depends on their mineral composition and crystallographic structure, the processes responsible for their formation, and the degree of pre- and post-depositional shape modification induced by weathering and diagenesis. The majority of quartz silt grains are blocky and angular or sub-angular (Fig. 9.17). Quartz grains derived from active chemical weathering environments, such as soils, often show edge rounding due to

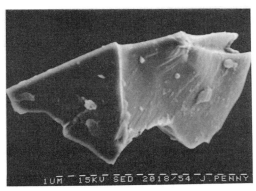

Figure 9.17. SEM micrograph showing an angular quartz silt grain in Holocene loess from Alaska. (Scale bars = 1 μm.)

solution and reprecipitation, whereas fresh grains formed by sub-glacial crushing or frost action often have sharp edges and display conchoidal and stepped fracture surfaces. Some authors have suggested that quartz grains finer than about 100 μm show a tendency to become more platy with decreasing size, due to some 'cleavage' control on quartz fracture (Hammond *et al.*, 1973; Krinsley and Smalley, 1973; Smalley, 1974), but such a tendency has not been observed in loess (Goudie *et al.*, 1984).

Mechanical breakage of feldspars is, however, strongly influenced by cleavage, and such grains are often equidimensional and blocky. Silt-sized grains of rhombohedral carbonates (calcite and dolomite) also frequently occur in loess as equidimensional rhombs bounded by flat crystal faces. Phyllosilicates occur predominantly as platy grains. Opal phytoliths and some heavy minerals such as zircon are easily recognized by their distinctive morphologies. Volcanic glass shards, if present, are often angular and have a smooth, glassy surface with conchoidal fractures.

9.4.3 Grain surface textures

The surfaces of loess grains are often covered by adhering clay-size particles and/or amorphous aluminosilicate coats (Figs. 9.18, 9.19 and 9.20). The adhering fine particles were initially interpreted by Smalley and Cabrera (1970) as glacial comminution debris, but similar particles have also been observed on grains from a wide range on non-glacial environments (Warnke, 1971; Whalley and Smith, 1981) and have been produced experimentally by non-glacial processes (Pye and Sperling, 1983; Pye, 1984c). Cegla *et al.* (1971), Vita-Finzi

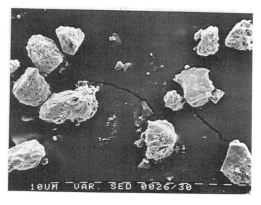

Figure 9.18. SEM micrograph showing typical grains of quartz and plagioclase feldspar in loess from the South Island of New Zealand. Note the large number of clay-size particles adhering to the grain surface. (Scale bars ± 10 μm.)

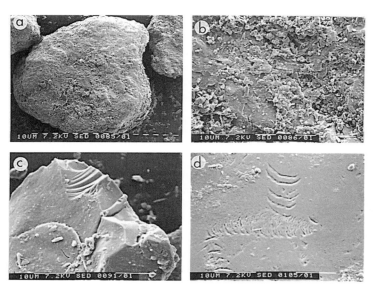

Figure 9.19. SEM micrographs of silt grains from late Pleistocene Tajikistan loess: (a) quartz silt grain with numerous adhering clay-size particles; (b) higher-magnification view of illite, kaolinite, quartz and calcite particles adhering to the surface of the grain shown in (a); (c) angular breakage features on a quartz silt grain; and (d) chattermarks of unknown origin on a quartz silt grain. (Scale bars = 10 μm.)

et al. (1973) and Smalley *et al.* (1973) suggested that some of the adhering particles on silt grains in European loess might represent secondary encrustations of calcium carbonate, possibly formed during post-depositional 'loessification' as suggested by Lozek (1965). However, Pye (1983b) found that many of the adhering particles on Polish and West German loess grains could be removed by sonic vibration, and that they consisted of a mixture of kaolinite, illite, clay-sized quartz and carbonate grains held in position by electrostatic charges or thin moisture films. In the case of grains from more weathered horizons in the loess, or grains derived from pre-weathered material, the particles appeared to be cemented together by amorphous silica or aluminosilicate material. Comparison of the surface textures of grains in the SEM before and after hydrochloric acid treatment to remove carbonate revealed no significant differences.

Minervin (1984) reported that Soviet Central Asian loess contains micro-aggregates or globules 10–100 μm in diameter which consist of a quartz (less commonly feldspar) core surrounded by concentric envelopes of amorphous silica gel, carbonate, and an outer jacket comprised of clay minerals, iron hydroxides, amorphous silica, finely dispersed quartz and carbonates. Accord-

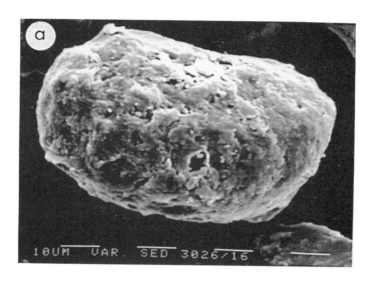

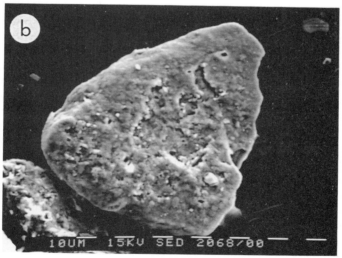

Figure 9.20. SEM micrographs of quartz silt grains in loess from: (a) Ariendorf, West Germany; and (b) Pegwell Bay, England, showing thick surface coatings composed of amorphous silica and clay minerals. (Scale bars = 10 μm.)

ing to Minervin, these structures are formed by cryogenic processes in fluvial environments during transport and deposition. Frost-induced breakage of quartz leaves broken Si–O bonds which increase the solubility of the surface layer and favour formation of a silica gel coating. Dissolved calcium and bicarbonate ions become absorbed on the hydrated gel surface. The patchy distribution of active centres on the quartz grain surfaces gives the carbonate envelope a perforated honeycomb structure. Subsequently, further amorphous silica, ferric hydroxide, carbonate and suspended clay is precipitated or becomes trapped on the surface.

An interesting question is whether the surface textural features of quartz silt and fine sand grains in loess can provide diagnostic evidence of their mode of formation and source environment. Scanning electron microscopy has been widely used to study the surface textures of quartz sand grains (Krinsley and Doornkamp, 1973), but much less work has been done on silt particles (Pye, 1984c; Wang Yong-yan et al., 1984). Three main problems arise in all attempts to use surface textural features to infer environmental histories. These are: (1) the problems of equifinality, the principle that similar textures may be produced by quite different processes; (2) indeterminacy, by which the same process may produce different textural features depending, for example, on crystallographic differences; and (3) superimposition of features formed in several different process environments. In some near-surface weathering environments, primary surface textural features can be almost completely overprinted within a few years.

Pye (1982, 1984c) examined the surface textures on quartz silt grains from a variety of natural environments, including sub-glacial, cold weathering, hot desert and humid tropical, and concluded that particle shape and surface texture are not clearly diagnostic of mode of origin or source environment, although some generalizations can be made. Frost action, salt weathering, glacial grinding and aeolian abrasion are all capable of producing blocky, angular or sub-angular quartz silt grains with sharp edges and conchoidal fracture surfaces (e.g. Fig. 9.19c). Description and interpretation of many distinctive but rare features is frequently problematical. For example, features described as 'chattermarks' or 'zip-like trails' (Fig. 9.19d) have been considered by some authors to be diagnostic of glacial abrasion, while others have maintained that they can also be produced by chemical action and fluvial abrasion (Bull et al., 1980).

Additional difficulties arise in studying the surface textural features of quartz silt grains in loess because the surfaces of the grains are typically obscured by coatings of clay minerals, carbonate and amorphous silica. If chemical alteration has not progressed too far, it may be possible to remove these coatings by ultrasonic vibration and acid etching. Experience has shown, however, that in many cases this does not provide a complete solution. It is also difficult to establish how much of the surface coating on quartz grains is pre-depositional and how much is post-depositional, since a great deal of redistribution of fine clastic and soluble material occurs in loess shortly after deposition.

9.4.4 Engineering properties and microfabric of loess

Unweathered loess is a relatively poorly consolidated sediment of medium to low compressive strength ($0 \cdot 5$–$2 \cdot 5$ kg cm^{-2}. The porosity typically is 40–55% and the moisture content above the water content 10–25%, the variation partly being related to the clay and organic matter content. The bulk density of typical loess lies in the range $1 \cdot 5$–$1 \cdot 8$ g cm^{-3} and the specific gravity is $2 \cdot 1$–$2 \cdot 75$. Dry density lies in the range $1 \cdot 1$–$1 \cdot 5$ g cm^{-3}. The voids ratio generally lies between $0 \cdot 7$ and $0 \cdot 9$. Weathered and diagenetically altered older loess usually has a higher dry density ($1 \cdot 5$–2 g cm^{-3}), higher specific gravity ($2 \cdot 6$–$2 \cdot 8$), and lower voids ratio ($0 \cdot 5$–$0 \cdot 7$). The Atterberg limits of unweathered and weathered loess show a considerable degree of overlap, however. Reported values for the Liquid Limit are 28–44%, for the Plastic Limit 12–20%, and for the Plasticity Index 16–29% (e.g. Watkins, 1945; Odell *et al.*, 1960; Liu Tung-sheng *et al.*, 1985*a*; Derbyshire and Mellors, 1987).

Loess is a metastable foundation material, i.e. it exhibits settlement when wetted. Loess formed in arid regions which has experienced limited pre-wetting may collapse under its own weight if it becomes saturated, but in more humid areas collapse on wetting normally occurs only if a load is applied. A number of subsidence criteria have been devised by engineering geologists based on the natural voids ratio (e_0), the voids ratio at the liquid limit (e_l), and the voids ratio at the plastic limit (e_p). Two of the most widely used criteria are the critical voids ratios (e_c) proposed by Denisov (1951) and Feda (1966):

$$\text{Denisov criterion: } e_c = e_l$$
$$\text{Feda criterion: } e_c = 0 \cdot 85 e_l + 0 \cdot 15 e_p$$

Another widely used parameter is the collapse factor, R, which is calculated from the results of oedometer tests to measure changes in the response of soils to flooding while subject to an applied load:

$$R = \frac{\text{change in thickness of specimen on flooding}}{\text{thickness of specimen immediately before flooding}} \times 100\%$$

Most unweathered loess collapses by up to 15% when flooded under low applied loads (<100 KN m^{-2}), but cemented loess, fluvially redeposited loess and weathered loess generally shows less collapse settlement and requires higher applied loads (Audric and Bouquier, 1976; Derbyshire and Mellors, 1987). The tendency for saturated loess soils to collapse is inversely related to

Figure. 9.21. Decalcified surface layer (approximately 2 m deep) developed in loess of late Last Glacial age, Vicksburg, Mississippi.

the clay content. Handy (1973) found in Iowa that loess containing <16% clay is highly prone to collapse; with 24% clay there is a 50% probability of collapse, and with 36% clay the chances of collapse are negligible. Some loess soils, notably those with a high smectite content, are capable of both swelling and collapse behaviour depending on the magnitude of the applied load. A common engineering solution to the problems posed by metastable loess soils is flooding and rolling (hydro-consolidation) before construction (Beles *et al.*, 1969; Lobdell, 1981).

The ability of loess to stand in vertical sections has been attributed to one or more of the following factors: (1) cementation of grain contacts and vertical root pipes by secondary calcium carbonate (Snowden and Priddy, 1968; Derbyshire and Mellors, 1987); (2) the presence of tensional moisture films and clay bridges at silt grain contacts (Lohnes and Handy, 1968; Snowden and Priddy, 1968; Derbyshire and Mellors, 1987); (3) the effect of van der Waals attractive forces between small particles (Smalley, 1966); (4) high permeability which prevents saturation above the water table (Snowden and Priddy, 1968); and (5) the nature of grain orientation (Matalucci *et al.*, 1969, 1970). Calcium carbonate cementation cannot be the main factor, since carbonate-free and leached loess can also stand in vertical sections (Fig. 9.21).

Matalucci *et al.* (1969, 1970) found that quartz grains showed imbrication

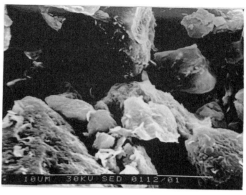

Figure 9.22. SEM micrograph showing typical open fabric of unweathered late Last Glacial loess, Vicksburg, Mississippi. (Scale bar = 10 μm.)

parallel to the palaeowind direction at Vicksburg, Mississippi, and that the shear strength of the loess is 12% greater in a direction perpendicular to the grain orientation than parallel to it. Failure planes developed parallel to the grain orientation produced offlapping, short, discontinuous fractures, whereas a uniform, continuous fracture developed normal to the grain orientation. However, grain orientation probably contributes little to the vertical stability of loess faces in Mississippi since these show a wide variety of orientations.

Several different mechanisms of collapse in loess soils have been proposed (Fookes and Best, 1969; Barden *et al.*, 1973; Lin Zaiguan and Liang Weiming, 1980; Lutenegger, 1981; Grabowska-Olszewska, 1975, 1982), but an important process is microshearing at points of grain contact; this may be related to dispersion of clay bridges at these points when the loess is wetted (Derbyshire and Mellors, 1986).

Scanning electron microscopy examination has shown that unweathered late Pleistocene loess is typified by a relatively open microfabric in which silt-sized grains are held together by meniscus-type clay bridges (Derbyshire, 1983*a*, *b*; Fig. 9.22). Softening or washing away of the clay bridges significantly reduces the cohesion between grains and may allow intergranular shearing to take place. Another process may also contribute to the hydro-sensitivity of loess. At the time of deposition, fine silt and clay particles are present both as coarse- and medium-sized aggregates and as individual platelets adhering to the surface of larger grains (Fig. 9.23a). After infiltration of rainwater and several wetting and drying cycles, some of the aggregates break down and the fines accumulate at silt grain contacts where meniscus moisture films are held. Disintegration of the primary aggregates creates secondary void spaces which enhance the

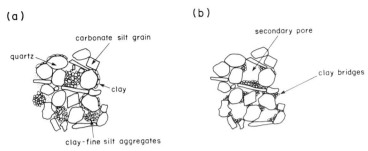

Figure 9.23. A suggested mechanism for the partial collapse of loess when wetted and subject to an applied load: (a) at the time of deposition clay-size particles are present both as coatings on larger silt grains and as silt-size aggregates; and (b) the loess microfabric is rapidly modified after deposition by percolating rainwater; aggregates of clay-size particles are partially broken-down and the individual particles redistributed. Most are deposited near the points of contact of silt grains, where moisture menisci are held. The breakdown of aggregates creates secondary pore spaces which facilitate rearrangement of the constituent grains when a load is applied. Settlement is enhanced by moisture which softens the intergranular clay bridges and lubricates the contact between quartz silt grains.

collapsing behaviour when the sediment becomes saturated (Fig. 9.23b). Additionally, on saturation some of the shearing resistance due to tensional moisture films is counteracted by a build-up of positive hydrostatic pore pressures.

With increased degree of weathering the loess microfabric becomes less open; silt grains become entirely coated with translocated clay which may partly fill the pores (Figs. 9.24 and 9.25). In addition to redistribution of fine and soluble material, changes in loess microfabric during weathering also result from dissolution of detrital carbonate grains, alteration of feldspars and heavy minerals, and formation of secondary minerals. The latter frequently include needle-calcite (Fig. 9.26; see also Šajgalík, 1979), authigenic K-feldspar, clay minerals and iron oxyhydroxides. As the silt/clay ratio, voids ratio and permeability are reduced during weathering, the opportunity for rearrangement of silt framework grains and migration of clays from points of silt-grain contact is also reduced, thereby restricting the tendency for the sediment to collapse when wetted.

Deep vertical fissures commonly develop sub-parallel to the face of high, steeply-inclined loess sections as a result of tensile stresses in the upper part of the slope (Terzaghi, 1943). In general, vertical faces in carbonate-bearing loess higher than about 5 m are liable to develop tensile cracks and are potentially unstable (Gwynne, 1950; Lohnes and Handy, 1968). Road cuts are frequently terraced at 5 m intervals to minimize the risk of slope failure (Fig. 9.8). The engineering geological problems associated with loess are discussed in more

Figure 9.24. SEM micrograph showing less-open fabric of slightly weathered loess which was deposited slowly between about 25 000 and 22 000 yr BP at Vicksburg, Mississippi. Translocated detrital and authigenic clay particles coat the surfaces of all grains and form bridges around pore throats. (Scale bar = 10 μm.)

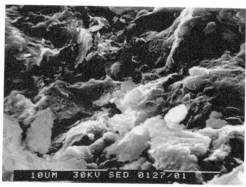

Figure 9.25. SEM micrograph of highly weathered early Last Glacial loess from Vicksburg, Mississippi. The clay content of this loess exceeds 50% and there are few visible pores. (Scale bar = 10 μm.)

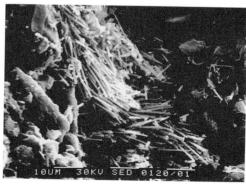

Figure 9.26. SEM micrograph showing authigenic needle-calcite infilling secondary pore in the loess shown in Fig. 9.24. (Scale bar = 10 μm.)

detail by Clevenger (1958) and Sheeler (1968), while many other useful references are contained in Lutenegger (1985).

9.4.5 Chemical and mineral composition

The major element composition of loess is related to its mineral composition and varies significantly between different regions. However, the majority of unweathered loess samples contain 50–75% SiO_2, 6–15% Al_2O_3, 2–7% Fe_2O_3, 0·5–15% CaO, 0–5% MgO, 1–6% K_2O, 1–4% Na_2O, 0·5–1.0% TiO_2, 0·02–0·2% MnO and 0·1–1·0% P_2O_5. In areas of humid climate, weathered loess is generally enriched in silicon, aluminium and iron relative to unweathered loess. Major element data for a number of unweathered and weathered loess samples are given in Table 9.2. The Timaru (New Zealand) and Alaska loesses are notably siliceous and low in calcium. By comparison, the Tajikistan loess is very rich in calcium (present mainly as carbonate). These regional differences reflect the composition of source rocks. The relatively sodium-rich loess at Timaru is derived largely from plagioclase-rich greywackes (Raeside, 1964; Griffiths, 1973), while the iron-rich Alaskan loess contains much biotite and chlorite derived from schists and gneissic rocks (Péwé, 1955). The Argentine loess is high in alkalis, iron and aluminium due to an abundance of volcanic feldspars and glass shards in the mineral assemblage (Teruggi, 1957). Much of the loess in Europe, North America, China and Soviet Central Asia contains abundant calcite and dolomite derived from marine limestones in addition to quartz and aluminosilicates derived from a variety of igneous and metamorphic rocks. In Europe an exception is provided by the highly siliceous *limon à doublets* of northern France which are considered to have experienced syn-depositional decalcification in a periglacial climate (Lautridou and Giresse, 1981; Lautridou *et al.*, 1984).

Pye and Johnson (1987) found that unleached late Wisconsinan loess in southern Mississippi contained less iron, aluminium, silicon, manganese and titanium than the leached Holocene soil developed at the top of the profile (Fig. 9.21) and much lower amounts of these elements than highly weathered early Wisconsinan and late Illinoian (?) loess (Table 9.3). A simple weathering index was devised to facilitate comparison between different stratigraphic units:

$$WI = Al_2O_3 + Fe_2O_3/Na_2O + K_2O$$

Typical values of the index were found to be 3–3·5 for unleached late Wisconsinan loess, 4–4·5 for the surface Holocene soil, 4·5–7·7 for early Wisconsinan loess, and 6·6–12 for late Illinoian (?) loess. A thin, clayey loess unit formed just before the onset of the main phase of late Wisconsinan loess

Table 9.2

Major element geochemistry of unweathered and weathered loess, determined by X-ray fluorescence spectrometry (weight % oxide). Determinations made on lithium tetraborate/lithium carbonate/lanthanum oxide fusion beads.

Unweathered loess

	Vicksburg 2K Mississippi	Vicksburg 2Q Mississippi	Vicksburg 2T Mississippi	Nurek Tajikistan	Karamaidan Tajikistan	Timaru New Zealand	La Plata Argentina[1]
SiO_2	61·72	61·79	61·87	49·39	53·10	71·31	63·34
Al_2O_3	7·92	7·73	7·52	11·28	12·08	13·21	16·13
TiO_2	0·62	0·65	0·64	0·58	0·62	0·62	0·74
Fe_2O_3	3·02	2·89	2·88	4·58	4·97	3·53	5·10
MgO	4·23	4·53	4·34	2·87	3·01	0·98	1·64
CaO	8·26	8·31	8·23	13·56	10·87	1·73	3·46
Na_2O	1·28	1·31	1·30	1·35	1·48	3·23	⎫ 3·68
K_2O	1·86	1·82	1·74	2·15	2·28	1·98	⎭
MnO	0·07	0·08	0·06	0·10	0·10	0·04	—
P_2O_5	0·13	0·14	0·12	0·14	0·16	0·14	—
LOI	11·51	11·81	12·28	13·74	11·12	2·78	3·80
Total	100·62	101·16	100·98	99·74	99·79	99·55	97·89

[1] Data of Teruggi (1957)

continued

Table 9.2 Continued.

Unweathered Loess

	Richardson, Alaska	Wallertheim, West Germany	Ariendorf, West Germany	Tyszowce, Poland	Mingtepe Uzbekistan	Yakutia Siberia[2]
SiO_2	72·67	60·69	63·56	73·02	56·28	59·01
Al_2O_3	11·93	8·44	8·53	8·69	10·18	11·84
TiO_2	0·67	0·59	0·59	0·62	0·56	0·72
Fe_2O_3	5·01	3·21	3·10	2·77	3·86	4·60
MgO	1·75	2·15	1·45	1·10	2·89	1·43
CaO	1·92	10·40	9·78	4·76	11·06	7·19
Na_2O	1·72	0·80	0·98	0·96	1·50	2·18
K_2O	1·94	1·79	1·69	2·04	2·03	2·56
MnO	0·07	0·09	0·08	0·04	0·07	0·16
P_2O_5	0·12	0·10	0·08	0·08	0·13	0·81
LOI	2·32	12·08	10·56	6·20	11·44	7·79
Total	100·12	100·34	100·40	100·28	100·10	98·29

[2] Data of Péwé and Journaux (1983)

Table 9.2 Continued.

	Unweathered loess		Weathered loess		
	limons à doublets, Villecartier, France[3]	Nebraska[4]	Vicksburg 2A Mississippi leached horizon of Holocene soil	Vicksburg 2E Mississippi leached horizon of Holocene soil	Karamaidan Tajikistan partially leached horizon of 9th pedocomplex
SiO_2	82·03	67·20	74·24	75·51	59·10
Al_2O_3	8·68	12·00	11·02	10·14	14·95
TiO_2	0·68	0·57	0·74	0·74	0·77
Fe_2O_3	3·38	3·36	4·30	3·95	6·24
MgO	0·65	2·08	0·76	0·89	2·45
CaO	0·61	4·76	0·69	0·85	4·72
Na_2O	1·09	1·43	1·17	1·38	1·17
K_2O	2·74	2·37	2·26	2·26	2·86
MnO	0·08	0·05	0·12	0·07	0·12
P_2O_5	–	0·16	0·18	0·14	0·14
LOI	–	5·50	3·57	2·76	7·23
Total	99·94	99·48	99·05	98·69	99·75

[3] Data of Lautridou et al. (1984)
[4] Data of Watkins (1945)

Table 9.3

Major and trace element composition of selected samples from Section V1, Vicksburg, Mississippi. Major elements reported as weight per cent oxide, trace elements reported as parts per million. Samples denoted 'T' are transitional zones. Loess stratigraphic units (column 2) are shown in Fig. 9.35. Data obtained by XRF analysis of lithium metaborate fusion beads (major elements) and pressed power pellets (trace elements).

Sample	Unit	Depth (m)	SiO_2	Al_2O_3	TiO_2	Fe_2O_3	MgO	CaO	Na_2O	K_2O	MnO	P_2O_5	LOI	Total	Weathering index
V1(i)	1	7.0	63.14	8.47	0.56	2.72	3.51	8.24	1.54	2.01	0.05	0.12	9.82	100.17	3.15
V1A	1	9.2	59.95	7.58	0.57	2.89	5.16	8.08	1.48	1.85	0.06	0.13	12.50	100.24	3.14
V1B	1	10.0	60.69	7.81	0.57	2.76	4.90	8.36	1.52	1.85	0.06	0.13	11.59	100.24	3.14
V1F	1	12.0	53.34	6.65	0.48	2.31	6.29	12.13	1.50	1.68	0.05	0.13	15.78	100.34	2.82
V1G	1	12.5	51.92	6.42	0.46	2.19	6.88	12.78	1.09	1.68	0.05	0.12	16.71	100.30	3.11
V1H	2	13.0	62.44	7.96	0.57	2.64	3.32	9.08	1.30	1.87	0.05	0.11	10.68	100.03	3.34
V1I	3	13.5	77.52	10.09	0.80	3.26	0.76	0.79	1.49	2.19	0.04	0.03	2.60	99.57	3.63
V1K	3	14.5	73.31	12.34	0.78	4.86	0.90	0.68	1.34	2.03	0.02	0.02	3.87	100.15	5.10
V1N	3	16.0	73.38	10.32	0.73	3.63	0.91	3.04	1.78	2.19	0.08	0.07	3.93	100.03	3.54
V1P	3T	17.0	79.15	9.96	0.95	3.51	0.48	0.70	0.40	1.35	0.01	0.04	3.71	100.28	7.70
V1R	4	18.0	69.16	15.48	0.87	6.17	0.76	0.46	0.37	1.47	0.01	0.07	5.56	100.36	11.77
V1S	4	18.5	70.21	13.77	0.81	4.83	0.85	0.83	0.44	1.76	0.03	0.04	5.95	99.63	8.45
V1T	4	19.0	70.31	15.30	0.85	5.15	0.71	0.39	0.52	1.78	0.01	0.04	5.36	100.40	8.89

Trace elements

Sample	As	Ba	Ce	Cl	Co	Cr	Cu	Ga	La	Ni	Nb	Pb	Rb	S	Sr	Th	U	V	Y	Zn	Zr
V1B	1	495	39	116	11	47	16	10	27	18	12	17	53	27	160	4	2	64	30	39	370
V1F	0	377	31	200	9	35	11	10	13	14	11	15	43	35	171	4	3	54	25	26	343
V1K	5	537	81	25	11	61	22	15	23	20	15	22	63	0	113	14	0	81	28	43	448
V1N	4	641	66	28	8	53	26	12	27	24	15	18	63	0	156	6	3	74	45	42	439
V1P	4	433	60	14	3	80	16	11	27	14	19	17	56	0	47	10	2	74	35	24	809
V1R	10	452	41	22	6	99	23	18	25	24	17	17	63	0	48	8	2	112	26	38	510
V1T	5	488	37	22	10	83	24	17	17	22	16	22	65	0	56	9	4	99	21	36	433

deposition had a weathering index of 3–4. In these sediments the sequence of elemental stability during weathering appears to be:

$$Ca < Mg < Na < K < Si < Ti < Fe < Al$$

The trace elements barium, cobalt, chromium, gallium, rubidium, thorium, uranium and zirconium appear to have been relatively stable during weathering, while arsenic, chlorine, copper, lanthanum, nickel, lead, strontium, vanadium and zinc decreased with depth (age) in the profiles, suggesting loss by leaching.

In late Wisconsinan loess of Missouri, Ebens and Connor (1980) found that aluminium, arsenic, cobalt, copper, fluorine, gallium, iron, lithium, scandium, strontium, vanadium and yttrium increased in abundance with downwind distance from the source while barium, calcium, carbon, magnesium, manganese, silicon, sodium, phosphorus, potassium and zirconium abundance decreased downwind. These trends were associated with an increase in clay content relative to quartz, feldspar, dolomite, zircon and apatite in the downwind direction. Wen Qi-zhong *et al.* (1982) reported a decrease in SiO_2 and increase in Al_2O_3 and Fe_2O_3 with distance from the source in the Chinese Central Loess Plateau. The total trace element concentration also increased in the direction of wind transport, although no systematic trends were observed in the abundance of rare earth elements (REE).

Wen Qi-zhong *et al.* (1982, 1984) also examined geochemical variations with depth in the Luochuan section, which records loess sedimentation since about 2·4 million years ago (Heller and Liu Tung-sheng, 1982, 1984). SiO_2/Al_2O_3, FeO/Fe_2O_3, CaO/MgO and $CaO + K_2O + Na_2O/Al_2O_3$ ratios were found to increase from the top to the bottom of the section, and were interpreted to indicate progressively more arid climatic conditions in the area since the beginning of the Pleistocene. The REE content decreased slightly from the top to the bottom of the section, but variations were small, suggesting that the character of the source sediments has not changed substantially.

Taylor *et al.* (1983) compared the composition of loess with that of the average upper crust. On the basis of analyses of a limited number of samples from four areas (China, New Zealand, Kansas and West Germany) they concluded that loess has a very similar major element composition, expressed on a carbonate-free basis, although some local provenance effects are evident. They suggested that the principal difference between the average composition of loess and the upper crust is that loess has a higher silica content, causing the other elements to fall below the crustal average (Fig. 9.27).

Typical loess contains 50–70% quartz, 5–30% feldspar, 5–10% mica, 0–30% carbonate, and 10–15% clay minerals. Markedly different mineralogies are sometimes found, however. In the Persian Gulf, pockets of loess contain up to 80% carbonate (calcite, dolomite and aragonite) and contain substantial

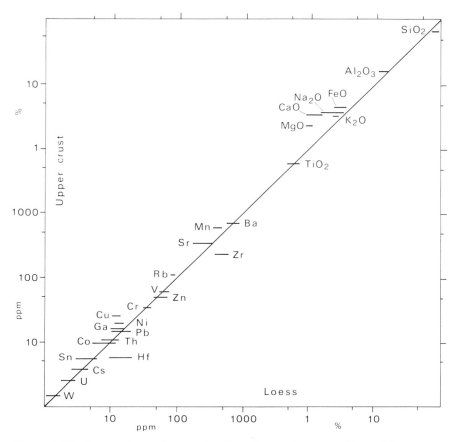

Figure 9.27. Comparison diagram for the chemical composition of the upper continental crust and the range of unweathered and non-carbonate loess samples from China, West Germany, Kansas and New Zealand. On average the loess is enriched in silica and depleted in most other major elements compared with the upper crust. Zirconium and hafnium are significantly enriched in the loess. (After Taylor et al. *(1983), Geochim. Cosmochim. Acta **47**, 322–332; ©*
Pergamon Press.)

amounts of gypsum (Doornkamp *et al.*, 1980). In Argentina, the Caucasus and parts of New Zealand, loess contains an unusually high proportion of glass shards, volcanic feldspars and (locally) tridymite (Terruggi, 1957; Kennedy, 1982; Galay *et al.*, 1982). Some of this material is airfall tephra which has been deposited on, and mixed with, the loess by bioturbation, but some owes its origin to aeolian reworking of tephra deposits.

The clay mineral assemblage in loess reflects both the nature of the source

Table 9.4
Bulk mineralogical composition of loess from several different parts of the world, determined by X-ray diffraction. Only those minerals comprising >5% of the sediment are shown.

Quartz	Feldspar	Calcite	Dolomite	Illite	Kaolinite	Smectite	Chlorite	Gibbsite	Biotite	Other¹	Unweathered loess
X	X	X	X	X	X	X					Vicksburg, Mississippi
X	X	X	X	X	X	X	X				Nurek, Tajikistan
X	X	X	X	X	X	X	X				Mingtepe, Uzbekistan
X	X	X		X	X						Karhlich, West Germany
X	X	X		X							Timaru, New Zealand
X	X	X	X	X	X						Pegwell Bay, Kent, UK
X	X	X	X	X	X						Tyszowce, Poland
X	X	X	X	X	X						Wallertheim, West Germany
X	X			X			X			X	Richardson, Alaska
X	X	X		X	X		X				Meng Xian, China (Derbyshire, 1983*b*)
X	X	X		X			X				Jiuzhoutai, China

Quartz	Feldspar	Calcite	Dolomite	Illite	Kaolinite	Smectite	Chlorite	Gibbsite	Biotite	Other¹	Leached loess
X	X			X	X	X	X				Vicksburg, Mississippi
X	X			X	X	X	X				Natchez, Mississippi
X	X	X		X	X		X				Karamaidan, Tajikistan
X	X			X	X						Karhlich, West Germany
X	X			X	X						Warneton, Belgium
X	X		X	X	X						Tongrinne, Belgium
X	X		X	X	X						Barton, Hampshire, UK
X	X			X				X		X	Nueva Braunau, Chile

¹ Includes tridymite and cristobalite.

material and weathering conditions experienced after deposition. Kaolinite, illite, smectite, mixed-layer clays, chlorite and vermiculite have frequently been identified in loess (Table 9.4). Halloysite and palygorskite have also been recorded (Schwaighofer, 1980). In the United States, differing clay mineral compositions of Peoria (late Wisconsinan) loess have been related to distinct valley sources which, in turn, are related to different up-valley provenances during glacial times (Frye *et al.*, 1962; Ruhe and Olson, 1980; Ruhe, 1984). In Illinois, for example, montmorillonite comprises 70% of the clay minerals in Peoria loess along the Mississippi Valley, 54% along the Illinois Valley, 26%

Table 9.5
Main clay mineral species and their relative amounts in Luochuan and Longxi sections, Chinese Loess Plateau.

Strata	Sediments	Amount	clay minerals and relative amount (%)			
			illite	chlorite	kao-linite	montmo-rilionite
Luochuan section						
Malan 1.	Loess	3	68	12	11	9
Lishi 1.	Loess	4	67	11	10	12
(Upper)	paleosol	6	72	8	8	13
Lishi 1.	Loess	9	63	11	11	16
(Lower)	paleosol	8	74	7	7	13
Wucheng 1.	Loess	12	60	11	10	19
	weathering B	10	61	10	9	20
Longxi section						
Holocene 1.	Loess	2	56	19	13	12
	paleosol	2	62	15	14	10
Malan 1.	Loess	1	51	22	17	10
Lishi 1.	Loess	7	62	11	12	15
	paleosol	10	64	10	11	15
Wucheng 1.	Loess	6	69	6	8	17
	weathering B	3	73	5	9	13

(After Zheng Hong-han, 1984, in M. Pecsi (ed.), *Lithology and Stratigraphy of Loess and Paleosols*, pp. 171–181; © Geographical Research Institute, Hungarian Academy of Sciences.)

along the Wabash Valley, and 9% along the Ohio Valley, whereas illite abundance is low along the Mississippi Valley (11%) and high along the Wabash Valley (71%) (Frye *et al.*, 1962).

Zheng Hong-han (1984) examined the clay mineral composition in two long Chinese loess sections at Luochuan and Longxi. Relatively little variation was found in the relative abundance of different clay minerals with increasing age, suggesting long-term uniformity of the source material. Illite was the dominant clay mineral throughout, with subsidiary amounts of chlorite, kaolinite and montmorillonite. In the palaeosols the relative abundance of illite was found to be even higher than in the loess units (Table 9.5).

In humid areas the detrital mineralogy of loess is frequently altered by post-depositional weathering and mineral authigenesis. In southern Mississippi, for example, fresh late-Wisconsinan loess is rich in dolomite and quartz, and has a relatively high plagioclase/K-feldspar ratio (Fig. 9.28a). The lower part of the Holocene soil and the slightly older, partially weathered loess unit

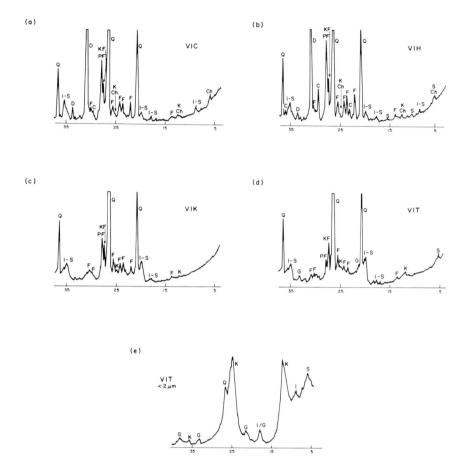

Figure 9.28. X-ray powder diffraction traces (CuK$_\alpha$ radiation) of loess from different litho-stratigraphic units at Vicksburg, Mississippi (see Fig. 9.33): (a) unleached loess from Unit 1, containing much detrital dolomite; (b) partially leached, slightly clayey loess from Unit 2; (c) leached and weathered loess from Unit 3; (d) leached and very highly weathered loess from Unit 4; and (e) <2 μm fraction from the Unit 4 loess. Q = quartz; PF = plagioclase feldspar; KF = K-feldspar; D = dolomite; C = calcite; K = kaolinite; I–S = illite–smectite plus mica; Ch = chlorite; G = goethite.

Figure 9.29. Loess dolls from Vicksburg loess.

underlying the unleached loess contain secondary calcite (partly as micro-nodules) in addition to dolomite, plus authigenic smectite and kaolinite (Fig. 9.28b). Early Wisconsinan loess in this area contains almost no dolomite or calcite while plagioclase and K-feldspar are present in approximately equal proportions (Fig. 9.28c). The oldest (late Illinoian ?) loess is also entirely decalcified and contains more K-feldspar than plagioclase (Fig. 9.28d). The <2 μm fractions of both the older loess units are enriched in authigenic smectite, kaolinite and goethite (Fig. 9.28e). While changes in the detrital mineralogy over time cannot be ruled out, the observed mineralogical changes are consistent with progressive leaching of carbonates and decomposition of

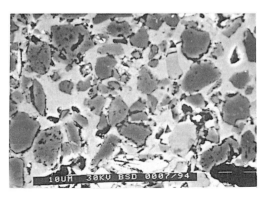

Figure 9.30. Back-scattered SEM micrograph of a polished thin section of a loess doll from Vicksburg, showing quartz and feldspar grains (dark grey) cemented by calcite (light grey). (Scale bar = 10 μm.)

less stable plagioclase feldspars and heavy minerals to form clay and iron hydroxides.

Secondary carbonate concretions, known as 'loess dolls' or 'lösskinchen' (Rosauer and Frechen, 1960; Chmielowiec, 1960; Leach, 1974), are common diagenetic features in partially leached loess (Fig. 9.29). The pore spaces between silt grains in such concretions are often completely filled by fine granular calcite (Figs. 9.30 and 9.31a). Many of the nodules form around plant roots and contain large internal voids partly filled by coarse rhombohedral calcite (Fig. 9.31b). Pedogenetic gypsum and iron–manganese concretions have also been recorded in loess (Schwertmann and Fanning, 1976; Childs and Leslie, 1977; Derbyshire, 1983*b*).

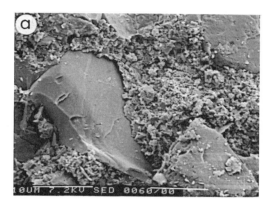

Figure 9.31. SEM micrographs showing (a) *fine granular calcite cementing quartz grains in a loess doll and* (b) *coarse, blocky calcite cement infilling a former root-tube within a Vicksburg loess doll. (Scale bars = 10 μm.)*

9.5 LOESS LANDFORMS

In areas of rolling or deeply dissected terrain loess typically forms a blanket of varying thickness which wholly or partly buries the existing topography (Norton, 1984). In mountainous areas, such as Tajikistan, loess is thin or absent on the steep upper mountain slopes and generally thickens in the downslope direction. In more flat-lying areas, the loess forms a flat surface and is of fairly uniform thickness.

Three main loess morphological types, referred to as 'yuan', 'liang' and 'mao', have been distinguished in Central China, where loess accumulated in a series of broad, shallow basins bounded by mountain ranges. The yuan, which is a flat or gently sloping loess plateau surface, is particularly well-developed in southern Shansi and westernmost Gansu (Zhang Zong-hu, 1984; Wu Zi-rhong *et al.*, 1982). Liang, or elongate ridges, occur extensively in northern Shansi and Shensi. Mao (hemispherical hills) are common in the liang areas but predominate in Shansi close to the Hwang Ho River. The three morphological types are regarded as sequential products of erosion (Wang Yong-yan and Zhang Zhong-hu, 1980). A fourth morphological type, consisting of residual cones and intervening plains (Derbyshire, 1983*a*) occurs in a broad region north of the Hwang Ho near Lanzhou. In this area, river incision has produced maximum relative relief of 500 m adjacent to the Hwang Ho Gorge, the valley systems being cut into the Miocene and Cretaceous rocks beneath the loess by as much as 200 m. The result is a series of thick residual loess bodies known as 'loess caps'. As the loess thins towards the north, the relative relief of the dissected spurs and cones decreases to 50–150 m, and flat plain areas become more extensive.

Due to its high porosity and permeability, infiltration rates in loess are relatively high, although on bare surfaces infiltration may be reduced by crusts formed during rainstorms (Duley, 1945; Williams and Allman, 1969). The drainage in areas of thick loess is almost entirely subterranean through pipes and fissures. Enlargement of pipes can lead to cavern formation and, ultimately, collapse (Fuller, 1922; Landes, 1933; Barbour, 1935). Features produced by collapse include dry valley systems, circular surface depressions, and active gullies (Fookes and Knill, 1969; Bariss, 1971; Volov and Orlovskii, 1977; Roloff *et al.*, 1981). Once formed, gullies increase in size by headwall erosion and collapse (Bradford *et al.*, 1978). Some gullies in the Chinese Loess Plateau are more than 40 km long. Soil erosion rates in the middle Hwang Ho loess lands are some of the highest in the world, being of the order of 10×10^3 t km^{-2} yr^{-1} and reaching $34 \cdot 5 \times 10^3$ t km^{-2} yr^{-1} in areas of severely gullied terrain (Derbyshire, 1978, 1983*a*).

Gullying and river erosion at the foot of loess slopes causes undercutting and collapse, forming steep bluffs with an angle of 70–85° and talus slopes composed of fallen slabs (Pecsi and Sheuer, 1979; Pecsi *et al.*, 1979). The process of collapse and bluff development is aided by the development of unloading joints parallel to the ground surface and tensile cracks parallel to the bluff face. The angle of the lower talus slopes depends on the values of ϕ' and c' (Lohnes and Handy, 1968), but is generally between 40 and 60°. During heavy rainstorms rotational slides and slurry flows sometimes occur on loess slopes, producing broad, flat-floored gullies and depositional lobes which may extend over distances of several hundred metres. Landslides and slurry flows can also be triggered by earthquakes. In China, the Haiyuan earthquake of 1920 which killed 230 000 people caused 650 loess collapses in Xiji County and formed 41 dammed lakes (Zhu Haizhi, 1982, cited in Liu Tung-sheng *et al.*, 1985*a*).

9.6 THEORIES OF LOESS ORIGIN

Throughout much of the last century, loess was widely believed to be a fluvial deposit (Lyell, 1834, 1847; Binney, 1846; Dumont, 1852; Todd, 1897; Fuller and Clapp, 1903). A lacustrine origin was favoured by Pumpelly (1866), Call (1882) and Warren (1878), while Kingsmill (1871) and Skertchley and Kingsmill (1895) favoured a marine origin. Howarth (1882) suggested loess is a volcanic deposit, while Keilhack (1920) concluded that it is of cosmic origin, being derived from dust rings similar to those of Saturn. Campbell (1889) suggested loess is derived from animal and vegetable matter.

The aeolian hypothesis advocated by von Richthofen (1877–85, 1882) was accepted by Udden (1897*a*, *b*) but initially received much criticism in North America (Todd, 1878; Child, 1881; Dana, 1895). Chamberlain (1897) proposed a compromise explanation of loess genesis in which he suggested that the Mississippi Valley loess is primarily water-deposited but has experienced some secondary reworking by the wind. As recently as 1944 Russell maintained that the Mississippi loess had formed mainly by weathering and downslope creep of backswamp river terrace silts (Russell, 1944*a*, *b*). Russell's interpretation was accepted by Fisk (1951) but rejected by Holmes (1944), Leighton and Willman (1949) and Doeglas (1949) in favour of an aeolian origin.

In the Soviet Union, the aeolian hypothesis was advocated in the early part of this century by Obruchev (1911, 1945). However, also in the Soviet Union, Berg (1916, 1964) developed an independent theory of loess formation which involved formation of loess and loess-like sediments by *in-situ* weathering and soil formation on carbonate-rich fine-grained rocks under a dry climate. Similar ideas were incorporated in the model proposed by Russell (1944*a*, *b*), who used the term 'loessification' to describe hypothesized changes in the properties of Mississippi alluvial silts due to weathering and downslope movement. Russell suggested that during loessification the carbonate content of loess increases due to precipitation from pore waters, while the sediment becomes better-sorted during downslope creep.

Elements of the aeolian theory and the loessification theory were combined by Lozek (1965), who concluded that loess does not originate only by accumulation of airborne dust, but also by a particular soil-forming process, found under a warm, dry, climatic regime, which is responsible for the typical structure, colour and calcareous character of loess. The presence of molluscs apparently indicative of such climatic conditions was cited as evidence for the applicability of the hypothesis to the Central European loess (see Smalley (1971) for further discussion of *in-situ* loess theories).

Syn-depositional weathering of dust is certainly rapid in some areas, but Lozek's hypothesis cannot be accepted as generally valid. Some loess deposits contain no calcium carbonate, while SEM examination and X-ray micro-analysis have shown that in others carbonate is present predominantly as detrital silt-size particles, including dolomite which is unlikely to be of post-depositional origin (Pye, 1983b). As discussed previously, the degree to which deposited dust becomes modified post-depositionally is dependent on the deposition rate and on climatic conditions at the deposition site (Flint, 1949; Frankel, 1957; Ray, 1967; Bruins and Yaalon, 1979). In humid areas, redistribution of carbonates in loess after deposition usually involves leaching of the 'A' soil horizon and carbonate redeposition, partly as nodules, in the 'B' horizon. The depth of leaching increases with time, although relict carbonate concretions may remain in the 'A' horizon as leaching progresses downward. Loessification as proposed by Lozek could only occur under an arid climate in areas of high water table.

The overwhelming majority of loess workers now accept that primary loess has an aeolian origin, but also recognize that in some areas the loess has been largely reworked and redeposited by slope processes and running water. Some of the processes involved in redeposition of loess have been investigated by Mucher and de Ploey (1977), Vreeken and Mucher (1981) and de Ploey (1984). (See Mucher (1986) for a review.) However, the following lines of evidence indicate that primary loess has an aeolian origin: (1) it typically forms a blanket over a variety of relief forms and extends over a wide altitudinal range; (2) the thickness of loess and mean grain size decrease in a regular manner with distance from the dust source; (3) belts of loess and associated deposits of windblown sand often show a clear geographical relationship to the prevailing wind or palaeowind direction; (4) loess is unstratified and free of pebble stringers which are common in sub-aqueous deposits; (5) the mineralogical composition of loess is often quite different from that of the underlying rock or sediment, making a loessification origin unlikely; (6) modern-day accretion of loess by deposition of windblown dust can be observed in some areas; and (7) the grain-size distribution of typical loess is identical to that of modern aeolian dusts which have not travelled great distances from the source.

9.7 STAGES IN THE FORMATION OF LOESS

Four main stages are involved in the formation of loess deposits (Smalley, 1966; Smalley and Smalley, 1983): (1) formation of loess-size particles (predominantly 20–60 μm); (2) transport of this material; (3) deposition; and (4) post-depositional modification. In simple cases, loess particles have been blown by the wind directly from the area where they were formed to the deposition site; in other cases silt has been transported and concentrated by glacial and/or fluvial processes before being blown by the wind. In very complex situations more than one phase of fluvial transport and aeolian transport may have been involved. Smalley (1980*b*) and Smalley and Smalley (1983) have discussed possible event sequences responsible for the formation of some of the world's major loess deposits.

As discussed in previous chapters, loess-size particles are formed by several different processes in a wide range of environments; the key requirements for a high, sustained rate of aeolian dust transport are: (1) the existence of bare, unstable geomorphic surfaces composed of poorly sorted sediments with a high silt/clay ratio; and (2) a fairly high frequency of strong, turbulent winds. Overwhelming evidence indicates an association of mid-latitude loess formation with glacial periods of the Pleistocene (Tutkovskii, 1900; Jahn, 1950; Dylik, 1954), when these conditions existed on a large scale. Today they occur only locally in high-latitude areas such as Alaska, Greenland, Spitzbergen and Antarctica, and in some high mountain belts such as the Andes, Himalayas and Tien Shan. The main sources of aeolian silt during glacial times were sandurs and braided river channels fed by seasonal glacial meltwaters. Some of the silt was produced by glacial grinding, some by frost action, some by fluvioglacial abrasion and some was reworked from existing weathering profiles or fine-grained rocks. Equally if not more important, however, was the role played by fluvioglacial meltwaters in separating the silt and clay fractions present in outwash sediment. New supplies of fine-grained material were provided by each annual melt. Deflation of fine material from sand and gravel bars at times of low flow was favoured in glacial times by strong winds blowing outwards from high-pressure centres located over the main ice sheets, as described by Hobbs (1933, 1942, 1943*a*, *b*). In areas of mountain glaciation, katabatic winds acted in a similar manner. In most instances dust was blown only relatively short distances from the source by low-level winds before being trapped by vegetation or topographic obstacles.

In Europe, loess was deposited in the unglaciated belt sandwiched between Scandinavian ice sheets in the north and Alpine glaciers to the south (Fig. 9.4). Pollen and faunal evidence indicates that the loess accumulated under relatively cold and dry conditions with steppe vegetation (Kukla, 1970; Kukla and Koči, 1972; Fink and Kukla, 1977). Ice-wedge casts and other geomorphological features indicate syn-depositional periglacial conditions (e.g. Haesaerts and van Vliet-Lanoë, 1973). During interglacial and interstadial periods, the ice-sheets retreated and the steppe vegetation in Central Europe was replaced by temperate broadleaf forests. Dust deflation from meltwater channels ceased and the loess experienced pedogenesis under relatively warm, humid conditions.

The source and transport history of loess in other parts of the world is less clear. This is particulary true of loess deposits which occur close to both glaciated mountain ranges and warm deserts, as in Soviet Central Asia and China. Smalley and Krinsley (1978) and Smalley (1980b) maintained that most of the loessic silt in these regions was formed by glaciation and cold weathering in the mountains and then carried out into the desert plains by rivers before being blown back on to the foothills. 'Without the cold particle origin the deposits could not form, and this is why the deserts which do not have the adjacent source of coarse silt do not have associated loess deposits' (Smalley and Krinsley, 1978, p. 59). In part, this view was based on the mistaken assumption that no effective silt-generating mechanisms are operative in deserts, and on earlier belief that the mountains and plateaux of China and Central Asia experienced much more extensive glaciation during cold stages of the Pleistocene (Sun Tien-ching and Yang Huan-jen, 1961; Smalley and Vita-Finzi, 1968). More recent work has tended to support the view that most of the Chinese loess, at least, is of desert rather than glacial origin. Pleistocene glaciation is now known to have been very limited in Tibet and southeast China (Zheng Ben-xing and Li Jijun, 1981; Derbyshire, 1983c), probably due to the arid nature of the glacial climate. In the Quilian Shan and Tien Shan, loess overlies, and hence post-dates, glacial moraines and terraces (Derbyshire, 1984c). Recent surveys have also indicated that significant amounts of silt are being produced in the cold, dry deserts of northern and northwest China by the combined action of frost and salt weathering. Clay mineralogical and geochemical evidence from the Central Chinese loess is also considered to be consistent with derivation mainly from desert source areas (Zhao Xi-tao and Qu Yong-xin, 1981). Dune patterns, loess grain-size trends, and the presence of loess in parts of the southern Gobi, point to large-scale aeolian transport from the deserts of Inner Mongolia towards the southeast (Fig. 9.32). Accumulation of thick loess in the area north of Xian appears to have been favoured by topographic obstacles to the west, south and east, and by the wetter climate and better vegetation cover towards the southeast. If the Tenger, Ordos and neighbouring deserts provided the main source of silt, the average distance of dust transport was about 500 km. This is greater than in the mid-latitude European and North American loess belts, and probably indicates more vigorous winds and the relatively inefficient nature of the silt-trapping vegetation on the southern edge of the Mongolian Desert source regions.

It is not certain how much of the silt-size material was actually formed in the

Figure 9.32. Distribution of loess and its source materials in northern China. (Modified after Liu Tung-sheng et al. (1982), in R. J. Wasson (ed.), Quaternary Dust Mantles of China, New Zealand and Australia, pp.1–17; © Australian National University.

Mountains and Bedrock

Dunes

Gravel

Sandy Loess

Loess

Clayey Loess

Redeposited loess
and alluvium

Wind Direction

Gobi and Shamo Deserts and how much was originally transported there by rivers flowing from the surrounding mountains during periods of wetter climate in the Pliocene and early Pleistocene (see Section 9.8 below). The possibility that a significant amount of silt, particularly in the Lanzhou area, where the total thickness of loess is considerably greater than in the Loess Plateau, was derived from the Nan Shan and Min Shan via the Hwang Ho valley during the late Pleistocene (Penck, 1930) has also not been disproven. Northwesterly winds may have transported some fine material to Lanzhou from the Quaidam Basin, but it is unlikely that they could have carried large amounts of medium and coarse silt across the mountain passes to the west of Lanzhou.

There is little firm evidence that the model proposed by Smalley and Krinsley (1978) and Smalley (1980b) is valid in relation to Soviet Central Asia. At the present day the Amu Darya and Syr Darya Rivers carry large loads of sand and silt from the Altai, Ghissars and Pamir mountains towards the Aral Sea, but it is unclear what proportion of this material is of glacial or cold weathering origin. In Central Asia there has been much fluvial incision associated with Quaternary tectonism, with reworking of older folded basinal sediments which contain fines.

At present fine dust is sometimes carried by northwesterly wind systems from the Karakum and Kyzylkum Deserts towards the foothills of Uzbekistan and Kazakhstan; this may also have occurred in the Pleistocene, when mid-latitude depression tracks were displaced further south. However, the loess flanking valleys in the more mountainous parts of Uzbekistan and Tajikistan is more likely to have been derived from sediments periodically exposed in local valleys than from distant desert sources. Long-distance transport of medium and coarse silt across the high and dissected terrain of South Tajikistan is unlikely to have occurred. The great thickness of some loess sections (up to 200 m) in Tajikistan also argues against long-distance transport.

Thus, although fluvial transport has certainly been involved in the formation of the Soviet Central Asian loess, the role of glacial and cold weathering processes has not yet been established, and the distance of aeolian dust transport between source and deposition site may have been much smaller than envisaged by some authors.

9.8 QUATERNARY HISTORY OF LOESS DEPOSITION

9.8.1 Europe

On the basis of palaeomagnetic, pollen, faunal and pedological studies at Krems, Austria, and Brno, Czechoslovakia, Fink and Kukla (1977) identified at least 17 episodes of loess deposition in Central Europe during the last 1·7 million years. At these sites windblown loess is interlayered with hillwash loams and forest soils which formed mainly during interglacial or interstadial times. Eight depositional cycles were recorded in the Brunhes palaeomagnetic epoch (approximately the last 700 000 years) and nine in the middle and Upper Matuyama epoch. The cycles are separated by marklines, levels of abrupt environmental change which were correlated with marked increases in global ice volume indicated by the oceanic oxygen isotope record. Each markline was interpreted by Fink and Kukla as indicating the boundary between full glacial and interglacial conditions.

Fink (1979) reported that the Central European loess record may extend further back in time at the Stranzendorf profile in Austria, where the basal loess is apparently older than the Matuyama–Gauss boundary (2·48 million years).

Multiple loess and interlayered soil horizons also occur at several sites in Hungary (Pecsi, 1982, 1985). At Paks a 60-m high section is exposed along bluffs eroded by the Danube River. The stratigraphic sequence here includes 12 loess layers, 13 fossil soils and two to three sand and sandy silt layers which have been divided into two younger and two older loess complexes (Pecsi, 1979b). The young loesses are intercalated with dark coloured forest soils, while in the older loesses red-coloured forest soils are characteristic and marked erosional hiatuses are evident. The base of the preserved older loess shows reversed magnetic polarity and is of later Matuyama age. Butrym and Maruszcak (1984) reported a thermoluminescence (TL) age of 150 000 years close to the top of the older loess and an age of 124 000 years for the base of the younger loess. According to these authors the top of the younger loess gave a TL age of 19 600 years, but the reliability of the TL datas is unknown since the analytical procedures were not described.

At Mende brickyard, five loess layers and intercalated chernozem-type soils are exposed (Pecsi *et al.*, 1979). The whole sequence shows normal polarity (Márton, 1979) and is equated with the younger loess as Paks. A TL date of 105 000 ± 17 000 was reported by Borsy *et al.* (1979) from soil below the lowermost loess. As in Austria and Czechoslovakia, the loess units at Mende and Paks accumulated during periods of relatively cold, dry climate.

In Poland, loess occurs in an east–west belt in the southern part of the country (Maruszcak, 1980). No Matuyama-age loess has been found, but the oldest loess pre-dates oxygen isotope stage 9 (approximately 350 000 yr BP; Maruszcak, 1985; Fig. 9.33). Palaeomagnetic studies of the loess have shown that periods of loess deposition coincided with periods of maximum ice volume (glacials) while soil formation occurred during interglacials and, to a lesser extent, interstadials (Tucholka, 1977; Dabrowska *et al.*, 1980; Maruszcak, 1980). Most of the thickness of Polish loess was deposited in the earlier and later parts of the last glacial period.

In northern France, Belgium and southern Britain, loess is relatively thin

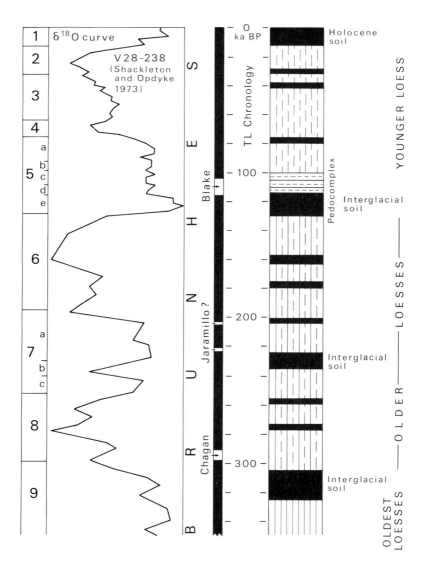

Figure 9.33. Generalized summary of loess stratigraphy in southern Poland. (After Maruszzcak (1985); © Marie Curie–Sklodowska University.

and does not provide a long-term continuous record of Quaternary environmental changes comparable with that in Central Europe and Central Asia. Most of the loess was deposited during the later part of the last glacial (Weischelian), but in many places, especially in northern France, there is also evidence of pre-Eemian (Saalian ?) loess (Wintle, 1981; Wintle *et al.*, 1984; Mees and Meijs, 1984; Wintle and Catt, 1985; Balescu and Haesaerts, 1984; Haesaerts, 1985; Juvigné, 1985; Lautridou *et al.*, 1985; Balescu *et al.*, 1986). Most of the loess in the Rhine Valley is also of later last glacial age (e.g. Wintle and Brunnacker, 1982). In parts of Normandy up to six pre-Weichselian loesses and one sandy loess are represented; all show normal magnetic polarity except the lowermost palaeosol in the Mesnil–Esnard profile (Lautridou, 1979).

9.8.2 North America

Radiocarbon dating of loess in the mid West and Mississippi Valley has shown that much of it was deposited during the later part of the last glacial period (Willman and Frye, 1970; Krinitzsky and Turnbull, 1967; Snowden and Priddy, 1968; McKay, 1979). Five major loess units have been recognized and named (from oldest to youngest): (1) pre Loveland; (2) Loveland; (3) Roxana; (4) Farmdale; and (5) Peorian. In parts of Nebraska a post-Peorian loess, termed the Bignell Loess (Schultz and Stout, 1945), has also been identified. The Peorian loess is geographically the most extensive and in many places comprises more than 90% of the total loess thickness. Numerous radiocarbon dates have shown it to be between 25 000 and 10 000 years old; in part it is diachronous, and in southern Mississippi only began to form about 21 000 ^{14}C yr BP. The ages of the older loess units are less well established. Pye and Johnson (1987) obtained TL ages of 21 000–22 000 years from the stratigraphic equivalent of the Farmdale loess in southern Mississippi. These dates were consistent with radiocarbon dates from contained gastropod shells which indicated ages of 20 000–23 000 ^{14}C yr BP. Weathered, reddish loess beneath this unit yielded TL ages of 74 000–77 000 years, suggesting deposition during the early part of the last glacial period. An even older, more highly weathered loess unit at the base of the sections studied (Figs. 9.34 and 9.35) gave TL ages of >130 000 years. Pye and Johnson concluded that this loess was probably deposited during the penultimate glacial period. Loess deposition in this area shows a close relationship with periods of glacial advance and early deglaciation in the upper Mississippi Valley. Weathering and soil formation occurred during the Holocene and Sangamon interglacials and during a long interstadial in the mid Wisconsinan.

Norton and Bradford (1985) attempted to date the pre Peorian loesses in Iowa using TL but the results obtained were not stratigraphically consistent.

In Nebraska, grain size and thickness trends indicate that loess was derived

Figure 9.34. Multiple loess units exposed in a roadcut near Natchez, southern Mississippi. Numbered units are those recognized by Pye and Johnson (1987). HS = Holocene soil.

both from fluvioglacial outwash and from the Nebraska Sandhills (Lugn, 1962, 1968). The last phase of major dune activity in the Nebraska Sand Hills occurred in the early to mid Holocene (Ahlbrandt and Fryberger, 1980), but radiocarbon dates obtained from peats deposited in interdune swamps indicate that sand movement and deposition occurred in different areas at different times, and that the dunefield has a complex history (Bradbury, 1980).

9.8.3 China

The stratigraphy and depositional history of two long loess sequences have been investigated in China, at Luochuan in Shaanxi Province (Heller and Liu

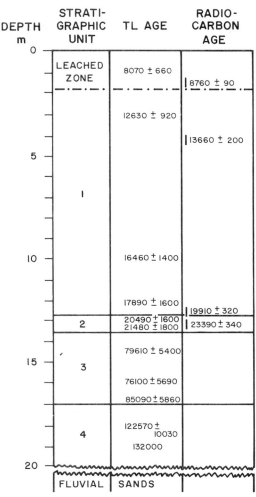

Figure 9.35. Summary of the loess stratigraphy at Section VI near Vicksburg, Mississippi. Radiocarbon ages were obtained from aragonitic gastropod shells. Thermoluminescence ages were obtained using the 4–11-μm size fractions. (After Pye and Johnson (1987); Earth Surf. Proc. Landf. (in press); © John Wiley.)

Tung-sheng, 1982, 1984), and at Lanzhou, 500 km further to the west in Gansu Province (Burbank and Li Jijun, 1985; Fig. 9.36). At Luochuan, the base of the loess was found to pre-date the Olduvai event (1·67–1·87 million years ago) and to post-date the Gauss–Matuyama boundary. Loess sedimentation probably began about 2·4 million years ago (Heller and Liu Tung-sheng, 1982). Eight loess units and eight major soils occur within the Brunhes epoch

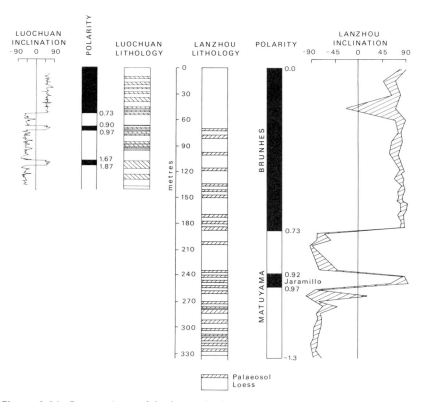

Figure 9.36. Comparison of the loess thickness, stratigraphy and age structure at Luochuan, Loess Plateau (after Heller and Liu Tung-sheng (1982), Nature **300**, 431–433); and Lanzhou, northwest China (after Burbank and Li Jijun (1985), Nature **316**, 429–431; © Macmillan Journals Ltd.)

(Fig. 9.36). The boundary between the basal Wucheng and overlying Lishih loesses was placed by Heller and Liu Tung-sheng (1982) at about 1·1 million years. The authors noted an apparent increase in the average loess deposition rate from 4·6 cm kyr^{-1} (0·046 mm yr^{-1}) below the Jaramillo event (0·9–0·93 million years ago) to 7·3 cm kyr^{-1} (0·073 mm yr^{-1}) in the upper 73 m of the section.

The apparent agreement between the timing of the ônset of loess sedimentation and the beginning of northern hemisphere glaciation, together with the similarity between the number of loess layers and the number of oxygen isotope stages in Atlantic deep-sea sediments (van Donk, 1976), led Heller and Liu Tung-sheng to suggest a correlation between loess depositional episodes and glacial stages.

In the 330-m thick section near Lanzhou, Burbank and Li Jijun (1985) obtained magnetostratigraphical results which indicate that the base of the loess succession dates from about 1·3 million years ago. This relatively young age, if correct, may reflect the fact that uplift along the northern fringe of the Tibetan Plateau precluded early Pleistocene loess accumulation in this area. Loess apparently accumulated during the later Matuyama epoch at an average rate of 37·5 cm kyr^{-1} (0·375 mm yr^{-1}) and during the Brunhes epoch at an average rate of 25 cm kyr (0·25 mm yr^{-1}).

The higher rate of deposition at Lanzhou compared with Luochuan was attributed by Li Jijun and Burbank to the proximity of the former site to the northern deserts and funnelling of dust-laden winds through the Gansu Corridor into the Lanzhou Basin where the winds expand, decelerate and deposit part of their sediment load.

The presence of interbedded palaeosols indicates that loess accumulation at Lanzhou was episodic. Palaeosols are more numerous in the basal 100 m of the section where they appear to have formed with an average periodicity of about 25 000 years. Li Jijun and Burbank suggested that palaeosol development in the earlier Pleistocene may have been modulated by orbital precession, though they did not elaborate on a possible causal connection.

It is not yet entirely clear whether the initiation of loess formation in China around 2·4 million yr BP was due to the onset of northern hemisphere cooling or to regional climatic and geomorphological changes associated with uplift of the Tibetan Plateau. Both probably played a part. In the early Pliocene, Tibet stood at an average altitude of only 1000 m above sea-level and northwest China experienced a moist tropical climate influenced by the Indian Ocean monsoon (Li Jijun *et al.*, 1979). By the end of the Pliocene the average altitude of Tibet was 2000 m and limited maritime-type glaciation had occurred. Continued uplift raised Tibet to an altitude of more than 4000 m by the end of the Pleistocene. Uplift created a barrier which impeded northward movement of the Indian monsoon into Central Asia, leading to progressive desiccation of basins in northwestern China and Mongolia. Lakes dried up, river systems were abandoned and dunefields initiated (Chao and Xing, 1982). Reduced precipitation also progressively changed the nature of glaciation to a more restricted continental type.

Pollen, faunal and pedological evidence suggests that superimposed on this general trend towards aridity were further fluctuations in temperature and precipitation related to changes in global ice volume. Loess accumulation episodes in Central China appear to have been drier and colder than at present with predominant cold desert steppe vegetation. According to Liu Tung-sheng *et al.* (1982), radiocarbon dates show that accumulation of the Malan loess in northern China began about 30 000 years ago and ceased about 10 000 years ago. At this time (coincident with the last glacial maximum) much of the continental shelf of the Yellow Sea was dry due to lowered sea-level; this would have enhanced the continentality of the climate in northern China and the increased aridity may have contributed to greater dust-blowing. The thickness of Malan loess at Luochuan is about 10 m, indicating an average deposition rate between 30 000 and 10 000 yr BP of 0·35 mm yr^{-1}. Although dustfalls have occurred during the Holocene, they appear not to have been sufficiently heavy or frequent to prevent development of a widespread surface soil. Few deposits of Holocene loess have been identified in China, although Liu Tung-sheng *et al.* (1981, 1982) reported a radiocarbon date of 5250 ± 250 yr BP from thin loess in the foothills of the Tien Shan.

Widespread soil formation also occurred in previous interglacials (and possibly some interstadials), when pollen and pedological evidence shows that the climate was relatively warm and humid with forest vegetation predominant (Liu Tung-sheng *et al.*, 1982). Climatic fluctuations appear to have been of smaller magnitude during accumulation of the Lower Pleistocene Wucheng loess than during accumulation of the Lishih and Malan loess in the mid and late Pleistocene.

According to Liu Tung-sheng *et al.* (1981), the higher rate of dust transport from the deserts during glacial periods may have been partly related to an increase in the frequency of cyclonic depressions (sandstorms) in the Gobi Desert and to more effective easterly transport of dust by a westerly jet stream centred north of the winter Tibetan anticyclone.

The apparent increase in the rate of loess sedimentation during the later Quaternary probably reflects the long-term increase in aridity and increasing frequency of strong winds, both due to the effect of Tibetan uplift of Tibet on atmospheric circulation systems.

9.8.4 Soviet Central Asia

Detailed studies have been made in the past two decades of loess stratigraphy and age structure in the area around Tashkent (Uzbekistan) and in the Tajik Depression (Tajikistan), where loess deposits are up to 200 m thick and up to two million years old (Dodonov *et al.*, 1977; Lazarenko *et al.*, 1980; Dodonov, 1979, 1984; Lazarenko, 1984; Fig. 9.37). These areas were characterized by

---

Figure 9.37. Map showing loess distribution and major loess sections in the areas around Tashkent and Dushanbe. Numbered loess sections in the Tashkent area are: (1) Gazelkent; (2) Orkutsai; (3) Yangiyul; (4) Keles; (5) Mingtepe; (6) Sarygach; (7) Pskent; (8) Charvak; and (9) Khumsan. Numbered sections in the Dushanbe area are: (1) Ak Jar; (2) Karamaidan; (3) Karatau; (4) Lakhuti; (5) Khonako; (6) Chashmanigar; and (7) Kuruksai. (Modified after Lazarenko et al. (1980) and Lazarenko (1984).)

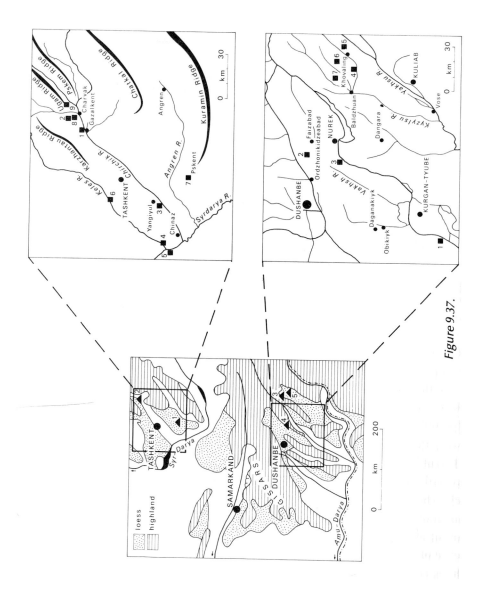

Figure 9.37.

Figure 9.38. Mid Pleistocene loess section at Mingtepe, Uzbekistan.

predominant downwarping in the late Cenozoic, and contain thick accumulations of late Tertiary–Quaternary molasse sediments. The Pliocene sediments which infill synclinal areas are mostly composed of fluvial silts, sands and gravels, in places modified by subaerial soil processes. The Lower Pleistocene deposits are also predominantly fluvial, but loess is an important component on the Middle and Upper Pleistocene sequences.

In the Tashkent area, loess occurs on many of the Quaternary terraces of the Keles and Chirchik Rivers (tributaries of the Syr Darya) and on the intervening foothills. The distribution of loess of different ages shows a close relationship with the pattern of river terrace development over time (Mavlyanov and Tetyukhin, 1982). A fairly complete record of Quaternary sedimentation is recorded at the Charvak and Orkutsai sections northeast of Tashkent, but elsewhere only incomplete Lower, Middle or Upper Pleistocene loess sections are found. At Charvak, nine major palaeosols and eight intervening loess layers occur above the Brunhes–Matuyama reversal. At Mingtepe, near the confluence of the Keles and Syr Darya Rivers (Fig. 9.38), grey Middle Quaternary loess containing a few weakly developed palaeosols overlies the fourth river

Figure 9.39. Upper Pleistocene loess at the top of the Orkutsai section, Uzbekistan, showing the well-developed Holocene soil (dark surface horizon).

terrace. Nearby, at Keles, late Pleistocene loess overlies the second terrace. Lazarenko *et al.* (1980) reported a TL date of 38 000 ± 17 000 near the top of this section. At Sarygach, loess overlying the fourth terrace belongs mainly to the Lower Quaternary. This loess is reddish in colour, well cemented, and is capped by a thick calcrete.

No loess has formed in the Tashkent area during the Holocene (Fig. 9.39). Pollen data suggest that during periods of Pleistocene loess deposition, and particularly during phases of soil formation, the climate was more humid than at present (Lazarenko *et al.*, 1980). Soils are better developed in the montane loess sections than in the desert steppe sections to the southwest of Tashkent.

The oldest loess (including loess soil) deposits in Soviet Central Asia are pre Olduvai in age and occur near the base of the Chashmanigar and Karamaidan sections (Dodonov, 1984; Lazarenko, 1984). Olduvai-age loess also occurs interstratified with fluvial deposits near the base of Kayrubak suite at Kuruksai. The 170-m thick Chashmanigar section records a complete subaerial Quaternary sedimentary sequence composed largely of loess and loess palaeosols. The Brunhes–Matuyama reversal is found beneath the ninth major palaeosol. Below this level the sequence consists of a series of superimposed palaeosols with thin interbedded loess horizons. According to Dodonov (1979, 1982, 1984), nine Brunhes-epoch palaeosols are also present at Kayrubak, Karamaidan, Lakhuti and Khonako (Fig. 9.40). At Karatau, at least seven soils are present (the Brunhes–Matuyama boundary is not exposed here).

Davis *et al.* (1980) reported TL dates from the Karatau section ranging from 21 000 years above the second palaeosol to 320 000 years below the seventh palaeosol. A date of 120 000 years was reported from just below the fourth

Figure 9.40. Loess just above the Brunhes–Matuyama magnetic reversal at the Karamaidan section, Tajikistan. N indicates normal polarity; IX$_B$ indicates the position of the lower member of the ninth pedocomplex.

palaeosol (approximately stratigraphically equivalent to the Blake event). At Lakhuti 1, a TL age of 530 000 years was reported from the seventh pedocomplex. No details of the TL dating procedure were given, however, and the reliability of the ages is uncertain. At Chashmanigar, TL dates extending as far back as 730 000 years were reported by Lazarenko (1984) for loess between the eighth and ninth palaeosols, but it is doubtful that reliable dates of this age can be obtained due to problems of electron trap saturation (Wintle, 1987).

Pollen and faunal evidence indicate that the soils were formed during periods of relatively moist and warm climate. Many of the soils are composite, reflecting irregular changes in the balance between sedimentation and pedogenesis. Most of the Upper Pleistocene soils are analogous to the 'chestnut' soils formed under semi-arid steppe conditions today. Pollen and snail species present in the loess indicate that it accumulated during periods when the climate was considerably cooler and drier than today. Archaeological evidence suggests that man inhabited the Afghan–Tajik depression mainly in inter-glacial periods when the climate was more hospitable (Ranov and Davis, 1979; Davis *et al.*, 1980). As in Uzbekistan, the rate of Holocene dust deposition in Tajikistan has been too low to form recognizable loess.

The loess soil record in Tajikistan suggests a long-term trend towards colder, drier conditions during the Quaternary (Fig. 9.41). As in northern China, this trend may be related to the progressive uplift of Tibet, the Tien Shan and Pamirs. The latter experienced rapid uplift totalling 2000–2500 m during the Quaternary, though there were marked regional differences in the rate of uplift at different times. The western Tien Shan showed marked differential movements, whereas the eastern Tien Shan developed as a single block during intense uplift (Krestnikov *et al.*, 1980).

As in China, superimposed on the general trend towards drier and cooler conditions during the Quaternary were up to 20 shorter-term climatic cycles. The loess record suggests that the magnitude of the fluctuations increased towards the late Quaternary (Fig. 9.41), with greater thicknesses of loess being deposited during colder periods of the Upper Pleistocene compared with the Lower Pleistocene. The fluctuations also appear to be more closely spaced in the Lower Pleistocene, though in part this may reflect compaction of the lower loess strata.

As noted previously, the distribution and geomorphological relationships of loess in Uzbekistan and Tajikistan strongly suggest derivation from local river channels which are fed partly by glacial meltwater. This does not, however, necessarily imply that all of the loessic silt was formed by glacial action; much may have been reworked during rapid fluvial incision into older late Cenozoic sediments, and some was probably formed by montane weathering processes. It is not known by how much the glaciated area in the Tien Shan, Pamirs, Altai and Ghissars increased during periods of global cooling. The effect of falling temperature would to some extent have been counterbalanced by reduced precipitation, and the areal extent of glaciation might have changed relatively little between warm and cold periods. However, in the upper Urumqui River valley and Bogdan Shan (Chinese Tien Shan) glacial tills deposited during the last Pleistocene glaciation indicate ice volumes of up to an order of magnitude greater than the present ones (Derbyshire, 1984*c*).

There is no simple correlation between the number of loess depositional units in Central Asia and the number of northern hemisphere cold stages as indicated by the oxygen isotope record and glacial sediments in Europe and North America. 'Wisconsinan' time in Central Asia is represented by four

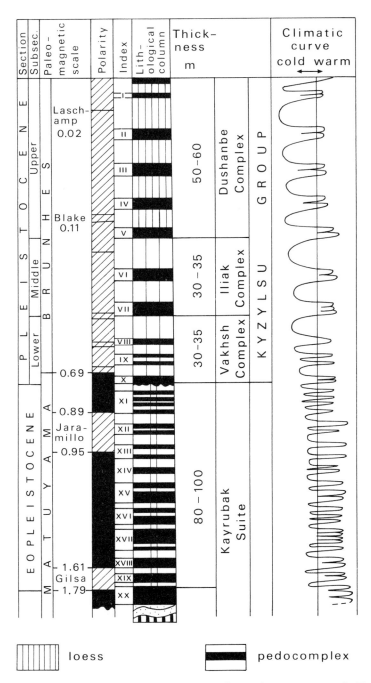

loess

pedocomplex

Figure 9.41. Summary of the loess stratigraphy and age structure in Tajikistan, together with a climatic curve based on pollen, faunal and pedological evidence. (Modified after Dodonov (1979), Acta Geol. Acad. Scient. Hung. **22**, 63–73; © Akademiai Kíado.

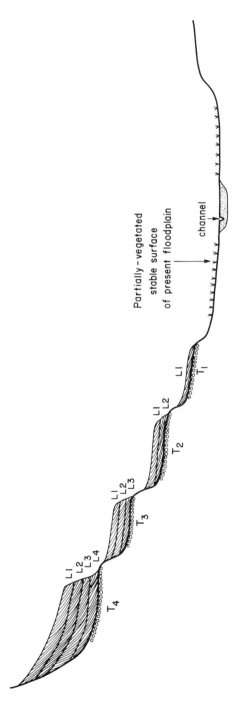

Figure 9.42 Schematic diagram showing the relationship between loess of different ages and river terraces in the piedmont areas of Soviet Central Asia. The surfaces of the present floodplains are mostly stable and have not provided large amounts of deflated dust for loess formation in the Holocene.

major palaeosols and intervening loess horizons, while the last interglacial appears to correspond with two loess units, the fifth and possibly the sixth pedocomplexes. If the interpretation of loess stratigraphy and age structure is correct, this suggests that general northern hemisphere cooling events were not the only driving mechanism behind loess sedimentation in Central Asia. An alternative possibility is that loess accumulation was controlled partly by changes in the rate of river incision and sediment load which in turn reflected episodic uplift. Uplift, in turn, would have affected the climate of the Tien Shan and Pamirs and might have increased the extent of glaciation.

In the river valleys of Tajikistan there are up to ten erosion-accumulation terraces between 10 and 220 m above the present valley floors. The terraces between 140 and 220 m (the Vakhsh complex) are classified as early Pleistocene, those between 80 and 140 m (Iliak complex) as mid Pleistocene, and those between 10 and 80 m (Dushanbe complex) as late Pleistocene (Dodonov, 1984). Buried soil horizons in the loess are correlated with the terraces.

In both Uzbekistan and Tajikistan the Holocene floodplain is in many places (where not cultivated) vegetated or has a surface crust of fine-grained sediment which cannot easily be deflated. Consequently, no loess is forming at present. Adjacent late Pleistocene terraces are capped by slightly younger loess; older terraces are capped by thicker and progressively more complex loess sequences, as shown schematically in Fig. 9.42. Lateral river erosion has removed some or all of the older terraces in places, forming steep bluffs, as at Mingtepe (Fig. 9.38).

The terrace sediments vary considerably but in many places consist of coarse, poorly sorted gravels overlain by sands and silts. The gravels suggest major increases in discharge, at least seasonally, during, or shortly after, episodes of river incision. These changes are consistent with an increase in seasonal meltwater flow related to an increase in the area of winter snowcover. The braided channels would have provided a much more suitable dust source at times of low flow than the present floodplain sediments, and it is possible that their development was responsible for the onset of each phase of loess accumulation.

There are currently insufficient data to allow firm conclusions about whether global cooling, tectonic uplift or a combination of the two was responsible for the changes in river regime. However, the poor correlation between the number and timing of terrace formation events and global cooling events in the late Pleistocene would suggest that tectonic movements in the Pamirs and Tien-Shan played a significant role.

9.8.5 Pakistan

Loess up to 11 m thick occurs in a belt extending through northern Pakistan from near Peshawar in the west to Gujar Khan in the east, a distance of about

200 km (Rendell, 1985). Preliminary TL results from loess in the Soan Valley, southeast of Rawalpindi, indicated a last glacial age (Rendell, 1985). In the Kashmir Valley, Pleistocene lake sediments (Karewas beds) are capped by 8–10 m of loess containing three weakly developed palaeosols. Agrawal *et al.* (1979) reported a radiocarbon date of 18 500 yr BP for the topmost palaeosol and ages of >40 000 ^{14}C yr BP for the lower two. The underlying lake sediments were found to show normal magnetization and probably belong to the Brunhes epoch. Formation of the loess was equated by Agrawal *et al.* with cold, dry glacial conditions while soil formation was equated with warmer and wetter episodes (interstadials) during the last glacial period (see also Pant *et al.*, 1983; Bronger and Pant, 1983).

9.8.6 Siberia

The central part of Yakutia remained unglaciated during late Pleistocene cold phases, although glaciers existed in the ranges to the north, east and south and a continental ice sheet lay to the west. Silt deflated from outwash plains, braided river channels and periglacial dune complexes was deposited on nearby river terraces and interfluves (Péwé and Journaux, 1983). Ice structures formed under periglacial conditions are numerous in the loess (Tomirdiaro, 1984). On the basis of radiocarbon and stratigraphic evidence, Péwé *et al.* (1977) recognized two major phases of loess deposition in Yakutia which they correlated respectively with the last glacial period and the penultimate glacial period. It is possible, however, that the older loess belongs to the earlier part of the last glacial period.

In southwestern Siberia, relatively thin loess which occurs in association with periglacial dunesands formed mainly during the later part of the last glacial period (Volkov and Zykina, 1984).

9.8.7 Ukraine

Loess up to 30 m thick covers more than 95% of the Ukraine lowlands and plains south of Poles'ye (Veklich, 1979; Balandin, 1984; Khalcheva, 1984; Veklich and Sirenko, 1984). Much of the loess sequence consists of superimposed palaeosols of different types, indicating a relatively low rate of deposition. The lithology, pedology, sedimentology and palynology of a considerable number of sections have been investigated, but no palaeomagnetic studies have been published and the age structure of the deposits is not well established. The record apparently extends back to the Lower Pleistocene (Sirenko, 1984), and some true loesses reportedly occur in Pliocene sediments (Veklich, 1979). Pollen and faunal data indicate the loess accumulated under periglacial steppe conditions, while the soil-forming episodes were relatively warm and humid. Climates were generally more humid in the north and drier in the south.

9.8.8 Australia and New Zealand

Although dunes cover extensive areas of the Central Australian deserts, no significant loess deposits occur on their margins. Deposits of aeolian clay (parna) occur mainly in western Victoria and the riverine plain of New South Wales, where the dominant Quaternary sediments are fluvial. Butler (1956) recognized that parna deposits occur both as discrete dunes (dune parna), usually on the downwind side of seasonally dry lakes and river beds, and as thin, discontinuous sheets (sheet parna). Virtually all parna was transported not as individual clay-size particles but as sand and coarse-silt size aggregates (Dare-Edwards, 1984). Sedimentological relationships and radiocarbon dates indicate that most of the parna is of late last glacial age, a period of widespread aridity in southeast Australia (Bowler, 1978).

The poor development of typical loess appears to be related to two main factors. First, Australia is a low-relief continent which has been dominated by stable landsurface conditions since the Mesozoic. Rates of weathering and surface stripping are low in comparison with the mountainous belts of Central Asia, and large areas of the continent are covered by soils and duricrusts of late Cenozoic age. Pleistocene glaciation and periglacial activity affected only small areas of southeast Australia and Tasmania (Bowler *et al.*, 1976). Large amounts of glacial and periglacial silt were not produced and concentrated by fluvio-glacial meltwaters, as happened in Europe, North America and the Soviet Union. Second, wind energy in the Australian deserts is lower than that of some other deserts, e.g. the Gobi and Takla Makan. At present, only the Lake Eyre Basin and parts of the Simpson Desert experience an average of more than five dust storms per year. This situation does not appear to have been radically different in the Pleistocene, for there is little evidence of long-distance movement of aeolian sand and large-scale dune migration.

By contrast, loess is widespread in both islands of New Zealand, though the deposits are thicker and more spectacular on the South Island. In the northern two-thirds of the North Island, loess material has been derived largely by aeolian reworking of tephra, whereas in the south it is derived mostly from quartzo-feldspathic sediments (McCraw, 1975). At Mount Curl and other sites in the southern part of North Island, loess is interstratified with tephra marker horizons. The age of the oldest loess has not been established exactly, but is estimated from dated tephra horizons to be 170 000–180 000 years old (Milne and Smalley, 1979). Five main loess units have been distinguished, separated by palaeosols and/or ash layers. The youngest (Ohakea) loess contains tephra which has been dated at 20 600 yr BP, and was clearly deposited in the last

Pleistocene cold stage. Two underlying loess units (the Rata Loess and Porewa Loess) also apparently belong to the last glacial period. The oldest two loesses may belong to the penultimate glacial period.

At Timaru, South Island, six loess members separated by palaeosols have been identified (Tonkin *et al.*, 1974; Runge *et al.*, 1973). Radiocarbon dates suggested accumulation of the youngest loess member between 9900 and 11 800 ^{14}C yr BP. The second loess member apparently began to accumulate before 31 000 ^{14}C yr BP. Tonkin *et al.* concluded that loess deposition occurred during periods of glacial recession, when broad areas of fluvioglacial deposits were exposed to the wind, while soil formation was initiated during warm interstadial conditions and continued throughout the cooling of the following stadial.

9.8.9 Israel

The best section of *in-situ* loess so far described from Israel occurs at Netivot on the northern desert fringe of the Negev (Bruins, 1976; Bruins and Yaalon, 1979; Fig. 9.15). With the exception of a thin (0·4 m) upper loess, the whole profile has been profoundly modified by syn- and post-depositional pedogenesis. Six carbonate-rich layers are interstratified with five intervening brown clayey layers (Fig. 9.43). The sequence was interpreted by Bruins and Yaalon as indicating continuous dust deposition under conditions of changing humidity. Clayey layers were equated with periods of wet to semi-arid climate in the late Quaternary, and calcareous layers with periods of dry to semi-arid climate. Only one radiocarbon date (27 100 ^{14}C yr BP) was obtained from the second carbonate-rich horizon down from the top of the section.

More recent work has suggested that, contrary to widespread opinion that it is forming today, the Negev loess was deposited mainly in glacial periods of the late Pleistocene (Pye and Tsoar, 1987). During the later part of the last glacial period, much of northern Sinai was covered not by dunes, as it is today, but by poorly-sorted fluvial deposits (Sneh, 1982) which could have provided a major source of dust. The late glacial climate of the Negev and Sinai was wetter than present (Issar and Bruins, 1983; Issar, 1985; Issar *et al.*, 1986), partly due to increased frequency of rainbearing cyclonic depressions during spring. Greater frequency of dust-transporting winds, more frequent wadi flow (which increased the available dust supply), and growth of thicker dust-trapping vegetation in the semi-arid fringe may have all favoured loess formation at this time (Tsoar and Pye, 1987). The Holocene saw a return to more arid conditions; silt supply diminished as wadi flow became less frequent, and vegetation cover in the loess areas was reduced, leading to widespread fluvial erosion and redeposition of the loess.

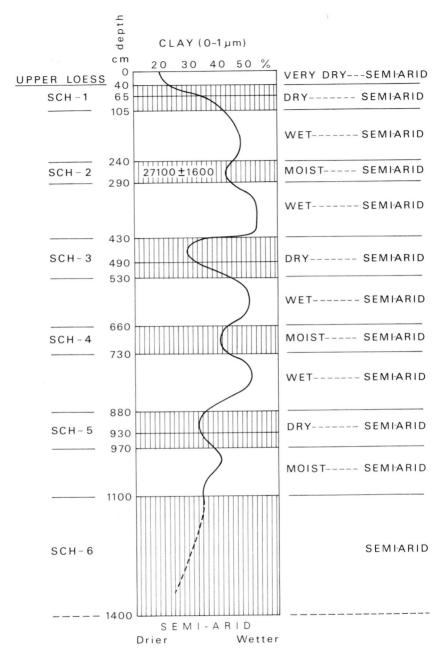

Figure 9.43. Stratigraphy of the loess at Netivot, southern Israel. (After Bruins and Yaalon (1979), Acta Geol. Acad. Scient. Hung. **22**, *161–170;* © *Akademiai Kíado.*

9.8.10 Alaska

Loess forms a blanket, ranging in thickness from a few millimetres to more than 60 m, which covers almost all parts of Alaska that lie below altitudes of 300–450 m. Aeolian dust is still being deposited in some areas close to outwash streams (Péwé, 1951). Rates of deposition of Holocene loess range from 0·78 to 2·07 mm yr^{-1} on the east side of the Delta River, from 0·15 to 0·27 mm yr^{-1} on the north side of the Tanana River, and from 1·40 to 1·57 mm yr^{-1} near Fairbanks (Péwé, 1975). However, most of the loess in central Alaska is of Wisconsinan and Illinoian glacial age. Wintle and Westgate (1986) reported TL dates of 86 000–91 000 yr BP for loess below the Old Crow Tephra near Fairbanks. At Cape Deceit on the south shore of Kotzebue Sound, several superimposed loess units have been recorded, including one which may be pre Illinoian in age (Péwé, 1975).

9.8.11 Summary of loess deposition during the Quaternary

The available evidence indicates that late Cenozoic loess accumulation began on a significant scale in China shortly after 2·4 million yr BP. In Central Asia and Europe loess began to form slightly later, between two million and 1·7 million yr BP. Loess may also have begun to accumulate in the Ukraine around this time. Loess sedimentation in all of these areas has clearly been episodic, with loess accumulating mainly in relatively cool, dry periods. During relatively warm, wet periods, dust deposition slowed or ceased, allowing soils to form.

Loess accumulation episodes in different regions were not exactly synchronous. In Europe, which was not significantly affected by Quaternary orogenesis, loess formation shows a fairly close relationship with northern hemisphere cooling and warming trends, although correlation with the oceanic oxygen isotope record is not perfect. In China and Soviet Central Asia, loess depositional episodes were also influenced by geomorphological and climatic changes related to rapid uplift of the Tibetan Plateau and associated mountain ranges. Although the number of loess–soil cycles in China and Soviet Central Asia is similar to that recorded in Europe, important differences are evident when the record is examined in detail. For example, although nine major soils have been recognized above the Brunhes–Matuyama reversal in Europe, Uzbekistan and Tajikistan, with eight reported in China, the number of soils recognized above the Blake event varies from one in China, two in Europe, three in Uzbekistan and four in Tajikistan.

No early and mid Pleistocene loess deposits have yet been identified on the margins of glaciated areas in North America, Argentina, Siberia and New Zealand. It is unknown whether loess was formed in these areas before the penultimate glacial period and subsequently eroded.

Table 9.6

Comparison of apparent loess accumulation rates in different parts of the world during the Quaternary.

Location	Period (yr BP)	Apparent average loess accumulation rate (mm yr^{-1})	Reference
Charvak, Uzbekistan	0–115 000	0·26	Lazarenko et al. (1980)
	115 000–700 000	0·05	Lazarenko et al. (1980)
Khumsan, Uzbekistan	0–700 000	0·064	Lazarenko et al. (1980)
Keles, Uzbekistan	38 000–60 000	0·45	Lazarenko et al. (1980)
	0–110 000	0·29	Lazarenko et al. (1980)
Pskent, Uzbekistan	0–1 250 000	0·184	Lazarenko et al. (1980)
Chashmanigar, Tajikistan	0–130 000	0·29	Lazarenko (1984)
	130 000–700 000	0·08	Lazarenko (1984)
	700 000–950 000	0·15	Lazarenko (1984)
Lakhuti 1, Tajikistan	0–700 000	0·164	Lazarenko (1984)
Khonako 2 Tajikistan	0–700 000	0·14	Dodonov (1979)
Kayrubak, Tajikistan	0–700 000	0·16	Dodonov (1979)
Karamaidan, Tajikistan	0–700 000	0·17	Dodonov (1979)
	700 000–200 000	0·06	Dodonov (1979)
Lanzhou, China	0–700 000	0·26	Burbank and Li Jijun (1985)
	700 000–1 300 000	0·25	Burbank and Li Jijun (1985)
Luochuan, China	0–700 000	0·07	Heller and Liu Tung-sheng (1982)
Červeny Kopeč, Czechoslovakia	700 000–1 670 000	0·05	Fink and Kukla (1977); Kukla (1978)
	0–700 000	0·09	Fink and Kukla (1977); Kukla (1978)
Krems, Austria	700 000–1 670 000	0·022	Fink and Kukla (1977); Kukla (1978)
Tyszowce, Poland	10 000–180 000	0·75	Wintle and Prószyńska (1983)
Vicksburg, Mississippi	10 000–210 000	2·73	Pye and Johnson (1987)
Timaru, New Zealand	9900–11 800	2·0	Tonkin et al. (1974)

There is no sedimentary evidence that significant loess deposits formed in Australia or Africa (though localized and relatively thin deposits of sandy loess do occur in Tunisia and northern Nigeria). The principal reasons for the limited development of loess in these areas appear to be: (1) large areas of bare, silt-rich sediment exposed to wind deflation were rare on these largely unglaciated and low-relief continents; and (2) silt deflated from alluvial deposits during periods of increased Quaternary aridity was not trapped by vegetation close to the dust source due to the gentle nature of the environmental gradients in these areas; instead, the dust was dispersed over a wide geographical area.

The average rate of loess accumulation in all parts of the world appears to have increased during the course of the Quaternary. Low rates (or infrequent periods) of loess deposition before one million yr BP are indicated by the fact that Lower Pleistocene loess deposits are typically highly weathered, with many sequences consisting entirely of superimposed pedocomplexes and reworked loessic silts (Pecsi, 1984b). Higher rates (or more frequent periods) of loess deposition in the last million years may be related to greater cooling and more extensive continental glaciation during cold periods of the Upper Pleistocene compared with the early Pleistocene. In China and Central Asia, the increase in the scale of dust-blowing during cold periods of the late Quaternary was probably enhanced by the trend towards drier (and possibly windier) conditions associated with the uplift of the Tibetan Plateau and neighbouring mountain ranges. Average loess accumulation rates in China, Soviet Central Asia and Europe were of the order of $0 \cdot 02 – 0 \cdot 06$ mm yr^{-1} during Matuyama time, and of the order of $0 \cdot 09 – 0 \cdot 26$ mm yr^{-1} during the Brunhes epoch (Table 9.6). These long-term average rates disguise the fact that deposition rates were up to two orders of magnitude higher during Pleistocene cold stages, and one or two orders of magnitude lower during warm periods when pedogenesis predominated.

Loess accumulated at a rate of $0 \cdot 5 – 3$ mm yr^{-1} around the time of the last glacial maximum, which was a period of exceptional worldwide dust activity. In addition to the loess record, evidence of very high dust deposition rates in the later part of the last glacial is provided by ocean sediments and ice cores. Dust-blowing on this scale was possibly unparalleled in previous Earth history. During the Holocene, dust deposition rates in most parts of the world have been too low for significant thicknesses of loess to accumulate, although aeolian additions to soils and ocean sediments have been significant. Dust transport and deposition rates in the latest Holocene have been artificially raised by human activities, but are still significantly lower than in the late Pleistocene. A situation similar to that in the early and mid Holocene probably prevailed for more than 80% of Quaternary time.

REFERENCES

Abdel-Salam, M. S. and Sowelim, M. A. (1967a) Dust deposits in the city of Cairo. *Atmos Env.* **1**, 211–220.

Abdel-Salam, M. S. and Sowelim, M. A. (1967b) Dust-fall caused by the spring Kamasin storms in Cairo. *Atmos. Env.* **1**, 221–226.

Adetungi, J. and Ong, C. K. (1980) Qualitative analysis of the Harmattan haze by X-ray diffraction. *Atmos. Env.* **14**, 857–858.

Adetungi, J., McGregor, J. and Ong, C. K. (1979) Harmattan haze. *Weather* **34**, 430–436.

Agrawal, D. P., Krishnamurthy, R. V., Kusumgar, S., Nautiyal, V., Athavale, R. N. and Radhakrishnamurthy, C. R. K. (1979) Chrono-stratigraphy of loessic and lacustrine sediments in the Kashmir Valley, India. *Acta Geol. Acad. Sci. Hung.* **22**, 185–196.

Ahlbrandt, T. S. and Fryberger, S. G. (1980) Eolian deposits in the Nebraska Sand Hills. *US Geol. Surv. Prof. Pap.* **1120A**, 1–24.

Akematsu, T., Dodson, R. F., Williams, M. G. and Hurst, G. A. (1982) The short-term effects of volcanic ash on the small airways of the respiratory system. *Env. Res.* **29**, 358–370.

Alexander, A. F. (1934) The dustfall of November 13, 1933, at Buffalo, New York. *J. Sed. Petrol.* **4**, 81–82.

Ali, Y. A. and West, I. (1983) Relationships between modern gypsum nodules in sabkhas of loess to compositions of brines in sediments in northern Egypt. *J. Sed. Petrol.* **53**, 1151–1168.

Allen, T. (1981) *Particle Size Measurement*, 3rd edn. London: Chapman & Hall.

Allen, C. C. (1978) Desert varnish of the Sonoran Desert – optical and electron microprobe analysis. *J. Geol.* **86**, 743–752.

Allen, H. D. (1986) *Late Quaternary of the Kopais Basin, Greece: Sedimentary and Environmental History*. Ph.D. thesis, Cambridge University, Cambridge.

Al-Najim, F. A. (1975) Dust storms in Iraq. *Bull. Coll. Sci.* **16**, 437–451.

Al-Sayeh, A. H. (1976) Geological and statistical study of the dust falling over Mosul town. *Bull. Coll. Sci.* **17**, 159–179.

Amit, R. and Gerson, R. (1986) The evolution of Holocene reg (gravelly) soils in deserts – an example from the Dead Sea region. *Catena* **13**, 59–79.

Anderson, G. E. (1926) Experiments on the rate of wear of sand grains. *J. Geol.* **34**, 144–258.

Anderson, R. S. and Hallet, B. (1986) Sediment transport by wind: toward a general model. *Geol. Soc. Am. Bull.* **97**, 523–535.

Anspaugh, L. R., Phelps, P. L., Kennedy, N. C. and Booth, H. G. (1973) Distribution and redistribution of airborne particulates from the Schooner Cratering event. *Lawrence Livermore Laboratory Report UCRL-74392.*

Archibold, O. W. (1985) The metal content of wind-blown dust from uranium tailings in northern Saskatchewan. *Water Air Soil Poll.* **24**, 63–76.

Arizona Department of Transportation (1974–75) *Soil Erosion and Dust Control on Arizona Highways.* Arizona Department of Transportation, Phoenix, 4 vols.

Armbrust, D. V. (1968) Windblown soil abrasive injury to cotton plants. *Agron. J.* **60**, 622–625.

Arvidson, R. E. (1972) Aeolian processes on Mars, erosive velocities, settling velocities, and yellow clouds. *Geol. Soc. Am. Bull.* **83**, 1503–1508.

Aspliden, C. J., Tourre, Y. and Sabine, J. B. (1976) Some climatological aspects of West African disturbance lines during GATE. *Mon. Weath. Rev.* **104**, 1029–1035.

Aston, S. R., Chester, R., Johnson, L. R. and Padgham, R. C. (1973) Eolian dust from the lower atmosphere of the Eastern Atlantic and Indian Oceans, China Sea and Sea of Japan. *Mar. Geol.* **14**, 15–28.

Atwater, M. A. (1970) Planetary albedo changes due to aerosols. *Science* **170**, 64–66.

Audric, T. (1973) Étude géologique et géotechnique des limons de plateau de la région parisienne. *Bull. Int. Ass. Eng. Geol.* **8**, 49–59.

Audric, T. and Bouquier, L. (1976) Collapsing behaviour of some loess soils from Normandy. *Q. J. Eng. Geol.* **9**, 265–277.

Augustinus, P. G. E. F. and Riezebos, H. T. (1971) Some sedimentological aspects of the fluvioglacial outwash plain near Soesterberg (The Netherlands). *Geol. Mijn.* **50**, 341–348.

Bach, W. (1986) Nuclear war: the effects of smoke and dust on weather and climate. *Prog. Phys. Geog.* **10**, 315–363.

Bagnold, R. A. (1937) The transport of sand by wind. *Geog. J.* **89**, 409–438.

Bagnold, R. A. (1941) *The Physics of Blown Sand and Desert Dunes.* Methuen, London. 241pp.

Bagnold, R. A. (1960) The re-entrainment of settled dusts. *Int. J. Air Poll.* **2**, 357–363.

Bain, D. C. and Tait, J. M. (1977) Mineralogy and origin of dust fall on Skye. *Clay Min.* **12**, 353–355.

Baker, G. (1959a) A contrast in the opal phytolith assemblages of two Victorian soils. *Aust. J. Bot.* **7**, 88–96.

Baker, G. (1959b) Opal phytoliths in some Victorian soils and 'red rain' residues. *Aust. J. Bot.* **7**, 64–87.

Baker, G. (1960) Phytoliths in some Australian dusts. *Proc. Roy. Soc. Vict.* **72**, 21–40.

Bal, L. and Buursink, J. (1976) An inceptisol formed in calcareous loess on the 'Dast-i-Esan Top' plain in Afghanistan. Fabric, mineral and trace element analysis. *Neth. J. Agric. Sci.* **24**, 17–42.

Balandin Y. G. (1984) Peculiarities of loess sequences of key sections in the Ukraine as reflections of the characteristics of the loess geosystem. In M. Pecsi (ed.) *Lithology and Stratigraphy of Loess and Paleosols*. Geographical Research Institute, Hungarian Academy of Sciences, Budapest, pp.61–69.

Balescu, S. and Haesaerts, P. (1984) The Sangatte raised beach and the age of the opening of the Strait of Dover. *Geol. Mijn.* **63**, 355–362.

Balescu, S., Dupuis, C. and Quinif, Y. (1986) La thermoluminescence du quartz: un marqueur stratigraphique et paléogeographique des loess Saaliens et Weichséliens du NW de l'Europe. *C. R. Acad. Sci. Paris* **302**, Ser. II, 779–784.

Barbour, G. B. (1927) The loess of China. *Smithsonian Inst. Rep. 1926*, 279–296.

Barbour, G. B. (1935) Recent observations on the loess of North China. *Geog. J.* **86**, 54–64.

Barden, L., McGown, A. and Collins, K. (1973) The collapse mechanism of partly saturated soil. *Eng. Geol.* **7**, 49–60.

Baris, Y. I., Sahin, A. A. and Ozemi, H. (1978) An outbreak of pleural mesothelioma and chronic fibrosing pleurisy in the village of Karain/Urgup in Anatolia. *Thorax* **33**, 181–192.

Bariss, N. (1971) Gully formation in the loesses of central Nebraska. *J. Rocky Mtn. Soc. Sci.* **8**, 47–59.

Barrie, L. A. (1986) Arctic air pollution: an overview of current knowledge. *Atmos. Env.* **20**, 643–663.

Bar-Ziv, J. and Goldberg, G. M. (1974) Simple siliceous pneumoconiosis in Negev Bedouins. *Arch. Env. Health* **29**, 121–126.

Beavington, F. and Cawse, P. A. (1979) The deposition of trace elements and major nutrients in the dust and rainwater in northern Nigeria. *Sci. Total Env.* **13**, 263–274.

Behairy, A. K., Chester, R., Griffiths, A. J., Johnson, L. R. and Stoner, J. H. (1975) The clay mineralogy of particulate material from some surface seawaters of the eastern Atlantic Ocean. *Mar. Geol.* **18**, M45–M56.

Beles, A. A., Stanculescu, I. I. and Scally, V. R. (1969) Pre-wetting of loess soil foundation for hydraulic structures. *Proc. 7th Int. Conf. Soil Mech. Found. Eng.* **2**, pp.17–25.

Belly, P. Y. (1964) Sand movement by wind. *US Army Coastal Eng. Res. Centre Tech. Mem. 1*, pp.1–38.

Beltagy, A. I., Chester, R. and Padgham, R. C. (1972) The particle-size distribution of quartz in North Atlantic deep-sea sediments. *Mar. Geol.* **13**, 297–310.

Bennett, J. G. (1980) Aeolian deposition and soil parent materials in northern Nigeria. *Geoderma* **24**, 241–255.

Berg, L. S. (1916) The origin of loess (in Russian). *Izv. Russ. Geog. Obsh.* **52**, 579–647.

Berg, L. S. (1964) *Loess as a Product of Weathering and Soil Formation.* Israel Program for Scientific Translations, Jerusalem, 207pp.

Berger, A. L. (1978) Long-term variations of caloric insolation resulting from the Earth's orbital elements. *Quat. Res.* **9**, 139–167.

Bertrand, J. J., Baudet, J. and Drochon, A. (1974) Importance des aérosols naturels en Afrique de l'Ouest. *J. Réch. Atmos.* **8**, 845–860.

Bhalotra, Y. P. R. (1963) Meteorology of the Sudan. *Sudan Met. Serv. Mem.* **6**, 113pp.

Binney, A. (1846) The bluff formation at Natchez, Mississippi. *Proc. Boston Nat. Hist. Soc.* **2**, 126–130.

Bisal, F. and Ferguson, W. (1970) Effect of non-erodible aggregates and wheat stubble on initiation of soil drifting. *Can. J. Soil Sci.* **50**, 31–34.

Bisal, F. and Hsieh, J, (1966) Influence of moisture on erodibility of soil by wind. *Soil Sci.* **102**, 143–146.

Bisal F. and Nielsen, K. F. (1962) Movement of soil particles in saltation. *Can. J. Soil Sci.* **42**, 81–86.

Biscaye, P. E. (1965) Mineralogy and sedimentation of Recent deep-sea clay in the Atlantic Ocean and adjacent seas and oceans. *Geol. Soc. Am. Bull.* **76**, 803–832.

Blackburn, G. (1981) Particle size analysis of Widgelli parna in southeast Australia. *Aust. J. Soil. Res.* **19**, 355–360.

Blacktin, J. C. (1934) *Dust.* Blackie, London, 296pp.

Blackwelder, E. (1925) Exfoliation as a phase of rock weathering. *J. Geol.* **33**, 793–806.

Blackwelder, E. (1934) Yardangs. *Geol. Soc. Am. Bull.* **45**, 159–166.

Blackwelder, E. (1933) The insolation hypothesis of rock weathering. *Am. J. Sci.* **26**, 97–113.

Blank, M. Leinen, M. and Prospero, J. M. (1985) Major Asian eolian inputs indicated by the mineralogy of aerosols and sediments in the western North Pacific. *Nature* **314**, 84–86.

Blatt, H. (1967) Original characteristics of clastic quartz grains. *J. Sed. Petrol.* **37**, 401–424.

Blatt, H. (1970) Determination of mean sediment thickness in the crust: a sedimentologic method. *Geol. Soc. Am. Bull.* **81**, 255–262.

Blong, R. J. (1985) *Volcanic Hazards.* Academic Press, Sydney, 424pp.

Blück, B. J. (1974) Structure and directional properties of some valley sandur deposits in southern Iceland. *Sedimentology* **21**, 533–554.

Blumel, W. D. (1982) Calcretes in Namibia and southeast Spain – relations to sub-stratum, soil formation and geomorphic factors. *Catena Supp.* **1**, 67–95.

Bolton, D. (1984) Generation and propagation of African squall lines. *Q. J. Roy. Met. Soc.* **110**, 695–721.

Bonatti, E. and Arrhenius, G. (1965) Eolian sedimentation in the Pacific off northern Mexico. *Mar. Geol.* **3**, 337–348.

Borsy, Z., Felszerfalvi, J. and Szabo, J. J. (1979) Thermoluminescence dating of several layers of the loess at Paks and Mende (Hungary). *Acta Geol. Acad. Sci. Hung.* **22**, 451–460.

Boulet, R. (1973) Toposéquence de sols tropicaux en Haute Volta. Équilibre et diséquilibre pédobioclimatique. *Mem. ORSTOM*, Paris **85**, 1–272.

Boulton, G. S. (1978) Boulder shapes and grain size distribution of debris as indicators of transport paths through a glacier and till genesis. *Sedimentology* **25**, 773–799.

Boulton, G. S. (1979) Processes of glacial erosion on different sub-strata. *J. Glaciol.* **23**, 15–38.

Bowden, L. W., Hunung, J. R., Hutchinson, C. F. and Johnson, C. W. (1974) Satellite photograph presents first comprehensive view of local wind: the Santa Ana. *Science* **184**, 1077–1078.

Bowen, H. J. M. (1966) *Trace Elements in Biochemistry.* Academic Press, New York.

Bowes, D. R., Langer, A. M. and Rohl, A. N. (1977) Nature and range of mineral dusts in the environment. *Phil. Trans. Roy. Soc. Lond.* **A 286**, 593–610.

Bowler, J. M. (1973) Clay dunes: their occurrence, formation and environmental significance. *Earth Sci. Rev.* **9**, 315–338.

Bowler, J. M. (1975) Deglacial events in southern Australia: their age, nature and palaeoclimatic significance. In R. P. Suggate and M. M. Cresswell (eds.), *Quaternary Studies.* Royal Society of New Zealand, Wellington, pp.75–82.

Bowler, J. M. (1978) Glacial-age events at high and low latitudes. In E. M. van-Zinderen Bakker (ed.), *Antarctic Glacial History and World Palaeoenvironments.* Balkema, Rotterdam, pp.149–172.

Bowler, J. M., Hope, G. S. Jennings, J. N., Singh, G. and Walker, D. (1976)

Late Quaternary climates of Australia and New Guinea. *Quat. Res.* **6**, 359–394.

Bowles, F. A. (1975) Paleoclimatic significance of quartz/illite variations in cores from the eastern equatorial Atlantic. *Quat. Res.* **5**, 225–235.

Bowman, D. (1982) Iron coating in recent terrace sequences under extremely arid conditions. *Catena* **9**, 353–359.

Bowman, D., Karnieli, A., Issar, A. and Bruins, H. J. (1986) Residual colluvio-aeolian aprons in the Negev Highlands (Israel) as a palaeoclimatic indicator. *Palaeogeog. Palaeoclimatol. Palaeoecol.* **56**, 89–101.

Braaten, D. A. and Cahill, T. A. (1986) Size and composition of Asian dust transported to Hawaii. *Atmos. Env.* **20**, 1105–1110.

Bradbury, J. P. (1980) Late Quaternary vegetation history of the Central Great Plains and its relationship to eolian processes in the Nebraska Sand Hills. *US Geol. Surv. Prof. Pap.* **1120A**, 29–36.

Bradford, J. M., Piest, R. F. and Spomer, R. G. (1978) Failure sequenced gully headwalls in Western Iowa. *Soil Sci. Soc. Am. J.* **42**, 323–327.

Brazel, A. J. and Idso, S. B. (1979) Thermal effects of dust on climate. *Ann. Ass. Am. Geog.* **69**, 432–437.

Brazel, A. J. and Hsu, S. (1981) The climatology of hazardous Arizona dust storms. *Geol. Soc. Am. Spec. Pap.* **186**, 293–303.

Brazel, A. J. and Nickling, W. G. (1986) The relationship of weather types to dust storm generation in Arizona (1965–1980). *J. Climatol.* **6**, 255–275.

Brazier, S., Sparks, R. S. J., Carey, S. N., Sigurdsson, H. and Westgate, J. G. (1983) Bimodal grain size distribution and secondary thickening in air-fall ash layers. *Nature* **301**, 115–119.

Breed, C. S. and Breed, W. J. (1979) Dunes and other windforms of central Australia (and a comparison with linear dunes on the Moenkopi Plateau, Arizona). In F. el-Baz and D. M. Warner (eds.), *Apollo–Soyuz Test Project Summary Science Report Volume 2, Earth Observations and Photography.* NASA, Washington DC, pp.319–358.

Briat, M., Royer, A., Petit, J. R. and Lorius, C. (1982) Late glacial input of eolian continental dust in the Dome C ice core: additional evidence from individual microparticle analysis. *Ann. Glaciol.* **3**, 27–31.

Bricker, O. P. and Mackenzie, F. T. (1971) Limestones and red soils of Bermuda: discussion. *Geol. Soc. Am. Bull.* **81**, 2523–2524.

Brittlebank, C. C. (1897) Red rain. *Vict. Nat.* **13**, 125.

Brockie, W. J. (1973) Experimental frost shattering. *Proc. 7th New Zealand Geog. Conf., Hamilton*, 177–186.

Bromfield, A. R. (1974) The deposition of sulphur in dust in northern Nigeria. *J. Agric. Sci.* **83**, 423–425.

Bronger, A. and Pant, R. K. (1983) Micromorphology and genesis of some selected loess profiles in the Kashmir Valley and their relevance to stratigraphy and paleoclimate. In *Current Trends in Geology*, vol. VI (*Climate and Geology of Kashmir*). Today and Tomorrow's Printers and Publishers, New Delhi, pp.131–140.

Brookfield, M. (1970) Dune trend and wind regime in Central Australia. *Z. Geomorph. Supp. Bd.* **10**, 121–158.

Brownlow, A. E., Hunter, W. and Parkin, D. W. (1965) Cosmic dust collections at various latitudes. *Geophys. J.* **9**, 337–368.

Browzin, B. S. (1985) Granular loess classification based on loessial fraction. *Bull. Ass. Eng. Geol.* **22**, 217–227.

Bruce, J. G. (1973*a*) Loessial deposits in southern South Island with a definition of Stewarts Claim Formation. *New Zealand J. Geol. Geophys.* **16**, 533–548.

Bruce, J. G. (1973*b*) A time stratigraphic sequence of loess deposits on near-coastal surfaces of the Balclutha district. *New Zealand J. Geol. Geophys.* **16**, 549–556.

Bruins, H. J. (1976) *The Origin, Nature and Stratigraphy of Paleosols in the Loessial Deposits of the N.W. Negev (Netivot, Israel)*. M.Sc. Thesis, Hebrew University of Jerusalem.

Bruins, H. J. and Yaalon, D. H. (1979) Stratigraphy of the Netivot section in the desert loess of the Negev (Israel). *Acta Geol. Acad. Sci. Hung.* **22**, 161–170.

Brunnacker, K. (1968) Loess stratigraphy at the Lower Rhine. In C. B. Schultz and J. C. Frye (eds.) *Loess and Related Eolian Deposits of the World*. University of Nebraska Press, Lincoln, pp.321–322.

Brunnacker, K. (1973) Einiges über löss-vorkommen in Tunisien. *Eiszeit. Gegen.* 23–24, 89–99.

Brunnacker, K. (1980) Young Pleistocene loess as an indicator for the climate in the Mediterranean area. *Paleoecol. Afr. Surround. Is.* **12**, 99–113.

Brunnacker, K. and Lozek, V. (1969) The presence of loess in southeast Spain. *Z. Geomorph.* **NF 13**, 297–316.

Bryant, I. D. (1982) Loess deposits in Lower Adventdalen, Spizbergen. *Polar Res.* **2**, 93–103.

Bryson, R. A. (1967) Is man changing the climate of the Earth? *Saturday Rev.* **50**, 52–55.

Bryson, R. A. and Baerreis, I. A. (1967) Possibilities of major climatic modifications and their implications: northwest India, a case for study. *Am. Met. Soc. Bull.* **48**, 136–142.

Bucher, A. and Lucas, C. (1984) Sédimentation éolienne intercontinentales, poussières sahariennes et géologie. *Bull. Centre Réch. Explor. Product. Elf Aquitaine* **8**, 151–165.

Bull, W. B. (1964) Geomorphology of segmented alluvial fans in western Fresno County, California. *US Geol. Surv. Prof. Pap.* **532F**, 87–128.

Bull, P. A., Culver, S. J. and Gardner, R. (1980) Chattermark trails as palaeoenvironmental indicators. *Geology* **8**, 318–322.

Buraczyński, J. (1979) Caracteristiques lithologiques des loess d'Achenheim (près de Strasbourg, France). *Acta Geol. Acad. Sci. Hung.* **22**, 229–253.

Buraczyński, J. (1982) Étude lithostratigraphique des loess d'Alsace (France). *Ann. Univ. M. Curie-Sklodowska, Lublin, Polon.* **37** (1), sect. B, 1–41.

Burbank, D. W. and Li Jijun, L. (1985) Age and palaeoclimatic significance of the loess of Lanzhou, North China. *Nature* **316**, 429–431.

Buritt, B. and Hyers, A. D. (1981) Evaluation of Arizona's highway dust warning system. *Geol. Soc. Am. Spec. Pap.* **186**, 281–292.

Burrin, P. J. (1981) Loess in the Weald. *Proc. Geol. Ass.* **92**, 87–92.

Butler, B. E. (1956) Parna – an aeolian clay. *Aust. J. Sci.* **18**, 145–151.

Butler, B. E. (1974) A contribution towards the better specification of parna and some other aeolian clays in Australia. *Z. Geomorph.* **NF 20**, 106–116.

Butler, B. E. (1982) The location of aeolian dust mantles in southeastern Australia. In R. J. Wasson (ed.), *Quaternary Dust Mantles of China, New Zealand and Australia*. ANU Press, Canberra, pp.141–143.

Butler, B. E. and Hutton, J. T. (1956) Parna in the Riverine Plain of southeastern Australia and the soils thereon. *Aust. J. Agric. Res.* **7**, 536–553.

Butrym, J. and Maruszcak, H. (1984) Thermoluminescent chronology of younger and older loesses. In M. Pecsi (ed.), *Lithology and Stratigraphy of Loess and Paleosols*. Geographical Research Institute, Hungarian Academy of Sciences, Budapest, pp.195–199.

Butzer, K. W. and Hansen, C. L. (1968) Towards a history of the Saharan Nile. In K. W. Butzer and C. L. Hansen, *Desert and River in Nubia*. University of Wisconsin Press, Madison, pp.431–457.

Cadle, R. D. (1975) *The Measurement of Airborne Particles*. Wiley, New York, 342pp.

Call, R. E. (1882) The loess of North America. *Am. Nat.* **23**, 369–381.

Cailleux, A. (1961) Sur une poussière transportée par le vent en Mer Rouge. *C. R. Acad. Sci. Paris* **252**, 905–907.

Cailleux, A. (1963) Sur une poussière transportée par le vent dans le Golfe Persique. *C. R. Acad. Sci. Paris* **256**, 2439–2440.

Campbell, J. T. (1889) Origin of the loess. *Am. Nat.* **23**, 785–792.

Carey, S. N. and Sparks, R. S. J. (1986) Quantitative models of fallout and dispersal of tephra from volcanic eruption columns. *Bull. Volcanol.* **48**, 109–125.

Carlson, T. N. and Prospero, J. M. (1972) The large-scale movement of Saharan air outbreaks over the equatorial North Atlantic. *J. App. Met.* **11**, 283–297.

Carter, J. (1979) *How Dust Fouled Up the Hostage Problem in Iran*. Oval Office Tapes Service.

Carter, L. J. (1977) Soil erosion: the problem persists despite the billions spent on it. *Science* **196**, 409–411.

Catt, J. A. (1977) Loess and coversands. In F. W. Shotton (ed.) *British Quaternary Studies: Recent Advances*. Oxford University Press, Oxford, pp.221–229.

Catt, J. A. (1978) The contribution of loess to soils in lowland Britain. In S. Limbrey and J. G. Evans (eds.) The effect of man on the landscape: the lowland zone. *CBA Res. Rep.* **21**, 12–20.

Catt, J. A. (1979) Distribution of loess in Britain. *Proc. Geol. Ass.* **90**, 93–95.

Catt, J. A. (1985) Soil particle size distribution and mineralogy as indicators of pedogenic and geomorphic history: examples from the loessial soils of England and Wales. In K. S. Richards, R. R. Arnett and S. Ellis (eds.), *Geomorphology and Soils*. Allen and Unwin, London, pp.202–218.

Catto, N. R. (1983) Loess in the Cypress Hills, Alberta, Canada. *Can. J. Earth Sci.* **20**, 1159–1167.

Cegla, J. (1969) Influence of capillary ground moisture on aeolian accumulation of loess. *Bull. Acad. Pol. Sci. Geol. Geog. Ser.* **17**, 25–27.

Cegla, J. (1972) Loess sedimentation in Poland. *Acta Univ. Wratislav. Stud. Geog.* **17**, 53–71.

Cegla, J., Buckley, T. and Smalley, I. J. (1971) Microtextures of particles from some European loess deposits. *Sedimentology* **17**, 129–134.

Chamberlain, A. C. (1975) The movement of particles in plant communities. In J. L. Monteith (ed.), *Vegetation and the Atmosphere*, vol. I. Academic Press, London, pp.155–203.

Chamberlain, T. C. (1897) Supplementary hypothesis respecting the origin of the loess of the Mississippi Valley. *J. Geol.* **5**, 795–802.

Chao, S. C. and Xing, J. M. (1982) Origin and development of the Shamo (sandy deserts) and the Gobi (stony deserts) of China. *Striae* **17**, 79–91.

Chapman, F. and Grayson, H. J. (1903) On 'red rain', with special reference to its occurrence in Victoria, with a note on Melbourne dust. *Vict. Nat.* **20**, 17–32.

Chapman, R. W. (1980) Salt weathering by sodium chloride in the Saudi Arabian desert. *Am. J. Sci.* **280**, 116–129.

Charba, J. (1974) Application of gravity current model to analysis of squall-line gust front. *Mon. Weath. Rev.* **102**, 140–156.

Charlson, R. J. (1969) Atmospheric visibility related to aerosol mass concentration. *Env. Sci. Tech.* **3**, 913–918.

Charlson, R. J. and Pilat, M. J. (1969) Climate: the influence of aerosols. *J. App. Met.* **8**, 1001–1002.

Charney, J. G. Stone, P. H. and Quirk, W. J. (1975) Drought in the Sahara: a biogeophysical feedback mechanism. *Science* **187**, 434–435.

Chen, Y, Tarchitzsky, J., Brouwer, J., Morin, J. and Banin, A. (1980) Scanning electron microscope observations on soil crusts and their formation. *Soil Sci.* **130**, 49–55.

Chepil, W. S. (1941) Relationship of wind erosion to the dry aggregate structure of a soil. *Sci. Agric.* **21**, 448–507.

Chepil, W. S. (1945) Dynamics of wind erosion. IV. The translocating and abrasive action of the wind. *Soil Sci.* **61**, 167–177.

Chepil, W. S. (1950) Properties of soil which influence wind erosion. I. The governing principle of surface roughness. *Soil Sci.* **69**, 149–162

Chepil, W. S. (1951) Properties of soil which influence wind erosion. V. Mechanical stability of structure. *Soil Sci.* **72**, 465–478.

Chepil, W. S. (1956) Influence of moisture on erodibility of soil by wind. *Proc. Soil Sci. Soc. Am.* **20**, 288–292.

Chepil, W. S. (1959) Equilibrium of soil grains at the threshold of movement by wind. *Proc. Soil Sci. Soc. Am.* **23**, 422–428.

Chepil, W. S. and Milne, R. A. (1941) Wind erosion of soil in relation to roughness of surface. *Soil Sci.* **52**, 411–432.

Chepil, W. S. and Woodruff, N. P. (1957) Sedimentary characteristics of dust storms. II. Visibility and dust concentration. *Am. J. Sci.* **255**, 104–114.

Chepil, W. S. and Woodruff, N. P. (1963) The physics of wind erosion and its control. *Adv. Agron.* **15**, 211–302.

Chepil, W. S., Siddoway, F. H. and Armbrust, D. V. (1962) Climatic factor for estimating wind erodibility of farm fields. *J. Soil Water Conserv.* **17**, 162–165.

Chepil, W. S., Siddoway, F. H. and Armbrust, D. V. (1963) Climatic Index of wind erosion conditions in the Great Plains. *Proc. Soil Sci. Soc. Am.* **27**, 449–451.

Chester, R. and Johnson, L. R. (1971a) Atmospheric dusts collected off the West African coast. *Nature* **229**, 105–107.

Chester, R. and Johnson, L. R. (1971b) Atmospheric dusts collected off the coast of North Africa and the Iberian Peninsula. *Mar. Geol.* **11**, 251–260.

Chester, R. and Johnson, L. R. (1971c) Trace element geochemistry of North Atlantic aeolian dusts. *Nature* **231**, 176–178.

Chester, R., Sharples, E. J. and Sanders, G. S. (1985) The concentrations of particulate aluminium and clay minerals in aerosols from the northern Arabian Sea. *J. Sedim. Petrol.* **58**, 37–41.

Chester, R. and Stoner, J. H. (1974) The distribution of Mn, Fe, Cu, Ni, Co, Ga, Cr, V, Ba, Sr, Sn, Zn, and Pb in some soil-sized particulates from the lower troposphere. *Mar. Chem.* **2**, 157–188.

Chester, R., Elderfield, H. and Griffin, J. J. (1971) Dust transported in the northeast and southest trade winds of the Atlantic Ocean. *Nature* **233**, 474–476.

Chester, R., Elderfield, J. J., Griffin, J. J., Johnson, L. R. and Padgham, R. C. (1972) Eolian dust along the eastern margins of the Atlantic Ocean. *Mar. Geol.* **13**, 91–106.

Chester, R., Baxter, G. G., Behairy, A. K. A., Connor, K., Cross, D., Elderfield, H. and Padgham, R. C. (1977) Soil-sized dusts from the lower troposphere of the eastern Mediterranean Sea. *Mar. Geol.* **24**, 201–217.

Chester, R., Griffiths, A. G. and Hirst, J. M. (1979) The influence of soil-sized atmospheric particulates on the elemental chemistry of deep-sea sediments of the northeastern Atlantic. *Mar. Geol.* **32**, 141–154.

Chester, R., Sharples, E. J., Sanders, G. S. and Saydam, A. C. (1984*a*) Saharan dust incursion over the Tyrrhenian Sea. *Atmos. Env.* **18**, 929–935.

Chester, R., Sharples, E. J., Sanders, G. S., Oldfield, F. and Saydam, A. C. (1984*b*) The distribution of natural and non-crustal ferrimagnetic minerals in soil-sized particulates from the Mediterranean atmosphere. *Water Air Soil Poll.* **23**, 25–35.

Chesworth, W. (1982) Late Cenozoic geology and the second oldest profession. *Geoscience Canada* **9**, 54–61.

Child, A. L. (1881) The loess of the western plains – subaerial or subaqueous. *Kansas City Rev. Sci. Ind.* **4**, 293–294.

Childs, C. and Leslie, D. (1977) Inter-element relationship in iron–manganese concretions from a catenary sequence of yellow grey earth soils in loess. *Soil Sci.* **123**, 369–376.

CHIQUA (1985) *International Symposium on Loess Research, Guidebook for Excursion from Xian to Ansai, Loess Plateau.* 33pp.

Chmielowiec, G. (1960) Calcareous concretions in the loess of Poland. *Ann. Univ. Marie Curie Sklodowska* **15B**, 39–49.

Choun, H. F. (1936) Dust storms in the southeastern Great Plains area. *Mon. Weath. Rev.* **64**, 195–199.

Chung, Y. S. (1986) Air pollution detection by satellites: the transport and deposition of air pollutants over oceans. *Atmos. Env.* **20**, 617–630.

Church, M. (1972) Baffin Island sandurs. *Geol. Surv. Can. Bull.* **216**, 208pp.

Clayton, R. N., Rex, R. W., Syers, J. K. and Jackson, M. L. (1972) Oxygen isotope abundance in quartz from Pacific pelagic sediments. *J. Geophys. Res.* **77**, 3907–3915.

Clements, T., Stone, R. O., Mann, J. F. and Eymann, J. L. (1963) A study of windborne sand and dust in desert areas. *US Army Natick Laboratories, Earth Sci. Div., Tech. Rep. ES8*, 61pp.

Clevenger, W. A. (1958) Experiences with loess as foundation material. *Trans.*

Am. Soc. Civ. Eng. **123**, 151–169. Discussion by H. R. Cedergren, R. B. Peck, H. O. Ireland and W. A. Clevenger, op. cit. 170–180.

Coffey, G. N. (1909) Clay dunes. *J. Geol.* **17**, 754–755.

Conca, J. L. and Rossman, G. R. (1982) Case hardening of sandstone. *Geology* **10**, 520–523.

Cooke, R. U. (1981) Salt weathering in deserts. *Proc. Geol. Ass.* **92**, 1–16.

Cooke, R. U., Brunsden, D., Doornkamp, J. C. and Jones, D. K. C. (1982) Problems of sand and dust movement by wind in dry lands. In R. U. Cooke, D. Brunsden, J. C. Doornkamp and D. K. C. Jones, *Urban Geomorphology in Dry Lands*, United Nations University, Oxford, pp.249–289.

Coque, R. (1955) Morphologie et croût dans le sud-Tunisien. *Ann. Géog.* **64**, 359–370.

Coudé-Gaussen, G. (1982) Les poussières éoliennes sahariennes. Mise au point. *Rév. Géomorph. Dyn.* **31**, 49–69.

Coudé-Gaussen, G. (1984) Le cycle des poussières éoliennes désertiques actualles et la sédimentation des loess peridésertiques quarternaires. *Bull. Centre Réch. Explor. Product. Elf-Aquitaine* **8**, 167–182.

Coudé-Gaussen, G. and Blanc, P. (1985) Présence de grains éolisés de palygorskite dans les poussières actuelles et les sediments récent d'origine désertique. *Bull. Soc. Geol. Franç.* **1**, 571–579.

Coudé-Gaussen, G., Mosser, C., Rognon, R. and Torenq, J. (1982) Une accumulation de loess Pleistocene superient dans le sud-Tunisien: la coupe de Techine. *Bull. Soc. Geol. France* **24**, 283–292.

Coudé-Gaussen, Rognon, P. and Fedoroff, N. (1984) Piégeage de poussières éoliennes dans les fissures de granitoides du Sinai oriental. *C. R. Acad. Sci. Paris* **298**, Ser. II, 369–374.

Covey, C., Schneider, S. H. and Thompson, S. L. (1984) Modelling the climatic effect of nuclear war. *Nature* **308**, 21–25.

Crook, K. A. W. (1968) Weathering and roundness of quartz sand grains. *Sedimentology* **11**, 171–182.

Crutzen, P. J. and Birks, J. W. (1982) The atmosphere after a nuclear war: twilight at noon. *Ambio* **11**, 115–125.

Cutts, J. A. (1973) Nature and origin of layered deposits of the Martian polar regions. *J. Geophys. Res.* **78**, 4231–4249.

Dabrowska, A., Tkacz, M. and Tucholka, P. (1980) Magnetostratigraphical elements of Vistulian glaciation in Poland. *Quat. Stud. Poland* **2**, 7–20.

Dan, J. and Yaalon, D. H. (1966) Trends of soil development in the Mediterranean environment of Israel. *Trans. Conf. Medit. Soils, Madrid*, pp.139–145.

Dan, J. Moshe, R. and Alperovitch, N. (1973) The soils of Sede Zin. *Israel J. Earth Sci.* **26**, 211–227.

Dana, J. D. (1895) *Manual of Geology*. American Book Company, New York.

Danin, A. and Yaalon, D. H. (1982) Silt plus clay sedimentation and decalcification during plant succession in sands of the Mediterranean coastal plain of Israel. *Israel J. Earth Sci.* **31**, 101–110.

Danin, A., Gerson, R. and Carty, J. (1983) Weathering patterns on hard limestone and dolomite by endolithic lichens and cyanobacteria: supporting evidence for eolian contribution to terra rossa soil. *Soil Sci.* **136**, 213–217.

Darby, D. A., Burckle, L. H. and Clark, D. L. (1974) Airborne dust on the Arctic pack ice, its composition and fallout rate. *Earth Planet. Sci. Lett.* **24**, 166–172.

Dare-Edwards, A. J. (1984) Aeolian clay deposits of southeastern Australia: parna or loessic clay. *Trans. Inst. Brit. Geog. N. S.* **9**, 337–344.

Darwin, C. (1846) An account of the fine dust which often falls on vessels in the Atlantic Ocean. *Q. J. Geol. Soc. Lond.* **2**, 26–30.

Darzi, M. and Winchester, J. W. (1982) Aerosol characteristics at Mauna Loa observatory, Hawaii, after East Asian dust storm episodes. *J. Geophys. Res.* **87**, 1251–1258.

Davis, R. S., Ranov, V. A. and Dodonov, A. E. (1980) Early man in Soviet Central Asia. *Sci. Am.* **243**, 92–102.

Davitaya, F. F. (1969) Atmospheric dust content as a factor affecting glaciation and climatic change. *Ann. Ass. Am. Geog.* **59**, 552–560.

Delany, A. C., Delany, A. C., Parkin, D. W., Griffin, J. J., Goldberg, E. D. and Reiman, B. E. F. (1967) Airborne dust collected at Barbados. *Geochim. Cosmochim. Acta* **31**, 885–909.

Denisov, N. Y. (1951) *The Engineering Properties of Loess and Loess Loams*. Gosstroiizdat, Moscow, 136pp. (in Russian).

Derbyshire, E. (1978) The middle Hwang Ho loess lands. *Geog. J.* **144**, 191–194.

Derbyshire, E. (1983*a*) On the morphology, sediments and origin of the Loess Plateau of Central China. In R. Gardner and H. Scoging (eds.), *Megageomorphology*. Oxford University Press, London, pp.172–194.

Derbyshire, E. (1983*b*) Origin and characteristics of some Chinese loess at two locations in China. In M. E. Brookfield and T. S. Ahlbrandt (eds.), *Eolian Sediments and Processes*. Elsevier, Amsterdam, pp.69–90.

Derbyshire, E. (1983*c*) The Lushan dilemma: Pleistocene glaciation south of the Chang Jiang (Yangtze River). *Z. Geomorph.* **NF 27**, 445–471.

Derbyshire, E. (1984*a*) Granulometry and fabric of the loess at Jiuzhoutai, Lanzhou, People's Republic of China. In M. Pecsi (ed.) *Lithology and Stratigraphy of Loess and Paleosols*. Geographical Research Institute, Hungarian Academy of Sciences, Budapest, pp.95–104.

Derbyshire, E. (1984*b*) Environmental change along the Old Silk Road: the record in the loess. *Geography* **69**, 108–118.

Derbyshire, E. (1984*c*) Till properties and glacier regime in parts of High Asia: Karakoram and Tien Shan. In R. O. Whyte (ed.), *The Evolution of the East Asian Environment*. Centre of Asian Studies, University of Hong Kong, pp.84–110.

Derbyshire, E. and Mellors, T. W. (1986) Loess. In P. G. Fookes and P. Vaughan (eds.) *A Handbook of Engineering Geomorphology*. Surrey University Press and Blackie, Edinburgh, pp.237–246.

Derbyshire, E. and Mellors, T. W. (1987) Geological and geotechnical characteristics of some loessic silts from China and Britain: a comparison. *Eng. Geol.* (in press).

Diester-Haas, L. (1976) Late Quaternary climatic variations in northwest Africa deduced from East Atlantic sediment cores. *Quat. Res.* **6**, 299–314.

Dietrich, R. V. (1977) Impact abrasion of harder by softer materials. *J. Geol.* **85**, 242–246.

Dincer, T., al-Mugrin, A. and Zimmerman, U. (1974) Study of the infiltration and recharge through sand dune in arid zones with special reference to the stable isotopes and thermonuclear tritium. *J. Hydrol.* **23**, 79–109.

Dobson, M. (1781) An account of the Harmattan, a singular African wind. *Phil. Trans. Roy. Soc. Lond.* **71**, 46-57.

Dodonov, A. E. (1979) Stratigraphy of the Upper Pliocene – Quaternary deposits of Tajikistan (Soviet Central Asia). *Acta Geol. Acad. Scient. Hung.* **22**, 63–73.

Dodonov, A. E. (ed.) (1982) *Guidebook for excursions A-11 and C-11, The Tajik SSR.* INQUA, Moscow, pp.31–66.

Dodonov, A. E. (1984) Stratigraphy and correlation of Upper Pliocene – Quaternary deposits of Central Asia. In M. Pecsi (ed.) *Lithology and Stratigraphy of Loess and Paleosols*. Geographical Research Institute, Hungarian Academy of Sciences, Budapest, pp.201–211.

Dodonov, A. E., Melamed, Y. R. and Nikiforova, K. V. (eds.) (1977) *Excursion Guide for an International Symposium on the Problem Boundary of the Neogene and Quaternary System*. Nauka, Moscow.

Doeglas, D. J. (1949) Loess; an eolian product. *J. Sed. Petrol.* **19**, 112–117.

Donk, J. van (1976) An [18]O record of the Atlantic Ocean for the entire Pleistocene. *Geol. Soc. Am. Mem.* **145**, 147–164.

Donkin, R. A. (1981) The 'manna lichen' *Lecanora esculenta. Anthropos* **76**, 562–576.

Doornkamp, J. C., Brunsden, D. and Jones, D. K. C. (1980) Aeolian landforms and deposits. In *Geology, Geomorphology and Pedology of Bahrain*, Chapter 10. GeoBooks, Norwich.

Douglas, R. G. and Savin, S. M. (1975) Oxygen isotope analyses of Tertiary and Cretaceous microfossils from Shatsky Rise and other sites in the North Pacific Ocean. *Init. Rep. DSDP* **32**, 509–520.

Dowgiallo, E. M. (1965) Mutual relation between loess and dune accumulation in southern Poland. *Geog. Polon.* **6**, 105–115.

Doyle, L. J., Hopkins, T. L. and Betzer, P. R. (1976) Black magnetic spherules fallout in the eastern Gulf of Mexico. *Science* **194**, 1157–1159.

Duce, R. A., Unni, C. K., Ray, B. J., Prospero, J. M. and Merrill, J. T. (1980) Long-range atmospheric transport of soil dust from Asia to the tropical North Pacific: temporal variability. *Science* **209**, 1522–1524.

Duley, F. L. (1945) Infiltration into loess soil. *Am. J. Sci* **243**, 278–282.

Dumanski, J., Pawluk, S., Vucetich, C. G. and Lindsay, J. D. (1980) Pedogenesis and tephrochronology of loess-derived soils, Hinton, Alberta. *Can. J. Earth Sci.* **17**, 52–59.

Dumont, A. H. (1852) Loess or lehm. *Q. J. Geol. Soc. Lond.* **8**, 278–281.

Dyer, A. J. and Hicks, B. B. (1967) Radioactive fallout from the French 1966 Pacific tests. *Aust. J. Sci.* **30**, 168–170.

Dylik, J. (1954) The problem of the origin of loess in Poland. *Biul. Peryglac.* **1**, 125–131.

Dymond, J., Biscaye, P. E. and Rex, R. W. (1974) Eolian origin of mica in Hawaiian soils. *Geol. Soc. Am. Bull.* **85**, 37–40.

Ebens, R. J. and Connor, J. J. (1980) Geochemistry of loess and carbonate residuum (Missouri). *US Geol. Surv. Prof. Pap. 954-G*, 32pp.

Eck, H. V., Hauser, V. L. and Ford, R. H. (1965) Fertilizer needs for restoring productivity on Pullman silty clay loam after various degrees of soil removal. *Proc. Soil Sci. Soc. Am.* **29**, 209–213.

Eden, D. (1980). The loess of northeast Essex. *Boreas* **9**, 165–177.

Eden, D. (1982) Starborough loess, Marlborough, New Zealand: statigraphy and correlation. In R. J. Wasson (ed.), *Quaternary Dust Mantles of China, New Zealand and Australia*. ANU Press, Canberra, pp.123–125.

Edwards, M. B. (1979) Late Precambrian loessites from North Norway and Svalbard. *J. Sed. Petrol.* **49**, 85–91.

Ehrenberg, C. G. (1847) The Sirocco dust that fell at Genoa on the 16th May 1846. *Q. J. Geol. Soc. Lond.* **3**, 25–26.

Ehrenberg, C. G. (1851) On the *infusoria* and other microscopic forms in dust showers and blood rain. *Am. J. Sci.* **11**, 372–389.

El Fandy, M. G. (1940) The formation of depression of the Khamasin type. *Q. J. Roy. Met. Soc.* **66**, 323–336.

El Fandy, M. G. (1949) Dust – an active meteorological factor in the atmosphere of northern Africa. *J. App. Phys.* **20**, 660–666.

El Fandy, M. G. (1953) On the physics of dusty atmospheres. *Q. J. Roy. Met. Soc.* **63**, 393–414.

Elmes, P. C. (1980) Fibrous minerals and health. *J. Geol. Soc. Lond.* **137**, 525–535.

Ernst, J. A. (1974) African dust layer sweeps into the southwest North Atlantic area. *Am. Met. Soc. Bull.* **5**, 1352–1353.

Eswaran, H. (1972) Micromorphological indicators of pedogenesis in some tropical soils derived from basalt. *Geoderma* **7**, 15–31.

Eswaran, H. and Stoops, G. (1979) Surface textures of quartz in tropical soils. *Soil Sci. Soc. Am. J.* **43**, 420–424.

Fahey, B. D. (1983) Frost action and hydration as rock weathering mechanisms on schist: a laboratory study. *Earth Surf. Proc. Landf.* **8**, 535–845.

Fahey, B. D. (1985) Salt weathering as a mechanism of rock break-up in cold climates: an experimental approach. *Z. Geomorph.* **NF 29**, 99–111.

Fairall, C. W., Davidson, K. L. and Schacher, G. E. (1983) An analysis of the surface production of sea-salt aerosols. *Tellus* **35 B**, 31–39.

Farquharson, J. S. (1937) Haboobs and instability in the Sudan. *Q. J. Roy. Met. Soc.* **63**, 393–414.

Feda, J. (1966) Structural stability of subsident loess from Praha Dejvice. *Eng. Geol.* **1**, 201–219.

Fennelly, P. F. (1976) The origin and influence of airborne particulates. *Am. Scient.* **64**, 46–56.

Ferguson, W. S., Griffin, J. J. and Goldberg, E. D. (1970) Atmospheric dusts from the North Pacific – a short note on a long-range eolian transport. *J. Geophys. Res.* **75**, 1137–1139.

Ferrar, H. T. (1914) Note on the occurrence of loess deposits in Egypt and its bearing on change of climate in recent geological times. *Rep. Brit. Ass. Adv. Sci.* p.363.

Fett, W. (1958) *Der atmosphärische Staub.* Deutches Verlag Wissen, Berlin, 309pp.

Fink, J. (1968) The loesses in Austria. In C. B. Schultz and J. C. Frye (eds.) *Loess and Related Eolian Deposits of the World.* University of Nebraska Press, Lincoln, pp.285–288.

Fink, J. (1979) Palaeomagnetic research in the northeastern foothills of the Alps and in the Vienna Basin. *Acta Geol. Acad. Scient. Hung.* **22**, 111–124.

Fink, J. and Kukla, G. J. (1977) Pleistocene climates of central Europe: at least seventeen interglacials after the Olduvai event. *Quat. Res.* **7**, 363–371.

Fisher, G. L., Chang, D. P. Y. and Brummer, M. (1976) Fly ash collected from electrostatic precipitator: microcrystalline structure and the mystery of spheres. *Science* **192**, 553–555.

Fisher, R. V. (1966) Textural comparison of John Day volcanic siltstone with loess and volcanic ash. *J. Sed. Petrol.* **36**, 706–718.

Fisher, R. V. and Schminke, H. U. (1984) *Pyroclastic Rocks*. Springer, Berlin, 473pp.

Fisk, H. N. (1951) Loess and Quaternary geology of the Lower Mississippi Valley. *J. Geol.* **59**, 333–356.

Fletcher, B. (1976*a*) The erosion of dust by an airflow. *J. Phys. D. App. Phys.* **9**, 913–924.

Fletcher, B. (1976*b*) The incipient motion of granular materials. *J. Phys. D. App. Phys.* **9**, 2471–2478.

Fletcher, J. E. and Martin, W. P. (1948) Some effects of algae and moulds in the rain crust of desert soils. *Ecology* **29**, 95.

Flint, R. F. (1949) Leaching of carbonates in glacial drift and loess as a basis for age correlation. *J. Geol.* **57**, 297–303.

Flohn, H. (1981) A hemispheric circulation asymmetry during late Tertiary. *Geol. Rund.* **70**, 725–736.

Foda, M. A. (1983) Dry fall of fine dust on sea. *J. Geophys. Res.* **88**, 6021–6026.

Foda, M. A., Khalaf, F. I. and al-Kadi, A. S. (1985) Estimation of dust fallout rates in the northern Arabian Gulf. *Sedimentology* **32**, 595–603.

Folger, D. W. (1970) Wind transport of land-derived mineral, biogenic and industrial matter over the North Atlantic. *Deep Sea Res.* **17**, 337–352.

Folger, D. W., Burckle, L. H. and Heezen, B. C. (1967) Opal phytoliths in a North African dust fall. *Science* **155**, 1243–1244.

Folk, R. L. (1978) Angularity and silica coatings of Simpson Desert sand grains, Northern Territory, Australia. *J. Sed. Petrol.* **48**, 611–624.

Folk, R. L. and Ward, W. C. (1957) Brazos River Bar: a study in the significance of grain size parameters. *J. Sed. Petrol.* **27**, 3–26.

Fookes, P. G. and Best, R. (1969) Consolidation characteristics of some late Pleistocene periglacial metastable soils of east Kent. *Q. J. Eng. Geol.* **2**, 103–128.

Fookes, P. G. and Knill, J. L. (1969) The application of engineering geology in the regional development of northern and central Iran. *Eng. Geol.* **3**, 81–120.

Foster, S. M. and Nicholson, T. H. (1980) Microbial aggregation of sand in a maritime dune succession. *Soil Biol. Biochem.* **13**, 205–208.

Frakes, L. A. (1979) *Climates Throughout Geologic Time*. Elsevier, Amsterdam, 310pp.

Francis, C. W., Chesters, G. and Erhardt, W. H. (1968) [210]Polonium entry into plants. *Env. Sci. Tech.* **2**, 690–695.

Frankel, L. (1957) Relative rates of loess deposition in Nebraska. *J. Geol.* **65**, 649–652.

Franzmeier, D. P. (1970) Particle size sorting of proglacial eolian materials. *Proc. Soil Sci. Soc. Am.* **34**, 920–924.

Fraudorf, P., Patel, R. I., Shirk, J., Walker, R. M. and Freeman, J. J. (1980)

Optical spectroscopy of interplanetary dust collected in the Earth's stratosphere. *Nature* **286**, 866–868.

Frazee, C. J., Fehrenbacher, J. B. and Krumbein, W. C. (1970) Loess distribution from a source. *Proc. Soil Sci. Soc. Am.* **34**, 296–301.

Free, E. E. (1911) The movement of soil material by the wind. *USDA Bur. Soil Bull.* **68**, 272pp.

Freeman, M. H. (1952) Dust storms of the Anglo-Egyptian Sudan. *Met. Rep. 11*, HMSO, London.

Friedlander, S. K. (1977) *Smoke, Dust and Haze*. Wiley, New York, 317pp.

Frye, J. C., Glass, H. D. and Willman, H. B. (1962) Stratigraphy and mineralogy of the Wisconsinan loess of Illinois. *Illinois Geol. Surv. Circ.* **334**, 1–55.

Frye, K. (1981) *The Encyclopedia of Mineralogy*. Hutchinson Ross, Stroudsberg, Pennsylvania, 794pp.

Fryrear, D. W. (1981) Long-term effect of erosion and cropping on soil productivity. *Geol. Soc. Am. Spec. Pap.* **186**, 253–260.

Fuller, M. L. (1922) Some unusual erosion features of the loess in China. *Geog. Rev.* **12**, 570–584.

Fuller, M. L. and Clapp, F. (1903) Marl-loess of the lower Wabash Valley. *Geol. Soc. Am. Bull.* **14**, 153–176.

Fullen, M. A. (1985) Wind erosion of arable soils in East Shropshire (England) during spring 1983. *Catena* **12**, 111–120.

Gabites, J. R. (1954) The drift of radioactive dust from the British nuclear bomb test in October 1953. *New Zealand J. Sci. Tech.* **36B**, 160–165.

Galay, B. F., Skorobogach, T. V. and Zhukov, Y. P. (1982) Volcanic material in loess of the Stavropol region. *Abs. 11th INQUA Cong., Moscow*, vol. II, p.76.

Game, P. M. (1964) Observations on a dustfall in the eastern Atlantic, February 1962. *J. Sed. Petrol.* **34**, 355–359.

Ganor, E. (1975) *Atmospheric Dust in Israel – Sedimentological and Meteorological Analysis of Dust Deposition*. Ph.D. Thesis, Hebrew University of Jerusalem (in Hebrew).

Ganor, E. and Mamane, Y. (1982) Transport of Saharan dust across the eastern Mediterranean. *Atmos. Env.* **16**, 581–587.

Gardner, R. A. M. (1977) Evidence concerning the existence of loess deposits at Tell Fasa, northern Negev, Israel. *J. Archaeol. Sci.* **4**, 377–386.

Gardner, R. A. M. and Pye, K. (1981) Nature, origin and palaeoenvironmental significance of red coastal and desert dune sands. *Prog. Phys. Geog.* **5**, 514–534.

Garratt, J. R. (1984) Cold fronts and dust storms during the Australian summer 1982/83. *Weather* **39**, 98–103.

Geikie, J. (1898) The tundras and steppes of prehistoric Europe. *Scot. Geog. Mag.* **14**, 281–294, 346–357.

Geis, J. W. (1973) Biogenic silica in selected species of deciduous angiosperms. *Soil Sci.* **116**, 113–130.

George, D. J. (1981) Dust fall and instability rain over Northern Ireland on the night of 28–29 January 1981. *Weather* **36**, 216–217.

Gerasimov, I. P. (1973) Chernozems, buried soils, and loesses of the Russian Plain; their age and genesis. *Soil Sci.* **116**, 202–210.

Gerson, R., Amit, R. and Grossman, S. (1985) *Dust Availability in Desert Terrains – A Study of the Deserts of Israel and Sinai*. Report to the US Army Research, Development and Standardization Group, UK, Contract No. DAJA 45-83-C-0041.

Gile, L. (1979) Holocene soils in eolian sediments of Bailey County, Texas. *Soil Sci. Soc. Am. J.* **43**, 940–1003.

Gile, L. H. and Grossman, R. B. (1979) *The Desert Project Soil Monograph*. USDA Soil Conservation Service, 984pp.

Gill, E. W. B. (1948) Frictional electrification of sand. *Nature* **162**, 568.

Gill, E. D. (1973) Loess in southeastern Australia. *Abs. 9th INQUA Cong., Christchurch*, 123–124.

Gill, E. D. and Reeckman, S. A. (1980) Loess in Western Victoria. *Loess Letter* **4**, 3–4.

Gill, E. D. and Segnitt, E. R. (1982) Strata of carbonate loess in southeastern Australia. *Loess Letter* **8**, 3–4.

Gillette, D. A. (1974) On the production of soil wind erosion aerosols having the potential for long-range transport. *J. Réch. Atmos.* **8**, 735–744.

Gillette, D. A. (1977) Fine particulate emissions due to wind erosion. *Trans. Am. Soc. Agric. Eng.* **20**, 890–897.

Gillette, D. A. (1978*a*) A wind tunnel simulation of the erosion of soil: effect of soil texture, sandblasting, wind speed, and soil consolidation on dust production. *Atmos. Env.* **12**, 1735–1743.

Gillette, D. A. (1978*b*) Tests with a portable wind tunnel for determining wind erosion threshold velocities. *Atmos. Env.* **12**, 2309–2313.

Gillette, D. A. (1980) Major contributions of natural primary continental aerosols: source mechanisms. *Ann. New York Acad. Sci.* **338**, 348–358.

Gillette D. A. (1981) Production of dust that may be carried great distances. *Geol. Soc. Am. Spec. Pap.* **186**, 11–26.

Gillette, D. A. and Goodwin, P. A. (1974) Microscale transport of sand-sized soil aggregates by wind. *J. Geophys. Res.* **79**, 4080–4084.

Gillette, D. A. and Walker, T. R. (1977) Characteristics of airborne particles produced by wind erosion of sandy soil, High Plains of West Texas. *Soil Sci.* **123**, 97–110.

Gillette, D. A., Blifford, I. H. and Fenster, C. R. (1972) Measurements of aerosol-size distributions and vertical fluxes of aerosols on land subject to wind erosion. *J. App. Met.* **11**, 977–987.

Gillette, D. A., Blifford, D. A. and Fryrear, D. W. (1974) The influence of wind velocity on the size distributions of aerosols generated by the wind erosion of soils. *J. Geophys. Res.* **79**, 4068–4075.

Gillette, D. A., Clayton, R. N., Mayeda, T. K., Jackson, M. L. and Sridhar, K. (1978) Tropospheric aerosols from some major dust storms of the southwestern United States. *J. App. Met.* **17**, 832–845.

Gillette, D. A., Adams, J., Endo, L. and Smith, D. (1980) Threshold velocities for input of soil particles into the air by desert soils. *J. Geophys. Res.* **C 85**, 5621–5630.

Gillette, D. A., Adams, J., Muhs, D. and Kihl, R. (1982) Threshold friction velocities and rupture moduli for crushed desert soils for input of soil particles into air. *J. Geophys. Res.* **87**, 9003–9015.

Ginzbourg, D. (1971) Note on the eolian deposits and loess of northern Sinai and the western Negev (Israel). *Proc. 8th INQUA Cong., Paris, 1969*, 757–758.

Ginzbourg, D. (1979) Loessic clay from the Beersheva Basin, Israel – potential raw material for the cement industry. *Acta Geol. Acad. Scient. Hung.* **22**, 319–326.

Ginzbourg, D. and Yaalon, D. H. (1963) Petrography and origin of the loess in the Be'er Sheva Basin. *Israel J. Earth Sci.* **12**, 68–70.

Giovanoli, R. (1982) Der Sahara-staubfall vom 8 Januar 1982 in der Schweiz. *Naturwissen.* **69**, 237–239.

Glaccum, R. A. and Prospero, J. M. (1980) Saharan aerosols over the tropical North Atlantic: mineralogy. *Mar. Geol.* **37**, 295–321.

Glasby, G. P. (1971) The influence of aeolian transport of dust particles on marine sedimentation in the southwest Pacific. *J. Roy. Soc. New Zealand* **1**, 285–300.

Glennie, K. W. (1970) *Desert Sedimentary Environments.* Elsevier, Amsterdam, 222pp.

Goh, K. M., Molloy, B. P. J. and Rafter, T. A. (1977) Radiocarbon dating of Quaternary loess deposits, Banks Peninsula, Canterbury, New Zealand. *Quat. Res.* **7**, 177–196.

Goldberg, E. D. (1971) Atmospheric dust, the sedimentary cycle and man. *Comments Earth Sci. Geophys.* **1**, 117–132.

Goldberg, E. D. and Griffin, J. J. (1970) The sediments of the northern Indian Ocean. *Deep Sea Res.* **17**, 513–537.

Golitsyn, G. S. and Ginsburg, A. S. (1985) Comparative estimates of climatic consequences of Martian dust storms and of possible nuclear war. *Tellus* **37B**, 173–181.

Goossens, D. (1985a) The granulometric characteristics of a slowly moving dust cloud. *Earth Surf. Proc. Landf.* **10**, 353–362.

Goossens, D. (1985b) A diffusion model for a settling non-consolidated dust mass. *Catena* **12**, 373–402.

Goudie, A. S. (1973) *Duricrusts*. Clarendon Press, Oxford, 174pp.

Goudie, A. S. (1974) Further experimental investigation of rock weathering by salt and other mechanical processes. *Z. Geomorph. Supp. Bd.* **21**, 1–12.

Goudie, A. S. (1977) Sodium sulphate weathering and the disintegration of Mohenjo Daro, Pakistan. *Earth Surf. Proc.* **2**, 75–86.

Goudie, A. S. (1978) Dust storms and their geomorphological implications. *J. Arid Env.* **1**, 291–310.

Goudie, A. S. (1983a) Dust storms in space and time. *Prog. Phys. Geog.* **7**, 502–530.

Goudie, A. S. (1983b) Calcrete. In A. S. Goudie and K. Pye (eds.) *Chemical Sediments and Geomorphology*. Academic Press, London, pp.93–132.

Goudie, A. S. (1984) Salt efflorescences and salt weathering in the Hunza Valley, Karakoram Mountains, Pakistan. In K. Miller (ed.) *Proceedings of the International Karakoram Project*, vol. II, Cambridge University Press, Cambridge, pp.607–615

Goudie, A. S. (1985) Salt weathering. *School of Geography, University of Oxford, Res. Pap. 33*, 31pp.

Goudie, A. S. (1986) Laboratory simulation of the 'wick effect' in salt weathering of rock. *Earth Surf. Proc. Landf.* **11**, 275–285.

Goudie, A. S. and Day, M. J. (1980) Disintegration of fan sediments in Death Valley, California, by salt weathering. *Phys. Geog.* **1**, 126–137.

Goudie, A. S. and Watson, A. (1984) Rock block monitoring of rapid salt weathering in southern Tunisia. *Earth Surf. Proc. Landf.* **9**, 95–99.

Goudie, A. S. and Thomas, D. S. G. (1985) Pans in southern Africa with particular reference to South Africa and Zimbabwe. *Z. Geomorph.* **NF 29**, 1–19.

Goudie, A. S., Cooke, R. U. and Evans, I. (1970) Experimental investigation of rock weathering by salts. *Area* **4**, 42–48.

Goudie, A. S., Allchin, B. and Hegde, K. T. M. (1973) The former extensions of the Great Indian Sand Desert. *Geog. J.* **139**, 243–257.

Goudie, A. S., Cooke, R. U. and Doornkamp, J. C. (1979) The formation of silt from quartz dune sand by salt processes in deserts. *J. Arid Env.* **2**, 105–112.

Goudie, A. S., Rendell, H. A. and Bull, P. A. (1984) The loess of Tajik S.S.R. In K. Miller (ed.) *Proceedings of the International Karakoram Project*, vol. I. Cambridge University Press, Cambridge, pp.399–412.

Goudie, A. S., Middleton, N. J. and Wells, G. L. (1986) The frequency and source areas of dust storms. In W. G. Nickling (ed.) *Binghampton Symposium on Eolian Processes*, Binghampton, New York (in press).

Grabowska-Olszewska, B. (1975) SEM analysis of microstructures of loess deposits. *Bull. Int. Ass. Eng. Geol.* **11**, 45–48.

Grabowska-Olszewska, B. (1982) Microstructural sensitivity of loesses. *Bull. Acad. Polon. Sci.* **30**, 181–187.

Graedel, T. E. and Franey, J. P. (1975) Field measurements of submicron aerosol washout by snow. *Geophys. Res. Lett.* **2**, 325–328.

Graf, J., Snow, R. H. and Draftz, R. G. (1976) *Aerosol sampling and analysis – Phoenix, Arizona.* Environmental Science Research Laboratory, US Environmental Protection Agency, Triangle Park, North Carolina, Report, 135pp.

Grant, C. C. (1949) Dust devils in the subarctic. *Weather* **4**, 402–403.

Greeley, R. (1979) Silt–clay aggregates on Mars. *J. Geophys. Res.* **84**, 6248–6254.

Greeley, R. and Iversen, J. D. (1985) *Wind as a Geological Process.* Cambridge University Press, Cambridge, 333pp.

Greeley, R. and Leach, R. N. (1979) A preliminary assessment of the effects of electrostatics on eolian processes. *NASA Tech. Mem. 79729,* 236–237.

Greeley, R., White, B. R., Pollack, J. B., Iversen, J. D. and Leach, R. N. (1981) Dust storms on Mars: considerations and simulations. *Geol. Soc. Am. Spec. Pap. 186,* pp. 101–122.

Green, F. H. Y., Vallyathan, V., Mentnech, M. S., Tucker, J. H., Merchant, J. A., Kiessling, P. J., Antonius, J. A. and Pashley, P. (1981) Is volcanic ash a pneumoconiosis risk? *Nature* **292**, 216–217.

Green, H. L. and Lane, W. R. (1964) *Particulate Clouds: Dusts, Smokes and Mists.* Spon, London.

Griffin, J. J., Windom, H. and Goldberg, E. D. (1968) The distribution of clay minerals in the world oceans. *Deep Sea Res.* **15**, 433–459.

Griffiths, E. (1973) Loess of Banks Peninsula. *New Zealand J. Geol. Geophys.* **16**, 657–676.

Griggs, D. (1936) The factor of fatigue in rock exfoliation. *J. Geol.* **44**, 781–796.

Grigoryev, A. A. and Kondratyev, K. J. (1980) Atmospheric dust observed from space. *WMO Bull.* **29**, 250–254.

Grove, A. T. (1960) The geomorphology of the Tibesti region. *Geog. J.* **126**, 18–31.

Gupta, J. P., Agrawal, R. K. and Raikhy, N. P. (1981) Soil erosion by wind from bare sandy plains in western Rajasthan. *J. Arid Env.* **4**, 15–20.

Gwynne, C. S. (1950) Terraced side slopes in loess, southwestern Iowa. *Geol. Soc. Am. Bull.* **61**, 1347–1354.

Haeberli, W. (1977) Sahara dust in the Alps – a short review. *Z. Gletscher.* **13**, 206–208.

Haesaerts, P. (1985) Les loess du Pléistocene supérieur en Belgique; comparisons avec les séquences d'Europe centrale. *Bull. Ass. Fr. Etude Quat.* **2–3**, 105–115.

Haesaerts, P. and van Vliet-Lanoë, B. (1973) Évolution d'un permafrost fossile dans les limons du dernier glaciare à Harmignies (Belgique). *Bull. Ass. Étude Quat.* **3**, 191–204.

Hagen, L. (1984) Soil aggregate abrasion by impacting sand and soil particles. *Trans. Am. Soc. Agric. Eng.* **27**, 805–808.

Hagen, L. and Woodruff, N. O. (1973) Air pollution in the Great Plains. *Atmos. Env.* **7**, 323–332.

Haldorsen, S. (1981) Grain size distribution of subglacial till and its relation to glacial crushing and abrasion. *Boreas* **10**, 91–105.

Haldorsen, S. (1983) Mineralogy and geochemistry of basal till and their relationship to till-forming processes. *Norsk Geol. Tidsskr.* **63**, 15–25.

Hall, F. F. (1981) Visibility reductions from soil dust in the western United States. *Atmos. Env.* **15**, 1929–1933.

Hall, K. (1986) Rock moisture content in the field and the laboratory and its relationship to mechanical weathering studies. *Earth Surf. Proc. Landf.* **11**, 131–142.

Hallett, J. (1969) A rotor-induced dust devil. *Weather* **24**, 133.

Hallett, J. and Hoffer, T. (1971) Dust devil systems. *Weather* **26**, 247–250.

Hammer, C. U., Clausen, H. B., Dansgaard, D., Gunderstrup, N., Johnsen, S. J. and Reeh, N. (1978) Dating of Greenland ice cores by flow models, isotopes, volcanic debris and continental dust. *J. Glaciol.* **20**, 3–26.

Hammond, C., Moon, C. F. and Smalley, I. J. (1973) High-voltage electron microscopy of quartz particles from post-glacial clay soils. *J. Mat. Sci.* **8**, 509–513.

Hand, I. F. (1934) The character and magnitude of the dust cloud which passed over Washington DC May 11, 1934. *Mon. Weath. Rev.* **62**, 157–158.

Handy, R. H. (1973) Collapsible loess in Iowa. *Proc. Soil Sci. Am.* **37**, 281–284.

Handy, R. L. (1976) Loess distribution by variable winds. *Geol. Soc. Am. Bull.* **87**, 915–927.

Handy, R. L. and Davidson, D. T. (1953) On the curious resemblance between fly ash and meteoric dust. *Proc. Iowa Acad. Sci.* **60**, 373–379.

Hattersley-Smith, G. (1961) Ablation effects due to wind-blown dust. *J. Glaciol.* **3**, 1153.

Hays, J. D. and Peruzza, A. (1972) The significance of calcium carbonate oscillations in eastern equatorial deep-sea sediments for the end of the Holocene warm interval. *Quat. Res.* **2**, 355–362.

Healy, T. R. (1970) Dust from Australia – a reappraisal. *Earth Sci. J.* **4**, 106–116.

Heathershaw, A. D. (1974) Bursting phenomena in the sea. *Nature* **248**, 394–395.

Hedin, S. (1896) A journey through the Takla Makan Desert, Chinese Turkestan. *Geol. J.* **8**, 264–278, 356–372.

Hedin, S. (1903) *Central Asia and Tibet*, vols. I and II. Charles Scribners and Sons, New York, 608pp.

Heiken, G. (1974) An atlas of volcanic ash. *Smithsonian Contrib. Earth Sci. 12*, 101pp.

Heiken, G. and Wohletz, K. (1985) *Volcanic Ash*. University of California Press, Berkeley, 246pp.

Heller, F. and Liu Tung-sheng (1982) Magnetostratigraphical dating of loess deposits in China. *Nature* **300**, 431–433.

Heller, F. and Liu Tung-sheng (1984) Magnetism of Chinese loess deposits. *Geophys. J. Roy. Astron. Soc.* **77**, 125–141.

Hess, S. L. (1975) Dust on Venus. *J. Atmos. Sci.* **32**, 1076–1087.

Hidy, G. M. (1984) *Aerosols*. Academic Press, Orlando, 774pp.

Hidy, G. M. and Brock, J. R. (1971) An assessment of the global sources of tropospheric aerosols. *Proc. 2nd Clean Air Cong.*, Washington DC, 1088–1097.

Hills, E. S. (1939) The lunette, a new landform of aeolian origin. *Aust. Geog.* **3**, 15–21.

Hinds, B. D. and Hoidale, G. B. (1975) *Boundary layer dust occurrences*, vol. II. *(Atmospheric dust over the Middle East, Near East and North Africa)*. Technical Report, US Army Electronics Command, White Sands Missile Range, New Mexico, 193pp.

Hirose, K. and Sugimura, Y. (1984) Excess [228]Th in the airborne dust: an indicator of continental dust from the East Asian deserts. *Earth Planet. Sci. Lett.* **70**, 110–114.

Hjulstrom, F. (1952) The geomorphology of the alluvial outwash plains (sandurs) of Iceland and the mechanics of braided rivers. *Proc. 17th Int. Geog. Conf., Washington DC*, 227–242.

Hobbs, W. H. (1931) Loess, pebble bands and boulders from glacial outwash of the Greenland continental glacier. *J. Geol.* **39**, 381–385.

Hobbs, W. H. (1983) The origin of loess associated with continental glaciation based upon studies in Greenland. *C. R. Int. Geol. Cong., Paris*, 1931, **2**, 408–411.

Hobbs, W. H. (1942) Wind – the dominant transportation agent within extramarginal zones to continental glaciers. *J. Geol.* **50**, 556–559.

Hobbs, W. H. (1943a) The glacial anticyclones and the European continental glacier. *Am. J. Sci.* **241**, 333–336.

Hobbs, W. H. (1943*b*) The glacial anticyclone and the continental glaciers of North America. *Proc. Am. Phil. Soc.* **86**, 368–402.

Hogan, A. W. (1975) Antarctic aerosols. *J. App. Met.* **14**, 550–559.

Hoidale, G. B. and Smith, C. M. (1968) Analysis of the giant particle component of the atmosphere over an interior desert basin. *Tellus* **20**, 251–268.

Hoidale, G. B., Smith, C. M., Blanco, A. J. and Barber, T. L. (1967) *A Study of Atmospheric Dust*. Atmospheric Sciences Laboratory Report ECOM-5067, US Army Electronics Command, White Sands Missile Range, New Mexico, 132pp.

Holm, K., Talbot, M. R., Williams, M. A. J. and Adamson, D. A. (1986) Sedimentation in low relief desert margin systems: the Triassic of south-west England and the Quaternary of New South Wales, Australia. Abstract, Conference on *Desert Sediments Ancient and Modern*, Geological Society of London, 20–21 May 1986, p.20.

Holmes, C. D. (1944) Origin of loess – a criticism. *Am. J. Sci.* **242**, 442–446.

Hooghiemstra, H. and Agwu, C. O. C. (1986) Distribution of palynomorphs in marine sediments: a record for seasonal wind patterns over northwest Africa and adjacent Atlantic. *Geol. Rund.* **75**, 81–95.

Hooke, R. le B. (1967) Processes on arid region alluvial fans. *J. Geol.* **75**, 438–460.

Ho Ping-ti (1969) The loess and the origin of Chinese agriculture. *Am. Hist. Rev.* **75**, 1–36.

Ho Ping-ti (1975) *The Cradle of the East*. Chinese University of Hong Kong and University of Chicago Press, Hong Kong, 440pp.

Horikawa, K. and Shen, H. (1960) Sand movement by wind action – on the characteristics of sand. *Beach Erosion Board Tech. Mem. 119*, 51pp.

Horowitz, A., Weinstein, M. and Ganor, E. (1975) Palynological determination of dust storm provenances in Israel. *Pollen et Spores* **17**, 223–232.

Houseman, J. (1961) Dust haze at Bahrain. *Met. Mag.* **90**, 50–52.

Houze, R. A. (1977) Structure and dynamics of a tropical squall-line system. *Mon. Weath. Rev.* **105**, 1540–1567.

Howarth, H. H. (1882) The loess – a rejoinder. *Geol. Mag.* (2nd ser.) **9**, 343–356.

Hudec, P. P. (1973) Weathering of rocks in arctic and sub-arctic environments. In J. D. Aitken and D. J. Glass (eds.) *Proceedings of a Symposium on the Geology of the Canadian Arctic*. Geological Association of Canada, pp.313–335.

Hudec, P. P. and Rigbey, S. G. (1976) The effect of sodium chloride on water absorption characteristics of rock aggregate. *Bull. Ass. Eng. Geol.* **13**, 199–211.

Huffman, G. G. and Price, W. A. (1949) Clay dune formation near Corpus Christi, Texas. *J. Sed. Petrol.* **19**, 118–127.

Humphreys, W. J. (1913) Volcanic dust and other factors in the production of climatic changes, and their possible relation to ice ages. *J. Franklin Inst.* **176**, 131–176.

Hunt, A. (1986) The application of mineral magnetic methods to atmospheric aerosol discrimination. *Phys. Earth. Planet. Int.* **42**, 10–21.

Hunt, J. C. R. and Nalpanis, P. (1985) Saltating and suspended particles over flat and sloping surfaces. I. Modelling concepts. In O. E. Barndorff-Nielsen, J. T. Møller, K. Romer-Rasmussen and B. B. Willetts (eds.), *Proceedings of International Workshop on the Physics of Blown Sand*, Aarhus, 28–31 May 1985. Department of Theoretical Statistics, Institute of Mathematics, University of Aarhus Mem. 8, vol. I, pp.9–36.

Hutton, J. T. (1982) Calcium carbonate in rain. In R. J. Wasson (ed.) *Quaternary Dust Mantles of China, New Zealand and Australia*. ANU Press, Canberra, pp.139–140.

Hutton, J. T. and Leslie, T. I. (1958) Accession of non-nitrogenous ions dissolved in rainwater to soils in Victoria. *Aust. J. Agric. Res.* **9**, 492–507.

Hyers, A. D. and Marcus, M. G. (1981) Land use and desert dust hazards in Central Arizona. *Geol. Soc. Am. Spec. Pap. 186*, pp.267–280.

Idso, S. B. (1972) Radiation fluxes during a dust storm. *Weather* **27**, 204–208.

Idso, S. B. (1973a) Haboobs in Arizona. *Weather* **28**, 154–155.

Idso, S. B. (1973b) Thermal radiation from a tropospheric dust suspension. *Nature* **241**, 448–449.

Idso, S. B. (1974a) Tornado or dust devil: enigma of desert whirlwinds. *Am. Scient.* **62**, 530–541.

Idso, S. B. (1974b) Thermal blanketing: a case for aerosol-induced climatic alteration. *Science* **186**, 50–51.

Idso, S. B. (1975) Low-level aerosol effects on Earth's surface energy balance. *Tellus* **27**, 318–320.

Idso, S. B. (1976) Dust storms. *Sci. Am.* **235**, 108–114.

Idso, S. B. (1981a) Climatic change: desert-forming processes. *Geol. Soc. Am. Spec. Pap.* 186, 217–222.

Idso, S. B. (1981b) Climatic change: the role of atmospheric dust. *Geol. Soc. Spec. Pap. 186*, pp.207–215.

Idso, S. B. and Brazel, A. J. (1977) Planetary radiation balance as a function of atmospheric dust: climatological consequences. *Science* **198**, 731–733.

Idso, S. B. and Brazel, A. J. (1978) Climatological effects of atmospheric particulate pollution. *Nature* **274**, 781–782.

Idso, S. B., Ingram, R. S. and Pritchard, J. M. (1972) An American haboob. *Am. Met. Soc. Bull.* **53**, 930–955.

Iler, R. K. (1979) *The Geochemistry of Silica*. Wiley, New York.

Ing, G. K. T. (1969) A dust storm over central China, April 1969. *Weather* 27, 136–145.

Ingram, R. S. (1973) Arizona 'eddy' tornadoes. *NOAA Tech. Mem. NWS WR91*, 9pp.

Issar, A. S. (1985) Fossil water under the Sinai–Negev Peninsula. *Sci. Am.* July 1985, 104–110.

Issar, A. S. and Bruins, H. J. (1983) Special climatological conditions in the deserts of Sinai and the Negev during the latest Pleistocene. *Palaeogeog. Palaeoclimatol. Palaeoecol.* 43, 63–72.

Issar, A. S., Karnieli, A., Bruins, H. J. and Gilead, I. (1984) The Quaternary geology and hydrology of Sede Zin, Negev, Israel. *Israel J. Earth Sci.* 33, 34–42.

Issar, A. S., Tsoar, H., Gilead, Y. and Zangvil, A. (1986) Geomorphological and palaeohydrological evidence of changes in the climate of the northern Negev and Sinai during the late Quaternary. In M. G. Wurtele and L. Berkofsky (eds.), *Progress in Desert Research*. Rowman and Allanheld, Totawa, New Jersey, (in press).

Itagi, K. and Koeunuma, A. (1962) Altitude distribution of fallout contained in rain and snow. *J. Geophys. Res.* 67, 3927–3933.

Ivanova, I. K. and Velichko, A. A. (1968) Short note about the loesses in the European part of the Soviet Union. In C. B. Schultz and J. C. Frye (eds.) *Loess and Related Eolian Deposits of the World*. University of Nebraska Press, Lincoln, pp.345–349.

Iversen, J. D. and Jensen, V. (1981) *Wind Transportation of Dust from Coal Stock Piles*. Skibsteknisk Laboratorium Report SL81054, Copenhagen, 82pp.

Iversen, J. D. and White, B. R. (1982) Saltation threshold on Earth, Mars and Venus. *Sedimentology* 29, 111–119.

Iversen, J. D., Pollack, J. B., Greeley, R. and White, B. R. (1976a). Saltation threshold on Mars; the effect of interparticle force, surface roughness, and low atmosphere density. *Icarus*, 29, 381–393.

Iversen, J. D., Pollack, J. B., Greeley, R. and White, B. R. (1976b) Windblown dust on Earth, Mars and Venus. *J. Atmos. Sci.* 33, 2425–2429.

Ives, R. L. (1947) Behaviour of dust devils. *Am. Met. Soc. Bull.* 28, 168–174.

Iwasaka, Y., Hiroaki, M. and Nagaya, K. (1983) The transport and spatial scale of Asian dust storm clouds: a case study of the dust storm event of April 1979. *Tellus* 35B, 189–196.

Jackson, M. L. (1981) Oxygen isotopic ratios in quartz as an indicator of provenance of dust. *Geol. Soc. Am. Spec. Pap.* 186, 27–36.

Jackson, M. L., Levelt, T. W. M., Syers, J. K., Rex, R. W., Clayton, R. N., Clayton, G. D., Sherman, G. D. and Uehara, G. (1971) Geomorphological relationships of tropospherically-derived quartz in soils on the Hawaiian Islands. *Soil Sci. Soc. Am. Proc.* **35**, 515–525.

Jackson, M. L., Gibbons, F. R., Syers, J. K. and Mokma, L. (1972) Eolian influence on soils developed on a chronosequence of basalts in Victoria, Australia. *Geoderma* **8**, 147–163.

Jackson, M. L., Gillette, D. A., Danielson, E. F., Blifford, I. H., Bryson, R. A. and Syers, J. K. (1973) Global dustfall during the Quaternary as related to environments. *Soil Sci.* **116**, 135–145.

Jackson, M. L., Clayton, R. N., Violante, A. and Violante, P. (1982) Eolian influence on terra rossa soils of Italy traced by quartz oxygen isotopic ratio. In H. van Olphen and F. Veniale (eds.) *International Clay Conference 1981*. Elsevier, Amsterdam, pp.293–300.

Jackson, P. S. and Hunt, J. C. R. (1975) Flow over a small hill. *Q. J. Roy. Met. Soc.* **101**, 929–955.

Jaenicke, R. and Schütz, L. (1978) Comprehensive study of physical and chemical properties of the surface aerosols in the Cape Verde Islands region. *J. Geophys. Res.* **83**, 3585–3599.

Jahn, A. (1950) Loess, its origin and connection with the climate of the Glacial Epoch. *Acta Geol. Polon.* **1**, 257–310.

Jamagne, M., Lautridou, J. P. and Sommé, J. (1981) Préliminaire a une synthèse sur les variations sédimentologiques des loess de la France du nord-ouest dans leur cadre stratigraphique et paléogéographique. *Bull. Soc. Géol. Fr.* **23**, 143–147.

Janecek, T. R. and Rea, D. K. (1983) Eolian deposition in the northwest Pacific Ocean: Cenozoic history of atmospheric circulation. *Geol. Soc. Am. Bull.* **94**, 730–738.

Janecek, T. R. and Rea, D. K. (1985) Quaternary fluctuations in the northern hemisphere trade winds and westerlies. *Quat. Res.* **24**, 150–163.

Jessup, R. W. (1960) An introduction to the soils of the southeastern portion of the Australian arid zone. *J. Soil Sci.* **11**, 92–105.

Johnson, L. R. (1976) Particle-size fractionation of eolian dusts during transport and sampling. *Mar. Geol.* **21**, M17–M21.

Johnson, L. R. (1979) Mineralogical dispersal patterns of North Atlantic deep-sea sediments with particular reference to eolian dusts. *Mar. Geol.* **29**, 335–345.

Jones, D. K. C., Cooke, R. U. and Warren, A. (1986) Geomorphological investigation, for engineering purposes, of blowing sand and dust hazard. *Q. J. Eng. Geol.* **19**, 251–270.

Jones, R. L. and Beavers, A. H. (1963a) Some mineralogical and chemical properties of plant opal. *Soil Sci.* **96**, 375–379.

Jones, R. L. and Beavers, A. H. (1963b) Sponge spicules in Illinois soils. *Proc. Soil Sci. Soc. Am.* **27**, 438–440.

Jones, R. L. and Hay, W. W. (1975) Bioliths. In G. E. Giesking (ed.) *Soil Components*, vol. II, (*Inorganic Components*). Springer, Berlin, pp.481–496.

Jones, R. L., Mackenzie, L. J. and Beavers, A. H. (1964) Opaline microfossils in some Michigan soils. *Ohio J. Sci.* **64**, 417–423.

Joseph, J. H., Manes, A. and Ashbel, D. (1973) Desert aerosols transported by khamsinic depressions and their climatic effects. *J. App. Met.* **12**, 792–797.

Joseph, P. W., Raipal, D. K. and Deka, S. N. (1980) 'Andhi'; the convective dust storm of northwest India. *Mausam* **31**, 431–432.

Junge, C. E. (1969) Comments on the concentration of size distribution measurements of atmospheric aerosols and a test of the theory of self-preserving size distributions. *J. Atmos. Sci.* **26**, 603–615.

Junge, C. E. (1979) The importance of mineral dust as an atmospheric constituent. In C. Morales (ed.) *Sahran Dust.* Wiley, Chichester, pp. 49–60.

Juvigné, E. (1985) The use of heavy mineral suites for loess stratigraphy. *Geol. Mijn.* **64**, 333–336.

Kadowaki (1979) Silicon and aluminium in urban aerosols for characterization of atmospheric soil particles in the Nagoya area. *Environ. Sci. Technol.* **13**, 1130–1133.

Kaimal, J. C. and Businger, J. A. (1970) Case studies of a convective plume and a dust devil. *J. App. Met.* **9**, 612–620.

Kalinske, A. A. (1943) Turbulence and the transport of sand and silt by wind. *Ann. N. Y. Acad. Sci.* **44**, 41–54.

Kalu, A. E. (1979) The African dust plume: its characteristics and propagation across West Africa in winter. In C. Morales (ed.) *Saharan Dust.* Wiley, Chichester, pp.95–118.

Karcz, I. (1972) Sedimentary structures formed by flashfloods in southern Israel. *Sedim. Geol.* **7**, 161–182.

Kármán, T. von (1934) Turbulence and skin friction. *J. Aeron. Sci.* **1**, 1–20.

Kastnelson, J. (1970) Frequency of dust storms at Be'er Sheva. *Israel J. Earth Sci.* **19**, 69–76.

Keilhack, K. (1920) The riddle of loess formation. *Zeit. Deutsch. Geol. Ges.* **72**, 146–161 (in German).

Kennedy, N. M. (1982) Tephric loess in the Rotorua – Bay of Plenty region, North Island, New Zealand. In R. J. Wasson (ed.) *Quaternary Dust Mantles of China, New Zealand and Australia.* ANU Press, Canberra, pp.119–122.

Kennett, J. P. (1977) Cenozoic evolution of the Antarctic glaciation, the circum-Antarctic Ocean, and their impact on global paleoceanography. *J. Geophys. Res.* **82**, 3843–3860.

Kerr, R. A. (1979) Global pollution: is the Arctic haze actually industrial smog? *Science* **205**, 290–293.

Kes, A. S. (1984) Zonation and faciality of loessic deposits. In M. Pecsi (ed.) *Lithology and Stratigraphy of Loess and Paleosols*. Geographical Research Institute, Hungarian Academy of Sciences, Budapest, pp.104–111.

Kessler, E., Alexander, D. and Rarick, J. (1978) Dust storms from the US High Plains in late winter 1977 – search for cause and implications. *Oklahoma Acad. Sci.* **58**, 116–128.

Khalaf, F. I. and al-Hashash, M. (1983) Aeolian sediments in the northwestern part of the Arabian Gulf. *J. Arid Env.* **6**, 319–332.

Khalaf, F. I., al-Kadi, A. and al-Saleh, S. (1985) Mineralogical composition and potential sources of dust fallout deposits in Kuwait, northern Arabian Gulf. *Sedim. Geol.* **42**, 255–278.

Khalcheva, T. (1984) Facial changes in the material composition of Upper Pleistocene loesses in western and eastern regions of European USSR: a case study of Volyno-Podolia and the Dneiper Basin. In M. Pecsi (ed.) *Lithology and Stratigraphy of Loess and Paleosols*. Geographical Research Institute, Hungarian Academy of Sciences, Budapest, pp.121–132.

Kidson, E. (1929) The meteorological conditions associated with the dust storm of October 1928. *New Zealand J. Sci. Technol.* **10**, 292–299.

Kidson, E. (1930) Dust from Australia. *New Zealand J. Sci. Technol.* **11**, 417–418.

Kidson, E. and Gregory, J. W. (1930) Australian origin of red rain in New Zealand. *Nature* **125**, 410–411.

Kimberlin, L. W., Hidlebaugh, A. L. and Grunewald, A. R. (1977) The potential wind erosion problem in the United States. *Trans. Am. Soc. Agr. Eng.* **20**, 873–879.

Kingsmill, T. W. (1871) The probable origin of deposits of 'loess' in North China and eastern Asia. *Q. J. Geol. Soc. Lond.* **27**, 376–384.

Kinsman, D. J. J. (1969) Modes of formation, sedimentary associations, and diagnostic features of shallow-water and supratidal evaporites. *Bull. Am. Ass. Petrol. Geol.* **53**, 830–840.

Kirchenheimer, F. (1969) Heidelberg und der Löss. '*Ruperto-Carola*' *Z. Verein. Freunde Student. Univ. Heidelberg XXI Jahr.* **46**, 3–7.

Klimenko, L. V. and Moskaleva, L. A. (1979) Frequency of occurrence of dust storms in the USSR. *Meteorol. Gidrol.* **9**, 93–97 (in Russian).

Knight, S. H. (1924) Eolian abrasion of quartz grains. *Geol. Soc. Am. Bull.* **35**, 107–108.

Knutson, E. O., Sood, S. K. and Stockham, J. D. (1977) Aerosol collection by snow and ice crystals. *Atmos. Env.* **11**, 395–402.

Kolbe, R. W. (1957) Freshwater diatoms from Atlantic deep-sea sediments. *Science* **126**, 1053–1056.

Kolla, V. and Biscaye, P. E. (1973) Clay mineralogy and sedimentation in the eastern Indian Ocean. *Deep Sea Res.* **20**, 727–738.

Kolla, V. and Biscaye, P. E. (1977) Distribution and origin of quartz in the sediments of the Indian Ocean. *J. Sedim. Petrol.* **47**, 642–649.

Kolla, V., Biscaye, P. E. and Hanley, A. F. (1979) Distribution of quartz in late Quaternary Atlantic sediments in relation to climate. *Quat. Res.* **11**, 261–277.

Kollmorgen, H. L. (1963) Isopachous map and study on thickness of Peorian loess in Nebraska. *Proc. Soil Sci. Soc. Am.* **27**, 445–458.

Konischev, V. N. (1982) Characteristics of cryogenic weathering in the permafrost zone of the European USSR. *Arc. Alp. Res.* **14**, 261–265.

Kraev, V. F. (1975) On the periglacial character of the Ukranian loess. *Biul. Peryglac.* **24**, 69–70.

Krestnikov, V. N., Nersesov, I. L. and Shtange, D. V. (1980) Quaternary tectonics and deep structure of the Pamir and Tien Shan. *Int. Geol. Rev.* **24**, 745–758.

Krieger, N. I. (1965) *Loess: Its Characteristics and Relation to the Geographical Environment.* Izd-vo Nauka, Moscow, 296pp.

Krigstrom, A. (1962) Geomorphological studies of sandur plains and their braided rivers in Iceland. *Geog. Ann.* **44**, 328–346.

Krinsley, D. H. and Doornkamp, J. C. (1973) *Atlas of Quartz Sand Grain Surface Textures.* Cambridge University Press, New York, 91pp.

Krinsley, D. H. and Greeley, R. (1986) Individual particles and Martian eolian action – a review. *Sedim. Geol.* **47**, 167–189.

Krinsley, D. H. and Leach, R. (1981) Properties of electrostatic aggregates and their possible presence on Mars. *Precamb. Res.* **14**, 167–178.

Krinsley, D. H. and McCoy, F. (1978) Aeolian quartz sand and silt. In W. B. Whalley (ed.) *Scanning Electron Microscopy in the Study of Sediments.* Geo Abstracts, Norwich, 249–260.

Krinsley, D. H. and Smalley, I. J. (1973) Shape and nature of small sedimentary quartz particles. *Science* **180**, 1277–1279.

Krinsley, D. H., Marshall, J. R. and McCauley, J. F. (1981) Production of fine silt and clay during natural eolian abrasion. *NASA Tech. Mem. 84211*, 251–255.

Krinitzsky, E. L. and Turnbull, W. J. (1967) Loess deposits in Mississippi. *Geol. Soc. Am. Spec. Pap. 94*, 64pp.

Kuenen, P. H. (1959) Experimental abrasion. 3. Fluviatile action on sand. *Am. J. Sci.* **257**, 172–190.

Kuenen, P. H. (1960) Experimental abrasion. IV. Eolian action. *J. Geol.* **68**, 427–449.

Kuenen, P. H. (1969) Origin of quartz silt. *J. Sed. Petrol.* **39**, 1631–1633.

Kuenen, P. H. and Perdok, W. G. (1962) Experimental abrasion. V. Frosting and defrosting of quartz grains. *J. Geol.* **70**, 648–658.

Kukal, Z. and Saadallah, A. (1970) Composition and rate of deposition of dust storm sediments in Central Iraq. *Cas. Min. Geol.* **15**, 227–234.

Kukal, Z. and Saadallah, A. (1973) Aeolian admixtures in the sediments of the northern Persian Gulf. In B. H. Purser (ed.) *The Persian Gulf*. Springer, Berlin, pp.115–121.

Kukla, G. J. (1970) Correlations between loesses and deep sea sediments. *Geol. Foren. Stock. Forhand.* **92**, 148–180.

Kukla, G. J. (1978) The classical European glacial stages: correlation with deep sea sediments. *Trans. Nebraska Acad. Sci.* **6**, 57–93.

Kukla, G. J. and Koči, A. (1972) End of the last interglacial in the loess record. *Quat. Res.* **2**, 374–382.

Kushelevsky, A., Shani, G. and Hacoun, A. (1983) Effect of meteorological conditions on total suspended particulate (TSP) levels and elemental concentration of aerosols in a semi-arid zone (Beer Sheva, Israel). *Tellus* **35B**, 55–64.

Kusumgar, S., Agrawal, D. P. and Krishnamurthy, R. V. (1980) Studies on the loess deposits of the Kashmir Valley and ^{14}C dating. *Radiocarbon* **22**, 757–762.

Kutzbach, J. E. (1981) Monsoon climate of the early Holocene: climate experiment with the Earth's orbital parameters for 9000 years ago. *Science* **214**, 59–61.

Kwaad, F. J. P. M. (1970) Experiments on the disintegration of granite by salt action. *Univ. Amsterdam Fys. Geog. Boden Kundig Lab. Pub.* **16**, 67–80.

Lamb, H. H. (1970) Volcanic dust in the atmosphere; with a chronology and assessment of its meteorological significance. *Phil. Trans. Roy. Soc. Lond.* A **266**, 425–533.

Lamb, H. H. (1971) Volcanic activity and climate. *Palaeogeog. Palaeoclimatol. Palaeoecol.* **10**, 203–230.

Lancaster, N. (1978a) Composition and formation of southern Kalahari pan margin dunes. *Z. Geomorph.* NF 22, 148–169.

Lancaster, N. (1978b) The pans of the southern Kalahari, Botswana. *Geog. J.* **144**, 81–98.

Landes, K. K. (1933) Caverns in loess. *Am. J. Sci.* **25**, 137–139.

Lange, H. (1982) Distribution of chlorite and kaolinite in eastern Atlantic sediments off North Africa. *Sedimentology* **29**, 427–431.

Lattman, L. H. (1973) Calcium carbonate cementation of alluvial fans in southern Nevada. *Geol. Soc. Am. Bull.* **84**, 3013–3028.

Lautridou, J. P. (1979) Lithostratigraphie et chronostratigraphie des loess de Haûte Normandie. *Acta Geol. Acad, Scient. Hung.* **22**, 125–132.

Lautridou, J. P. and Giresse, P. (1981) Genèse et signification climatique des limons à doublets de Normandie. *Biul. Periglac.* **28**, 175–188.

Lautridou, J. P. and Ozouf, J. C. (1982) Experimental frost shattering: fifteen years of research at the Centre de Geomorphologie du CNRS. *Prog. Phys. Geog.* **6**, 215–232.

Lautridou, J. P., Sommé, J. and Jamagne, M. (1984) Sedimentological, mineralogical and geochemical characteristics of the loess in northwest France. In M. Pecsi (ed.) *Lithology and Stratigraphy of Loess and Paleosols.* Geographical Research Institute, Hungarian Academy of Sciences, Budapest, pp.123–132.

Lautridou, J. P., Sommé, J., Hein J., Puisségur, J.-J. and Rousseau, D.-D. (1985) La stratigraphie des loess et formations fluviatiles d'Achenheim (Alsace): nouvelles dounées bioclimatiques et corrélations avec des séquences Pléistocènes de la France du Nord-Ouest. *Ball. Ass. Franç Étude Quat.* **2–3**, 125–132.

Lawson, T. J. (1971); Haboob structure at Khartoum. *Weather* **26**, 105–112.

Lazarenko, A. A. (1984) The loess of Central Asia. In A. A. Velichko (ed.) *Late Quaternary Environments of the Soviet Union.* Longman, London, pp.125–131.

Lazarenko, A. A., Bolikhovskaya, N. S. and Semenov, V. V. (1980) An attempt at a detailed stratigraphic sub-division of the loess association of the Tashkent region. *Int. Geol. Rev.* **23**, 1335–1346.

Leach, J. A. (1974) Soil structures preserved in carbonate concretions in loess. *Q. J. Eng. Geol.* **7**, 311–314.

Leathers, C. R. (1981) Plant components in desert dust in Arizona and their significance for man. *Geol. Soc. Am. Spec. Pap.* **186**, 191–206.

Ledbetter, J. O. (1972) *Air Pollution.* Dekker, New York, 424pp.

Leighton, M. M. and Willman, H. B. (1949) Loess formations of the Mississippi Valley. *Geol. Soc. Am. Bull.* **60**, 1904–1905.

Leinen, M. and Heath, G. R. (1981) Sedimentary indicators of atmospheric circulation in the northern hemisphere during the Cenozoic. *Palaeogeog. Palaeoclimatol. Palaeoecol.* **36**, 1–21.

Leonhard, K. C. von (1823–24) *Charakteristik der Falsarten.* Engelman, Heidelberg, 3 vols.

Lepple, F. K. and Brine, C. J. (1976) Organic constituents in eolian dust and surface sediments from northwest Africa. *J. Geophys. Res.* **81**, 1141–1147.

Lever, A. and McCave, I. N. (1983) Eolian components in Cretaceous and Tertiary North Atlantic sediments. *J. Sed. Petrol.* **53**, 811–832.

Lidstrom, L. (1968) Surface and bond-forming properties of quartz and silicate minerals and their application in mineral processing techniques. *Acta Poly. Scand. Chem. Metall. Ser.* **75**, 1–149.

Li Jijun, Wen, S., Zhang, Q., Wang, F., Zheng, B. and Li, B. Y. (1979) A discussion on the period, amplitude and type of the uplift of the Qinghai–Xizang Plateau. *Scient. Sinica* **22**, 1314–1328.

Lill, G. O. and Smalley, I. J. (1978) Distribution of loess in Britain. *Proc. Geol. Ass.* **89**, 57–65.

Lim, C. H. and Jackson, M. L. (1984) Mineralogy of soils developed in periglacial deposits of southwestern Canada. *Soil Sci. Soc. Am. J.* **48**, 684–687.

Lindsay H. A. (1933) A typical Australian line-squall dust storm. *Q. J. Roy. Met. Soc.* **59**, 350.

Lin Zaiguan and Liang Weiming (1980) Distribution and engineering properties of loess and loess-like soils in China. *Bull. Int. Ass. Eng. Geol.* **21**, 112–117.

Liu Tung-sheng (ed.) (1985) *Quaternary Geology and Environment of China.* Ocean Press, Beijing.

Liu Tung-sheng and Chang Tsung-yu (1964) The 'huangtu' (loess) of China. *Rep. 6th INQUA Cong.* IV pp.503–524.

Liu Tung-sheng, Gu Xiong-fei, An Zhi-sheng and Fan Yong-xiang (1981) The dustfall in Beijing, China, on April 18, 1980. *Geol. Soc. Am. Spec. Pap. 186*, pp.149–158.

Liu Tung-sheng, An Zhi-sheng and Yuan Bao-yin (1982) Aeolian processes and dust mantles (loess) in China. In R. J. Wasson (ed.) *Quaternary Dust Mantles of China, New Zealand and Australia.* ANU Press, Canberra, pp.1–17.

Liu Tung-sheng *et al.* (un-named) (1985*a*) *Loess and the Environment.* China Ocean Press, Beijing, xii + 251pp.

Liu Tung-sheng, Chen Ming-yang and Li Xiu-fang (1985*b*) A satellite images study of the dust storm at Beijing on April 17–21, 1980. In Liu Tung-sheng (ed.) *Quaternary Geology and Environment of China.* Ocean Press, Beijing, pp.17–21.

Liversidge, A. (1902) Meteoric dusts, New South Wales. *J. Proc. Roy. Soc. NSW* **36**, 241–285.

Lobdell, G. T. (1981) Hydroconsolidation potential of Palouse loess. *Am. Soc. Civ. Eng. J. Geotech. Eng. Div.* **107**, 733–742.

Lockertz, W. (1978) The lessons of the dust bowl. *Am. Scient.* **66**, 560–569.

Loewe, F. (1943) Dust storms in Australia. *Comm. Bur. Met. Bull.* **28**.

Logie, M. (1982) Influence of roughness elements and soil moisture of sand to wind erosion. *Catena Supp.* **1**, 161–173.

Lohnes, R. A. and Handy, R. L. (1968) Slope angles in friable loess. *J. Geol.* **76**, 247–258.

Lourensz, R. S. and Abe, K. (1983) A dust storm over Melbourne. *Weather* **38**, 272–275.

Löye-Pilot, M. D., Martin, J. M. and Morelli, J. (1986) Influence of Saharan dust on the rain acidity and atmospheric input to the Mediterranean. *Nature* **321**, 427–428.

Lozek, V. (1965) Das problem des Lössbildung und die Lössmollusken. *Eiszeit. Gegen.* **16**, 61–75.

Lozek, V. (1968*a*) The loess environment in Central Europe. In C. B. Schultz and J. C. Frye (eds.) *Loess and Related Eolian Deposits of the World.* University of Nebraska Press, Lincoln, pp.67–80.

Lozek, V. (1968*b*) The loesses of Czechoslovakia. In C. B. Schultz and J. C. Frye (eds.) *Loess and Related Eolian Deposits of the World.* University of Nebraska Press, Lincoln, pp.305–308.

Lugn, A. L. (1960) The origin and sources of loess in the Great Plains of North America. *Proc. 21st Int. Geol. Cong.* pp.223–235.

Lugn, A. L. (1962) The origin and sources of loess in the central Great Plains and adjoining areas of the Central Lowland. *Nebraska Univ. Stud.* **26**, 105pp.

Lugn, A. L. (1968) The origin of loesses and their relation to the Great Plains in North America. In C. B. Schultz and J. C. Frye (eds.) *Loess and Related Eolian deposits of the World.* University of Nebraska Press, Lincoln, pp.139–182.

Lukashev, K. I., Lukashev, V. K. and Dobrovolskaya, I. A. (1968) Lithogeochemical properties of loess in Byelorussia and Central Asia. In C. B. Schultz and J. C. Frye (eds.) *Loess and Related Eolian Deposits of the World.* University of Nebraska Press, Lincoln, pp.223–232.

Lumley, J. L. and Panofsky, H. A. (1964) *The Structure of Atmospheric Turbulence.* Wiley, New York, 139pp.

Lundquist, J. and Bentsson, K. (1970) The red snow – a meteorological and pollen analytical study of long transported material from snow falls in Sweden. *Geol. Foren. Stock. Forhand.* **92**, 288–301.

Lutenegger, A. J. (1981) Stability of loess in light of the inactive particle theory. *Nature* **291**, 360.

Lutenegger, A. J. (1985) Bibliographic review of geotechnical investigations of loess in North America. *Loess Letter Supp.* 7, 12pp.

Lyell, C. (1834) Observations on the loamy deposit called 'loess' of the Basin of the Rhine. *Edin. New Phil. J.* **17**, 110–113, 118–120.

Lyell, C. (1847) On the delta and alluvial deposits of the Mississippi River, and other points of the geology of North America, observed in the years 1845, 1846. *Am. J. Sci.* **3**, 34–39, 267–269.

Lyles, L. (1975) Possible effects of wind erosion on soil productivity. *J. Soil Water Conserv.*, **30**, 279–283.

Lyles, L. (1977) Wind erosion: processes and effect on soil productivity. *Trans. Am. Soc. Agr. Eng.* **20**, 880–884.

Lyles, L. and Allison, B. (1976) Wind erosion: the protective role of simulated standing stubble. *Trans. Am. Soc. Agr. Eng.* **19**, 61–64.

Lyles, L. and Krauss, R. K. (1971) Threshold velocities and initial particle motion as influenced by air turbulence. *Trans. Am. Soc. Agr. Eng.* **14**, 563–566.

Lyles, L., Schrandt, R. L. and Schneidler, N. F. (1974) How aerodynamic roughness elements control sand movement. *Trans. Am. Soc. Agr. Eng.* **17**, 134–139.

Macleod, D. A. (1980) The origin of the red Mediterranean soils in Epirus, Greece. *J. Soil Sci.* **31**, 125–136.

Mainguet, M. (1968) Le Borkou, aspects d'un modelé éolien. *Ann. Géog.* **77**, 296–322.

Maley, J. (1977) Palaeoclimates of central Sahara during the early Holocene. *Nature* **269**, 573–577.

Maley, J. (1980) Les changements climatiques de la fin du tértiare en Afrique: leur consequence sur l'apparition du Sahara et de sa végetation. In M. A. J. Williams and H. Faure (eds.) *The Sahara and the Nile.* Balkema, Rotterdam, pp.63–86.

Maley, J. (1982) Dust, clouds, rain types and climatic variations in tropical North Africa., *Quat. Res.* **18**, 1–16.

Malina, F. J. (1941) Recent developments in the dynamics of wind erosion. *Trans. Am. Geophys. Un. 1941*, pp.262–284.

Markovic-Marjanovic, J. (1968) Loess sections in the Danube Valley in Yugoslavia and their importance for the Quaternary stratigraphy of southeastern Europe. In C. B. Schultz and J. C. Frye (eds.) *Loess and Related Eolian Deposits of the World.* University of Nebraska Press, Lincoln, Nebraska, pp.261–280.

Marshall, J. R., Krinsley, D. H. and Greeley, R. (1981) An experimental study of the behaviour of electrostatically-charged fine particles in atmospheric suspension. *NASA Tech. Mem. 84211*, pp.208–210.

Marshall, P. (1903) Dust storms in New Zealand. *Nature* **68**, 233.

Marshall, P. and Kidson, E. (1929) The dust storm of October 1928. *New Zealand J. Sci. Tech.* **10**, 291–299.

Marsland, P. S. and Woodruff, J. G. (1937) A study of the effects of wind transportation on grains of several minerals. *J. Sed. Petrol.* **7**, 18–30.

Martin, R. J. (1936) Dust storms of February and March 1936 in the United States. *Mon. Weath. Rev.* **64**, 87–88.

Martin, T. R., Wehner, A. P. and Butler, J. (1983) Pulmonary toxicity of Mt. St. Helens volcanic ash. *Am. Rev. Respir. Dis.* **128**, 158–162.

Márton, P. (1979) Palaeomagnetism of the Mende Brickyard exposure. *Acta Geol. Acad. Scrent. Hung.* **22**, 403–407.

Maruszczak, H. (1965) Development conditions of the relief of loess areas in East-Middle Europe. *Geog. Polon.* **6**, 93–104.

Maruszczak, H. (1980) Stratigraphy and chronology of the Vistulian loess in Poland. *Quat. Stud. Poland* **2**, 57–76.

Maruszczak, H. (ed.) (1985) *Guidebook of the International Symposium, Problems of the Stratigraphy and Palaeogeography of Loesses, Poland, 6th–10th September, 1985.* Marie Curie-Sklodowska University, Lublin, 156pp.

Mason, B. (1966) *Principles of Geochemistry*, 3rd ed. Wiley, New York.

Mason, B. J. (1971) *The Physics of Clouds.* Oxford University Press, London, 671pp.

Matalucci, R. V., Shelton, J. W. and Abdel-Hardy, M. (1969) Grain orientation in Vicksburg loess. *J. Sed. Petrol.* **39**, 969–979.

Matalucci, R. V., Abdel-Hardy, M. and Shelton, J. W. (1970) Influence of grain orientation on direct shear strength of a loessial soil. *Eng. Geol.* **4**, 121–132.

Mavlyanov, G. A. (1958) *Genetic Types of Loesses and Loess-like Rocks in the Central and Southern Parts of Middle Asia and Their Engineering–Geological Properties.* Akademia Nauka Uzbekistan SSR, Institute of Geology, Tashkent, 609pp.

Mavlyanov, G. A. and Tetyukhin, G. F. (eds.) (1982) The Uzbek SSR. In *Guidebook for Excursions A-11 and C-11, XI INQUA Congress, Moscow.* INQUA, Moscow, pp.1–31.

Mayer, L., Gerson, R. and Bull, W. B. (1984) Alluvial gravel production and deposition: a useful indicator of Quaternary climatic changes (a case study in southwestern Arizona). *Catena Supp.* **5**, 137–151.

Mazzullo, J., Sims, D. and Cunningham, D. (1986) The effects of eolian sorting and abrasion upon the shapes of fine quartz sand grains. *J. Sed. Petrol.* **56**, 45–56.

McCauley, J. F., Grolier, M. J. and Breed, C. S. (1977) Yardangs. In D. O. Doehring (ed.) *Geomorphology in Arid Regions.* State University of New York, Binghampton, pp.233–269.

McCauley, J. F., Breed, C. S., Grolier, M. J. and MacKinnon, D. J. (1981) The US dust storm of February 1977. *Geol. Soc. Am. Spec. Pap. 186*, pp. 123–148.

McCormick, R. A. and Ludwig, J. H. (1967) Climate modification by atmospheric aerosols. *Science* **156**, 1358–1359.

McCraw, J. D. (1975) Quaternary airfall deposits in New Zealand. In R. P. Suggate and M. M. Cresswell (eds.) *Quaternary Studies.* Royal Society of New Zealand, Wellington, pp. 35–44.

McDonald, W. F. (1938) *Atlas of Climatic Charts of the Oceans*. USDA Weather Bureau, Washington DC, 60pp.

McGreevy, J. P. (1981) Some perspectives on frost shattering. *Prog. Phys. Geog.* **4**, 56–75.

McGreevy, J. P. (1982) Frost and salt weathering: further experimental results. *Earth Surf. Proc. Landf.* **7**, 475–488.

McGreevy, J. P. and Whalley, W. B. (1982) The geomorphic significance of rock temperature variations in cold environments: a discussion. *Arc. Alp. Res.* **14**, 157–162.

McGreevy, J. P. and Smith, B. J. (1982) Salt weathering in hot deserts: observations on the design of simulation experiments. *Geog. Ann.* **64A**, 161–170.

McKay, E. D. (1979) Wisconsinan loess stratigraphy of Illinois. *Illinois State Geol. Surv. Guidebook 13*, pp.95–108.

McTainsh, G. (1980) Harmattan dust deposition in northern Nigeria. *Nature* **286**, 587–588.

McTainsh, G. (1984) The nature and origin of aeolian mantles in central northern Nigeria. *Geoderma* **33**, 13–37.

McTainsh, G. (1985a) Dust processes in Australia and West Africa: a comparison. *Search* **16**, 104–106.

McTainsh, G. (1985b) Desertification and dust modelling in West Africa. *Desert. Control Bull., UNEP.* **12**, 26–33.

McTainsh, G. (1986) A dust monitoring programme for desertification control in West Africa. *Env. Conserv.* **13**, 17–25.

McTainsh, G. and Walker, P. H. (1982) Nature and distribution of Harmattan dust. *Z. Geomorph.* **NF 26**, 417–436.

Mees, R. P. R. and Meijs, E. P. M. (1984) Note on the presence of pre-Weichselian loess deposits along the Albert canal near Kesselt and Vroenhoven (Belgian Limbourg) *Geol. Mijn.* **63**, 7–11.

Melia, M. B. (1984) The distribution and relationship between palynomorphs in aerosols and deep-sea sediments off the coast of northwest Africa. *Mar. Geol.* **58**, 345–371.

Membery, D. A. (1983) Low-level wind profiles during the Gulf Shamal. *Weather.* **38**, 18–24.

Mensching, H. (1964) Zur Geomorphologie sud-Tunesiens. *Z. Geomorph.* **NF 8**, 424–439.

Middleton, N. J. (1984) Dust storms in Australia: frequency, distribution and seasonality. *Search* **15**, 46–47.

Middleton, N. J. (1985) Effect of drought on dust production in the Sahel. *Nature* **316**, 431–434.

Middleton, N. J. (1986a) Dust storms in the Middle East. *J. Arid Env.* **10**, 83–96.

Middleton, N. J. (1986b) A geography of dust storms in southwest Asia. *J. Climatol.* **6**, 183–196.

Middleton, N. J. Goudie, A. S. and Wells, G. L. (1986) The frequency and source areas of dust storms. In W. G. Nickling (ed.) *Aeolian Geomorphology*, Allen and Unwin, New York, pp.237–259.

Mill, H. R. and Lempfert, R. G. K. (1904) The great dust fall of February 1903 and its origins. *Q. J. Roy. Met. Soc.* **30**, 57–73.

Milne, D. J. G. and Smalley, I. J. (1979) Loess deposits in the southern part of the North Island of New Zealand: an outline stratigraph. *Acta Geol. Acad. Scient. Hung.* **22**, 197–204.

Minervin, A. V. (1984) Cryogenic processes in loess formation in Central Asia. In A. A. Velichko (ed.) *Late Quaternary Environments of the Soviet Union.* Longman, London, pp.133–140.

Mitchell, J. M. (1971) The effect of atmospheric aerosols on climate with special reference to temperature near the Earth's surface. *J. App. Met.* **10**, 703–714.

Miyake, Y., Gugiura, Y. and Katsuragi, Y. (1956) Radioactive fallout at Asahikawa, Hokkaido, in April, 1955. *J. Met. Soc. Japan* **34**, 50–54.

Mockma, D. L., Syers, J. K., Jackson, M. L., Rex, R. W. and Clayton, R. N. (1972) Aeolian additions to soils and sediments in the South Pacific area. *J. Soil Sci.* **23**, 147–162.

Moharram, M. A. and Sowelim, M. A. (1980) Infrared study of minerals and compounds in atmospheric dustfall in Cairo. *Atmos. Env.* **14**, 853–856.

Mojski, J. E. (1968a) Outline of loess stratigraphy in Poland. *Biul. Periglac.* **18**, 149–170.

Mojski, J. E. (1968b) Loesses in Poland. In C. B. Schultz and J. C. Frye (eds.) *Loess and Related Eolian Deposits of the World.* University of Nebraska Press, Lincoln, pp.325–328.

Molina-Cruz, A. (1977) The relation of the southern tradewinds to upwelling processes during the last 75 000 years. *Quat. Res.* **8**, 324–339.

Morales, C. (ed.) (1979) *Saharan Dust – Mobilization, Transport, Deposition.* Wiley, New York, 316pp.

Morey, G. W., Fournier, R. O. and Rowe, J. J. (1962) The solubility of quartz in water in the temperature interval from 25° C to 300° C. *Geochim. Cosmochim. Acta* **22**, 1029–1043.

Morton, B. R., Taylor, G. I. and Turner, J. S. (1956) Turbulent gravitational convection from maintained and instantaneous sources. *Phil. Trans. Roy. Soc. Lond.* **A 234**, 1–23.

Mosley-Thompson, E. and Thompson, L. G. (1982) Nine centuries of microparticle deposition at the South Pole. *Quat. Res.* **17**, 1–13.

Moss, A. J. (1966) Origin, shaping and significance of quartz sand grains. *J. Geol. Soc. Aust.* **13**, 97–136.

Moss, A. J. (1972) Initial fluviatile fragmentation of granitic quartz. *J. Sed. Petrol.* **42**, 905–916.

Moss, A. J. (1973) Fatigue effects in quartz sand grains. *Sedim. Geol.* **10**, 239–247.

Moss, A. J. and Green, P. (1975) Sand and silt grains: predetermination of their formation and properties by microfractures in quartz. *J. Geol. Soc. Aust.* **22**, 485–495.

Moss, A. J., Walker, P. H. and Hutka, J. (1973) Fragmentation of granitic quartz in water. *Sedimentology* **20**, 489–511.

Moss, A. J., Green, P. and Hutka, J. (1981) Static breakage of granitic detritus by ice and water in comparison with breakage by flowing water. *Sedimentology* **28**, 261–272.

Mucher, H. J. (1986) *Aspects of loess and loess-derived slope deposits: an experimental and micromorphological approach.* Fys. Geog. Boden. Lab. Univ. Amsterdam.

Mucher, H. J. and de Ploey, J. (1977) Experimental and micromorphological investigation of erosion and redeposition of loess by water. *Earth Surf. Proc. Landf.* **2**, 117–124.

Muhs, D. R. (1983) Airborne dustfall on the California Channel Islands, USA *J. Arid Env.* **6**, 223–228.

Murty, A. S. and Murty, B. V. (1973) Role of dust on rainfall in northwest India. *Pure App. Geophys.* **104**, 614–622.

Nagelschmidt, G. (1960) The relation between lung dust and lung pathology in pneumoconiosis. *Br. J. Ind. Med.* **17**, 247–259.

Nagelschmidt, G. (1965) Some observations of the dust content and composition in lungs with asbestosis, made during work on coal miners' pneumoconionsis. *Ann. N. Y. Acad. Sci.* **132**, 64–76.

Nahon, D. and Trompette, R. (1982) origin of siltstones: glacial grinding versus weathering. *Sedimentology* **29**, 25–35.

Nakata, J. K., Wilshire, H. G. and Barnes, C. G. (1976) Origin of Mojave Desert dust plumes photographed from space. *Geology* **4**, 644–648.

Nalivkin, D. V. (1982) *Hurricanes, Storms and Tornadoes.* Amerind, New Delhi.

Nalpanis, P. (1985) Saltating and suspended particles over flat and sloping surfaces. II. Experiments and numerical simulations. In O. E. Barndorff-Nielsen, J. T. Møller, K. Romer-Rasmussen, and B. B. Willetts (eds.) *Proceedings of International Workshop on the Physics of Blown Sand, Aarhus, 28–31 May 1985.* Department of Theoretical Statistics, Institute of Mathematics, University of Aarhus Mem. 8, vol. I, 37–66.

Nalpanis, P. and Hunt, J. C. R. (1986) *Suspension, Transport and Deposition of Dust from Stockpiles.* Dust and Materials Handling Report 27, Warren Spring Laboratory, 83pp. + 57 figs.

Newell, R. E. and Kidson, J. W. (1979) The tropospheric circulation over Africa and its relation to the global tropospheric circulation. In Morales, C. (ed.) *Saharan Dust*. Wiley, New York, pp.133–170.

Nicholson, S. E. (1982) Pleistocene and Holocene climates in Africa. *Nature* **296**, 779–780.

Nicholson, S. E. and Flohn, H. (1980) African environmental and climatic changes and the general atmospheric circulation in the late Pleistocene and Holocene. *Climatic Change* **2**, 313–348.

Nickling, W. G. (1978) Eolian sediment transport during dust storms: Slims River Valley, Yukon territory. *Can. J. Earth Sci.* **15**, 1069–1084.

Nickling, W. G. (1983) Grain-size characteristics of sediments transported during dust storms. *J. Sed. Petrol.* **53**, 1011–1024.

Nickling, W. G. (1984) The stabilizing role of bonding agents on the entrainment of sediment by wind. *Sedimentology* **31**, 111–117.

Nickling, W. G. and Brazel, A. J. (1984) Temporal and spatial characteristics of Arizona dust storms (1965–80). *J. Climatol.* **4**, 645–660.

Nickling, W. G. and Ecclestone, M. (1981) The effect of soluble salts on the threshold shear velocity of fine sand. *Sedimentology* **28**, 505–510.

Nicol, E. R., Eng Seng Lee and Murrow, P. J. (1985) Respirable crystalline silica in ash erupted from Galunggung volcano, Indonesia, 1982. *Atmos. Env.* **19**, 1027–1028.

Nihlén, T. and Solyom, Z. (1986) Dust storms and eolian deposits in the Mediterranean area. *Geol. Foren. Stock. Forhand.* **108**, 235–242.

Norberg, P. (1980) Mineralogy of a podzol formed in sandy materials in northern Denmark. *Geoderma* **24**, 25–43.

Norgren, A. (1973) *Opal Phytoliths as Indicators of Soil Age and Vegetative History*. Ph.D. Thesis, Oregon State University.

Norton, L. D. (1984) The relationship of present topography to pre-loess deposition topography in east-central Ohio. *Soil Sci. Soc. Am. J.* **48**, 147–151.

Norton, L. D. and Bradford, J. M. (1985) Thermoluminescence dating of loess from western Iowa. *Soil Sci. Soc. Am. J.* **49**, 708–712.

Nur, A. and Simmons, G. (1970) The origin of small cracks in igneous rocks. *Int. J. Rock Mech. Min. Sci. Geomech. Abs.* **7**, pp.307–314.

O'Brien, M. P. and Rindlaub, B. D. (1936) The transportation of sand by wind. *Civ. Eng.* **6**, 325–327.

Obruchev, V. A. (1911) The question of the origin of loess – in defence of the aeolian hypothesis. *Izv. Tonsk. Tekhol. Inst.* **23** (3) (in Russian).

Obruchev, V. A. (1945) Loess types and their origin. *Am. J. Sci.* **243**, 256–262.

Odell, R. T., Thornburn, T. H. and McKenzie, L. J. (1960) Relationship of Atterberg Limits to some other properties of Illinois soils. *Proc. Soil Sci. Soc.* **24**, 297–300.

Oke, T. R. (1978) *Boundary Layer Climates*. Methuen, London, 372pp.

Oldfield, F., Hunt, A., Jones, M. D. H., Chester, R., Dearing, J. A., Olsson, L. and Prospero, J. M. (1985) Magnetic differentiation of atmospheric dusts. *Nature* **317**, 516–518.

Oliver, F. W. (1945) Dust storms in Egypt and their relation to the war period, as noted in Maryut, 1939–45. *Geog. J.* **106**, 26–49.

Orev, Y. (1984) Sand is greener. *Teva va-Aretz* **26**, 15–16 (in Hebrew).

Orgill, M. M. and Sehmel, G. A. (1976) Frequency and diurnal variations of dust storms in the contiguous USA. *Atmos. Env.* **10**, 813–825.

Orombelli, G. (1970) I depositi loessici di Copreno (Milano). *Bull. Soc. Geol. Ital.* **89**, 529–546.

Owen, P. R. (1960) Dust deposition from a turbulent airstream. *J. Air Poll.* **3**, 8–25.

Padu, S. M. and Kelly, C. D. (1954) Aerobiological studies of fungi and bacteria over the Atlantic Ocean. *Can. J. Bot.* **32**, 202–212.

Pant, R. K., Krishnamurthy, R. V. and Tandon, S. K. (1983) Loess lithostratigraphy of the Kashmir Basin, India. In *Current Trends in Geology*, vol. VI (*Climate and Geology of Kashmir*). Today and Tomorrow's Printers and Publishers, New Delhi, pp.123–129.

Parkin, D. W. (1974) Trade winds during the glacial cycles. *Proc. Roy. Soc. Lond.* **A 337**, 73–100.

Parkin, D. W. and Padgham, R. C. (1975) Further studies on the trade winds during the glacial cycles. *Proc. Roy. Soc. Lond.* **A 346**, 245–260.

Parkin, D. W. and Shackleton, N. J. (1973) Trade wind and temperature correlations down a deep-sea core off the Sahara coast. *Nature* **245**, 455–457.

Parkin, D. W., Delany, A. C. and Delany, A. C. (1967) A search for airborne dust on Barbados. *Geochim. Cosmochim. Acta* **31**, 1311–1320.

Parkin, D. W., Phillips, D. R., Sullivan, R. A. L. and Johnson, L. R. (1970) Airborne dust collections over the North Atlantic. *J. Geophys. Res.* **75**, 1782–1795.

Parmenter, C. and Folger, D. W. (1974) Eolian biogenic detritus in deep-sea sediments: possible index of equatorial Ice Age aridity. *Science* **185**, 695–698.

Parrington, J. R., Zoller, W. H. and Aras, N. K. (1983) Asian dust: seasonal transport to the Hawaiian Islands. *Science* **220**, 195–197.

Pasquill, F. and Smith, F. B. (1983) *Atmospheric Diffusion*, 3rd edn. Ellis Horwood, Chichester, 437pp.

Pastouret, L., Chamley, H., Delibrias, G., Duplessy, J. C. and Thiede, J. (1978) Late Quaternary climatic changes in western tropical Africa deduced from deep-sea sedimentation off the Niger delta. *Oceanol. Acta* **1**, 217–232.

Patterson, E. M. (1977) Atmospheric extinction between 0.55 μm and 10.6 μm due to soil-derived aerosols. *App. Optics* **16**, 2414–2418.

Patterson, E. M. and Gillette, D. A. (1977) Measurements of visibility vs. concentration for airborne soil particles. *Atmos. Env.* **11**, 193–196.

Pecsi, M. (1968*a*) Loess. In R. W. Fairbridge (ed.) *The Encyclopedia of Geomorphology*. Reinhold, New York, pp.674–678.

Pecsi, M. (1968*b*) The main genetic types of Hungarian loesses and loess-like sediments. In C. B. Schultz and J. C. Frye (eds.) *Loess and Related Eolian Deposits of the World*. University of Nebraska Press, Lincoln, pp.317–320.

Pecsi, M. (ed.) (1979*a*) *Studies on Loess*. Akademiai Kiádo, Budapest, 555pp.

Pecsi, M. (1979*b*) Lithostratigraphical subdivision of the loess profiles at Paks. *Acta Geol. Acad. Scient. Hung.* **22**, 409–418.

Pecsi, M. (1982) The most typical loess profiles in Hungary. In *Quaternary Studies in Hungary*. INQUA Hungarian National Committee, Budapest, pp.145–169.

Pecsi, M. (ed.) (1984*a*) *Lithology and Stratigraphy of Loess and Paleosols*. Geographical Research Institute, Hungarian Academy of Sciences, Budapest, 325pp.

Pecsi, M. (1984*b*) Is typical loess older than one million years? In M. Pecsi (ed.) *Lithology and Stratigraphy of Loess and Paleosols*. Geographical Research Institute, Hungarian Academy of Sciences, Budapest, pp.215–224.

Pecsi, M. (1985) Chronostratigraphy of Hungarian loesses and the underlying subaerial formation. In M. Pecsi (ed.) *Loess and the Quaternary*. Akademiai Kiádo, Budapest, pp.33–49.

Pecsi, M. and Sheuer, G. (1979) Engineering geological problems of the Dunaujvaros loess bluff. *Acta Geol. Acad. Scient. Hung.* **22**, 345–354.

Pecsi, M., Schweitzer, G. and Sheuer (1979) Engineering geological and geomorphological investigation of landslides in the loess bluffs along the Danube in the Great Hungarian Plain. *Acta Geol. Acad. Scient. Hung.* **22**, 327–344.

Pecsi-Donath, E. (1985) On the mineralogical and petrological properties of the Younger Loess in Hungary. In M. Pecsi (ed.) *Loess and the Quaternary*. Akademiai Kiádo, Budapest, pp.93–104.

Pedgley, D. E. (1972) Desert depression over northeast Africa. *Met. Mag.* **101**, 228–244.

Peel, R. F. (1974) Insolation weathering: some measurements of diurnal temperature changes in exposed rocks in the Tibesti region, central Sahara. *Z. Geomorph. Supp. Bd.* **21**, 19–28.

Penck, A. (1930) Central Asia. *Geog. J.* **76**, 477–487.

Perry, R. and Young, R. J. (1977) *Handbook of Air Pollution Analysis*. Chapman and Hall, London, 506pp.

Perry, R. S. and Adams, J. B. (1978) Desert varnish: evidence for cyclic deposition of manganese. *Nature* **276**, 489–491.

Peterson, J. T. and Junge, C. E. (1971) Source of particulate matter in the atmosphere. In W. Matthews, W. Kellog and G. D. Robinson (eds.) *Man's Impact on Climate.* MIT Press, Cambridge, Mass., pp.310–320.

Petit, J. R., Briat, M. and Royer, A. (1981) Ice-age aerosol content from east Antarctic ice core samples and past wind strength. *Nature* **293**, 391–394.

Petrov, M. P. (1968) Composition of eolian dust in southern Turkmenia, as in dust storm of January 1968. *Int. Geol. Rev.* **13**, 1178–1182.

Péwé, T. L. (1951) An observation on wind blown silt. *J. Geol.* **59**, 399–401.

Péwé, T. L. (1955) Origin of the upland silt near Fairbanks, Alaska. *Geol. Soc. Am. Bull.* **66**, 699–724.

Péwé, T. L. (1968) Loess deposits in Alaska. *23rd Int. Geol. Cong.* VIII, pp.297–309.

Péwé, T. L. (1975) Quaternary geology of Alaska, *US Geol. Surv. Prof. Pap. 835*, 145pp.

Péwé, T. L. (ed.) (1981*a*) Desert Dust. *Geol. Soc. Am. Spec. Pap. 186*, 303pp.

Péwé, T. L. (1981*b*) Desert dust: an overview. *Geol. Soc. Am. Spec. Pap. 186*, pp.1–10.

Péwé, T. L. and Journaux, A. (1983) Origin and character of loess-like silt in unglaciated south-central Yakutia, Siberia, USSR. *US Geol. Surv. Prof. Pap. 1262*, 45pp.

Péwé, T. L., Journaux, A. and Stuckenrath, R. (1977) Radiocarbon dates and late Quaternary stratigraphy from Mamontova Gora, unglaciated central Yakutia, Siberia, USSR. *Quat. Res.* **8**, 51–63.

Péwé, T. L., Péwé, E. A., Péwé, R. H., Journaux, A. and Slatt, R. M. (1981) Desert dust: characteristics and rates of deposition in central Arizona, USA. *Geol. Soc. Am. Spec. Pap. 186*, pp.169–190.

Pias, J. (1971) Les loess en Afghanistan oriental et leurs pédogenèses successives au Quaternaire récent. *C. R. Acad. Sci. Paris* 272D, 1602–1605.

Pissart, A., Vincent, J. S. and Edlund, S. A. (1977) Dépôts et phénomènes éoliens sur Île de Banks, Térritoires du Nord-Ouest, Canada. *Can. J. Earth Sci.* **14**, 2462–2480.

Pitcher, W. S., Shearman, D. J. and Pugh, D. C. (1954) The loess of Pegwell Bay, Kent, and its associated frost soils. *Geol. Mag.* **91**, 308–314.

Pitty, A. (1968) Particle size of the Saharan dust which fell in Britain in July 1968. *Nature* **220**, 364–365.

Ploey, J. de (1984) Hydraulics of runoff and loess loam deposition. *Earth Surf. Proc. Landf.* **9**, 533–539.

Pokras, E. M. and Mix, A. C. (1985) Eolian evidence for spatial variability of late Quaternary climates in tropical Africa. *Quat. Res.* **24**, 137–149.

Pollack, J. B., Toon, O. B., Ackerman, T. R., McKay, C. P. and Turco, R. P. (1983) Environmental effects of an impact-generated dust cloud: implications for the Cretaceous–Tertiary extinctions. *Science* **219**, 287–289.

Potter, R. M. and Rossman, G. R. (1977) Desert varnish: the importance of clay minerals. *Science* **196**, 1446–1448.

Prandtl, L. (1935) The mechanics of viscous fluids. In F. Durand (ed.) *Aerodynamic Theory*, Vol. III. Julius Springer, Berlin, pp.57–109.

Priddy, R. R., Lewand, R. L. and McGee, E. H. (1964) Pseudoanticlines in Vicksburg loess. *J. Miss. Acad. Sci.* **10**, 178–179.

Pringle, A. W. and Bain, D. C. (1981) Saharan dust falls on northwest England. *Geog. Mag.* **53**, 729–732.

Prodi, F. and Fea, G. (1979) A case of transport and deposition of Saharan dust over the Italian Peninsula and southern Europe. *J. Geophys. Res.* **84C**, 6951–6960.

Prospero, J. M. (1979) Mineral and sea salt aerosol concentrations in various ocean regions. *J. Geophys. Res.* **84**, 725–731.

Prospero, J. M. (1981a) Arid regions as sources of mineral aerosols in the marine atmosphere. *Geol. Soc. Am. Spec. Pap. 186*, pp.71–86.

Prospero, J. M. (1981b) Eolian transport to the world ocean. In C. Emiliani (ed.) *The Sea*, vol. VII, *(The Oceanic Lithosphere)*. Wiley, New York, pp.801–874.

Prospero, J. M. and Bonatti, E. (1969) Continental dust in the atmosphere of the eastern equatorial Pacific. *J. Geophys. Res.* **74**, 3362–3371.

Prospero, J. M. and Carlson, T. N. (1972) Vertical and aerial distribution of Saharan dust over the western equatorial North Atlantic Ocean. *J. Geophys. Res.* **77**, 5255–5265.

Prospero, J. M. and Nees, R. T. (1977) Dust concentration in the atmosphere of the equatorial North Atlantic; possible relationship to Sahelian drought. *Science* **196**, 1196–1198.

Prospero, J. M. and Nees, R. T. (1986) Impact of the North African drought and el Niño on mineral dust in the Barbados trade winds. *Nature* **320**, 735–738.

Prospero, J. M., Bonatti, E., Schubert, C. and Carlson, T. N. (1970) Dust in the Caribbean atmosphere traced to an African dust storm. *Earth Planet. Sci. Lett.* **9**, 287–293.

Prospero, J. M., Glaccum, R. A. and Nees, R. T. (1981) Atmospheric transport of soil dust from Africa to South America. *Nature* **289**, 570–572.

Prospero, J. M., Charlson, R. J., Mohnen, V., Jaenicke, R., Delany, A. C., Moyers, J., Zoller, W. and Rahn, K. (1983) The atmospheric aerosol system: an overview. *Rev. Geophys. Space Phys.* **21**, 1607–1629.

Pumpelly, R. (1866) Geological researches in China, Mongolia and Japan during the years 1862 to 1865. *Smithsonian Inst. Contrib.* **202**, 162pp.

Punjrath, J. S. and Heldman, D. R. (1972) Mechanism of small particle re-entrainment from flat surfaces. *J. Aerosol Sci.* **3**, 429–440.

Pye, K. (1980) Beach salcrete and eolian sand transport: evidence from North Queensland. *J. Sed. Petrol.* **50**, 257–261.

Pye, K. (1982) SEM observations of natural and experimentally-formed quartz silt particles in relation to the problem of loess origin. *Abs. 11th INQUA Cong., Moscow*, vol. I, p. 260.

Pye, K. (1983*a*) Formation of quartz silt during humid tropical weathering of dune sands. *Sedim. Geol.* **34**, 267–282.

Pye, K. (1983*b*) Grain surface textures and carbonate content of late Pleistocene loess from West Germany and Poland. *J. Sed. Petrol.* **53**, 973–980.

Pye, K. (1983*c*) Red beds. In A. S. Goudie and K. Pye (eds.) *Chemical Sediments and Geomorphology.* Academic Press, London, pp.227–263.

Pye, K. (1984*a*) Loess. *Prog. Phys. Geog.* **8**, 176–217.

Pye, K. (1984*b*) Some perspectives on loess accumulation. *Loess Letter* **11**, 5–10.

Pye, K. (1984*c*) SEM investigations of quartz silt micro-textures in relation to the source of loess. In M. Pecsi (ed.) *Lithology and Stratigraphy of Loess and Paleosols.* Geographical Research Institute, Hungarian Academy of Sciences, Budapest, pp.139–151.

Pye, K. (1985) Granular disintegration of gneiss and migmatites. *Catena* **12**, 191–199.

Pye, K. and Johnson, R. (1987) Stratigraphy, geochemistry and thermoluminescence ages of Lower Mississippi Valley loess. *Earth Surf. Proc. Landf.* (in press).

Pye, K. and Paine, A. D. M. (1984) Nature and source of aeolian deposits near the summit of Ben Arkle, northwest Scotland. *Geol. Mijn.* **63**, 13–18.

Pye, K. and Sperling, C. H. B. (1983) Experimental investigation of silt formation by static breakage processes: the effect of temperature, moisture and salt on quartz dune sand and granitic regolith. *Sedimentology* **30**, 49–62.

Pye, K. and Tsoar, H. (1987) The mechanics and geological implications of dust transport and deposition in deserts, with particular reference to loess formation and dune sand diagenesis in the northern Negev, Israel. In I. Reid and L. Frostick (eds.) *Desert Sediments Ancient and Modern.* Blackwell, Oxford (in press).

Pye, K., Goudie, A. S. and Watson, A. (1985) An introduction to the physical geography of the Kora area of central Kenya. *Geog. J.* **151**, 168–181.

Rabenhorst, M. C., Wilding, L. P. and Girdner, C. L. (1984) Airborne dusts in the Edwards Plateau region of Texas. *Soil Sci. Soc. Am.* **48**, 621–627.

Radczewski, O. E. (1939) Eolian deposits in marine sediments. In P. D. Trask (ed.) *Recent Marine Sediments.* American Association of Petroleum Geologists, Tulsa, pp.496–502.

Raeside, J. D. (1964) Loess deposits of the South Island, New Zealand, and the soils formed on them. *New Zealand J. Geol. Geophys.* **7**, 811–838.

Rahn, K. A. (1976) Silicon and aluminium in atmospheric aerosols: crustal fractionation? *Atmos. Env.* **10**, 597–601.

Rahn, K. A. (1985) Progress in Arctic air chemistry, 1980–84. *Atmos. Env.* **19**, 1987–1994.

Rahn, K. A., Borys, R. A. and Shaw, G. E. (1977) The Asian source of Arctic haze bands. *Nature* **268**, 712–714.

Rahn, K. A., Borys, R. D. and Shaw, G. E. (1981) Asian dust over Alaska: anatomy of an Arctic haze episode. *Geol. Soc. Am. Spec. Pap. 186*, pp.37–70.

Ranov, V. A. and Davis, R. S. (1979) Toward a new outline of the Soviet Central Asian Paleolithic. *Current Anthropol.* **20**, 249–270.

Rapp, A. (1984) Are terra rossa soils in Europe eolian deposits from Africa? *Geol. Foren. Stock. Forhand.* **105**, 161–168.

Rapp, A. and Nihlén, T. (1986) Dust storms and eolian deposits in North Africa and the Mediterranean. *Geodynamik* **7**, 41–62.

Rathjens, C. (1928) Löss in Tripolitanien. *Z. Gesell. Erk. Berlin* **5/6**, 211–228.

Ravikovitch, S. (1953) The aeolian soils of the northern Negev. *Res. Council Israel Pub. 2*, pp.404–433.

Ray, L. L. (1967) An interpretation of profiles of weathering of the Peorian loess of western Kentucky. *US Geol. Surv. Prof. Pap. 575D*, pp.D221–D227.

Rea, D. K. and Bloomstine, M. K. (1986) Neogene history of the South Pacific tradewinds: evidence for hemispherical asymmetry of atmospheric circulation. *Palaeogeog. Palaeoclimatol. Palaeoecol.* **55**, 55–64.

Rea, D. K. and Janecek, T. R. (1981) Late Cretaceous history of eolian accumulation in the mid-Pacific Mountains, central North Pacific Ocean. *Palaeogeog. Palaeoclimatol. Palaeoecol.* **36**, 55–67.

Rea, D. K. and Janecek, T. R. (1982) Late Cenozoic changes in atmospheric circulation deduced from North Pacific eolian sediments. *Mar. Geol.* **49**, 149–167.

Rea, D. K., Leinen, M. and Janecek, T. R. (1985) Geologic approach to the long-term history of atmospheric circulation. *Science* **227**, 721–725.

Reck, R. A. (1974a) Aerosols in the troposphere: calculation of the critical absorption/backscatter ratio. *Science* **186**, 1034–1036.

Reck, R. A. (1974b) Influence of surface albedo on the change in the atmospheric radiation balance due to aerosols. *Atmos. Env.* **8**, 823–833.

Reed, R. J., Norquist, D. C. and Recker, E. E. (1977) The structure and properties of African wave disturbances as observed during phase III of GATE. *Mon. Weath. Rev.* **105**, 317–373.

Reist, P. C. (1984) *Introduction to Aerosol Science.* Macmillan, New York, 299pp.

Rendell, H. M. (1985) The precision of water content estimates in the thermoluminescence dating of loess from northern Pakistan. *Nuclear Tracks* **10**, 763–768.

Rendell, H. M., Gamble, I. J. A. and Townsend, P. D. (1983) Thermoluminescence dating of loess from the Potwar Plateau, northern Pakistan. *PACT* **9**, 555–562.

Rex, R. W. and Goldberg, E. D. (1958) Quartz contents of pelagic sediments of the Pacific Ocean. *Tellus* **10**, 153–159.

Rex, R. W., Syers, J. K., Jackson, M. L. and Clayton, R. N. (1969) Eolian origin of quartz in soils of Hawaiian Islands and in Pacific sediments. *Science* **163**, 277–279.

Rice, A. R. (1976) Insolation warmed over. *Geology* **4**, 61–62.

Richthofen, F. von (1877–85) *China: Ergebnisse eigner Reisen und darauf gegründeter Studien.* Reimer, Berlin, 5 vols.

Richthofen, F. von (1882) On the mode of origin of the loess. *Geol. Mag.* **9**, 293–305.

Riezebos, P. A. and van der Waals, L. (1974) Silt-sized quartz particles: a proposed source. *Sedim. Geol.* **12**, 279–285.

Riggi, J. C. (1968) El loess de Rio Tercero y al probable origen de los mallinis (Cordoba). *J. Geol. Argent.* 3rd ser. **2**, 67–77.

Risebrough, R. W., Hugget, R. J., Griffin, J. and Goldberg, E. D. (1968) Pesticides: transatlantic movement in the Northeast Trades. *Science* **159**, 1233–1236.

Robinson, G. D. (1980) Possible quartz synthesis during weathering of quartz-free mafic rock, Jasper County, Georgia. *J. Sed. Petrol.* **50**, 193–203.

Robinson, S. G. (1986) The late Pleistocene palaeoclimatic record of North Atlantic deep-sea sediments revealed by mineral-magnetic measurements. *Phys. Earth Planet. Int.* **42**, 22–47.

Robock, A. and Matson, M. (1983) Circumglobal transport of the el Chichon volcanic dust cloud. *Science* **221**, 195–197.

Rognon, P. and Williams, M. A. J. (1977) Late Quaternary climatic changes in Australia and North Africa: a preliminary interpretation. *Palaeogeog. Palaeoclimatol. Palaeoecol.* **21**, 285–327.

Roloff, G., Bradford, J. M. and Scrivner, C. L. (1981) Gully development in the deep loess hills region of central Missouri. *Soil Sci. Soc. Am. J.* **45**, 119–123.

Rosauer, E. and Frechen, J. (1960) Carbonate concretions in the Karlicher loess profile, Rheinland. *Proc. Iowa Acad. Sci.* **67**, 346–356.

Roth, E. S. (1965) Temperature and water content as factors in desert weathering. *J. Geol.* **73**, 454–468.

Rouse, H. (1937) Modern conceptions of the mechanics of fluid turbulence. *Trans. Am. Soc. Civ. Eng.* **102**, 403–543.

Ruddiman, W. F. and McIntyre, A. (1976) Northeast Atlantic palaeoclimate changes over the past 600 000 years. *Geol. Soc. Am. Bull.* **145**, 111–146.

Rudge, W. A. D. (1914) On the electrification produced during the raising of a cloud of dust. *Proc. Roy. Soc. Lond.* **A 90**, 256–272.

Ruegg, G. H. J. (1977) Features of middle Pleistocene sandur deposits in The Netherlands. *Geol. Mijn.* **56**, 5–24.

Ruhe, R. V. (1969) *Quaternary Landscapes of Iowa*. Iowa State University Press, Ames, Iowa.

Ruhe, R. V. (1976) Stratigraphy of mid-continent loess, USA. In W. C. Mahaney (ed.) *Quaternary Stratigraphy of North America*. Dowden, Hutchinson and Ross, Stroudsberg, pp.197–211.

Ruhe, R. V. (1983) Clay minerals in thin loess, Ohio River Basin, USA. In M. E. Brookfield and T. S. Ahlbrandt (eds.) *Eolian Sediments and Processes*. Elsevier, Amsterdam, pp.91–102.

Ruhe, R. V. (1984) Clay-mineral regions in Peoria loess, Mississippi River Basin. *J. Geol.* **92**, 339–343.

Ruhe, R. V. and Olson, C. G. (1980) Clay mineral indicators of glacial and non-glacial sources of Wisconsinan loesses in southern Indiana, USA. *Geoderma* **24**, 283–297.

Runge, E. C. A., Goh, K. M. and Rafter, T. A. (1973) Radiocarbon chronology and problems in its interpretation for Quaternary loess deposits – South Canterbury, New Zealand. *Proc. Soil Sci. Soc. Am.* **37**, 742–746.

Russell, R. J. (1944*a*) Lower Mississippi Valley loess. *Geol. Soc. Am. Bull.* **55**, 1–40.

Russell, R. J. (1944*b*) Origin of loess – a reply. *Am. J. Sci.* **242**, 447–450.

Ryan, J. A. and Carroll, J. J. (1970) Dust devil wind velocities: mature state. *J. Geophys. Res.* **75**, 531–541.

Safar, M. I. (1980) *Frequency of dust in day-time summer in Kuwait*. Meteorological Department, Kuwait, 107pp.

Sagan, C. (1975) Windblown dust on Venus. *J. Atmos. Sci.* **32**, 1079–1083.

Sagan, C. and Bagnold, R. A. (1975) Fluid transport on Earth and aeolian transport on Mars. *Icarus* **26**, 209–218.

Sagan, C. and Pollack, J. B. (1969) Windblown dust on Mars. *Nature* **223**, 791–794.

Šajgalik, J. (1979) Dependence of microstructure of loesses on their genesis. *Acta Geol. Acad. Scient. Hung.* **22**, 255–266.

Samways, J. (1975) A synoptic account of an occurrence of dense Harmattan haze at Kano in February 1974. *Savanna* **4**, 187–190.

Samways, J. (1976) Ill wind over Africa. *Geog. Mag.* 1976, 218–220.

Sarnthein, M. (1978) Sand deserts during glacial maximum and climatic optimum. *Nature* **272**, 43–46.

Sarnthein, M. and Diester-Haas, L. (1977) Eolian-sand turbidites. *J. Sed. Petrol.* **47**, 868–890.

Sarnthein, M. and Koopman, B. (1980) Late Quaternary deep-sea record of northwest African dust supply and wind circulation. *Palaeoecol. Afr. Surround. Is.* **12**, 239–253.

Sarnthein, M., Tetzlaaf, G., Koopman, B., Wolter, K. and Pflaumann, U. (1981) Glacial and interglacial wind regimes over the eastern sub-tropical Atlantic and northwest Africa. *Nature* **293**, 193–196.

Sayed-Ahmed, M. M. (1949) Khamasin and Khamasin conditions. *Meteorol. Dept. Pap. 1*, Government Press, Cairo.

Scheidegger, K. F. and Krissek, L. A. (1982) Dispersal and deposition of eolian and fluvial sediments off Peru and northern Chile. *Geol. Soc. Am. Bull.* **93**, 150–162.

Scheidig, A. (1934) *Der Löss und sein geotechniscene Eigenschafter.* Steinkopf, Dresden, 233pp.

Schneider, S. H. (1971) A comment on 'Climate: The influence of aerosols'. *J. App. Met.* **10**, 840–841.

Schoorl, RF. (1973) Mineralogical analysis of the 'brown rain' of February 7th 1972. *Geol. Mijn.* **52**, 37.

Schroeder, J. H. (1985) Eolian dust in the coastal desert of the Sudan: aggregates cemented by evaporites. *J. Afr. Earth Sci.* **3**, 370–386.

Schroeder, J. H., Kachholz, K. D. and Heuer, M. (1984–85) Eolian dust in the coastal desert of Sudan: aggregates cemented by evaporites. *Geo. Marine Lett.* **4**, 139–144.

Schultz, C. B. and Frye, J. C. (eds.) (1968) *Loess and Related Eolian Deposits of the World.* University of Nebraska Press, Lincoln, 369pp.

Schultz, C. B. and Stout, T. M. (1945) Pleistocene loess deposits of Nebraska. *Am. J. Sci.* **243**, 231–244.

Schütz, L. (1980) Long range transport of desert dust with special emphasis on the Sahara. *Ann. N. Y. Acad. Sci.* **338**, 515–532.

Schütz, L., Jaenicke, R. and Pietrer, H. (1981) Saharan dust transport over the North Atlantic Ocean. *Geol. Soc. Am. Spec. Pap. 186*, pp.87–100.

Schwaighofer, B. (1980) Pedogenetischer palygorskit in einem lössprofil bei Stillfried an der March (Niederösterreich). *Clay Min.* **15**, 283–289.

Schwertmann, U. and Fanning, D. S. (1976) Iron–manganese concretions in hydrosequences of soils in loess in Bavaria. *Soil Sci. Soc. Am. J.* **40**, 731–738.

Schweisow, R. L. and Cupp, R. E. (1976) Remote Doppler velocity measurements of atmospheric dust devil vortices. *App. Optics* **15**, 1–2.

Sehmel, G. A. and Lloyd, F. D. (1976) Resuspension of plutonium at Rocky Flats. *ERDA, Oak Ridge, Tennessee, Symp. Ser. 38*, pp.757–779.

Seignolis, C. and Arago, F. (1846) Pluie colorée en rouge dans le département de l'Ardèche. *C. R. Acad. Sci. Paris* **23**, 832.

Selby, M. J. (1976) Loess. *New Zealand J. Geog.* **61**, 1–18.

Sergius, L. A. (1952) The Santa Ana. *Weatherwise* **5**, 66–68.

Sergius, L. A., Ellis, G. R. and Ogden, R. M. (1962) The Santa Ana winds of southern California. *Weatherwise* **15**, 102–105.

Servant, M. and Servant-Vildary, S. (1980) L'environment quaternaire du bassin du Tchad. In M. A. J. Williams and H. Faure (eds.) *The Sahara and the Nile*. Balkema, Rotterdam, pp.133–162.

Sevink, J. and Kummer, E. A. (1984) Eolian dust deposition on the Giara di Gesturi Basalt Plateau, Sardinia. *Earth Surf. Proc. Landf.* **9**, 357–364.

Shackleton, N. J. and Opdyke, N. D. (1977) Oxygen isotope and palaeomagnetic evidence for early northern hemisphere glaciation. *Nature* **270**, 216–218.

Shadfan, H, Mashhady, A. S., Dixon, J. B. and Hussen, A. A. (1985) Palygorskite from Tertiary formations of eastern Saudi Arabia. *Clays Clay Min.* **33**, 451–457.

Sharp, M. and Gomez, B. (1985) Processes of debris comminution in the glacial environment and implications for quartz sand grain micromorphology. *Sedim. Geol.* **46**, 33–47.

Shaw, G. E. (1980) Transport of Asian desert aerosol to the Hawaiian Islands. *J. App. Met.* **19**, 1254–1259.

Sheeler, J. B. (1968) Summarization and comparison of engineering properties of loess in the United States. *Highway Res. Rec.* **212**, 1–9.

Shikula, N. K. (1981) Prediction of dust storms from meteorological observations in the South Ukraine, USSR. *Geol. Soc. Am. Spec. Pap. 186*, pp.261–266.

Sidwell, R. (1938) Sand and dust storms in the vicinity of Lubbock, Texas. *Econ. Geog.* **14**, 99–102.

Siever, R. (1962) Silica solubility 0°–200° C and the diagenesis of siliceous sediments. *J. Geol.* **70**, 127–150.

Simmons, G. and Richter, D. (1976) Microcracks in rocks. In R. G. J. Strens (ed.) *The Physics and Chemistry of Minerals and Rocks*. Wiley, London, pp.105–137.

Simonson, R. W. and Hutton, C. E. (1954) Distribution curves for loess. *Am. J. Sci.* **252**, 99–105.

Simpson, J. E. (1969) A comparison between laboratory and atmospheric density currents. *Q. J. Roy. Met. Soc.* **95**, 758–765.

Simpson, J. E. and Britter, R. E. (1979) The dynamics of the head of a gravity current advancing over a horizontal surface. *J. Fluid Mech.* **94**, 477–495.

Sinclair, P. C. (1964) Some preliminary dust devil measurements. *Month. Weath. Rev.* **92**, 363–367.

Sinclair, P. C. (1965) On the rotation of dust devils. *Am. Met. Soc. Bull.* **46**, 388–391.

Sinclair, P. C. (1969) General characteristics of dust devils. *J. App. Met.* **8**, 32–45.

Sinclair, P. C. (1976) Vertical transport of desert particulates by dust devils and clear thermals. In R. Engelman and G. Sehmel (eds.) *Atmosphere–Surface Exchange of Particulate and Gaseous Pollutants*. ERDA, Oak Ridge, Tennessee, pp.497–527.

Singer, A. (1967) Mineralogy of the non-clay fractions from basaltic soils in the Galilee, Israel. *Israel J. Earth Sci.* **16**, 215–228.

Singer, A. and Galan, E. (eds.) (1985) *Palygorskite–Sepiolite: Occurrence, Genesis and Uses*. Elsevier, Amsterdam.

Sirenko, N. A. (1984) Pliocene and Pleistocene soil formations of the Ukraine. In M. Pecsi (ed.) *Lithology and Stratigraphy of Loess and Paleosols*. Geographical Research Institute, Hungarian Academy of Sciences, Budapest, pp.27–32.

Skertchley, S. B. J. and Kingsmill, T. W. (1895) On the loess and other superficial deposits of Shantung (North China). *Q. J. Geol. Soc. Lond.* **51**, 238–254.

Skidmore, E. L., Fisher, P. S. and Woodruff, N. P. (1970) Wind erosion equation: computer solution and application. *Soil Sci. Soc. Am. Proc.* **34**, pp.931–935.

Slatt, R. M. and Eyles, N. (1981) Petrology of glacial sand: implications for the origin and mechanical durability of lithic fragments. *Sedimentology* **28**, 171–183.

Smalley, I. J. (1966) The properties of glacial loess and the formation of loess deposits. *J. Sed. Petrol.* **36**, 669–676.

Smalley, I. J. (1970) Cohesion of soil particles and the intrinsic resistance of simple soil systems to wind erosion. *J. Soil Sci.* **21**, 154–161.

Smalley, I. J. (1971) 'In situ' theories of loess formation and the significance of the calcium-carbonate content of loess. *Earth Sci. Rev.* **7**, 67–85.

Smalley, I. J. (1972) The interaction of great rivers and large deposits of primary loess. *Trans. N. Y. Acad. Sci.* **34**, 534–542.

Smalley, I. J. (1974) Fragmentation of granitic quartz in water: discussion. *Sedimentology* **21**, 633–635.

Smalley, I. J. (ed.) (1975) *Loess: Lithology and Genesis*. Benchmark Papers in Geology 26. Dowden, Hutchinson and Ross, Stroudsberg, 430pp.

Smalley, I. J. (1980*a*) *Loess: a partial bibliography*. Geo Abstracts, Norwich, 103pp.

Smalley, I. J. (1980*b*) The formation of loess materials and loess deposits: some observations on the Tashkent loess. *Geophys. Geol. Geophys. Ver. KMU Leipzig Bd.* **II, H2**, 247–257.

Smalley, I. J. and Cabrera, J. G. (1970) The shape and surface texture of loess particles. *Geol. Soc. Am. Bull.* **81**, 1591–1595.

Smalley, I. J. and Davin, J. E. (1980) The first hundred years – a historical bibliography of New Zealand loess, 1878–1978. *New Zealand Soil Bur. Bibliog. Rep.* **28**, 166pp.

Smalley, I. J. and Krinsley, D. H. (1978) Loess deposits associated with deserts. *Catena* **5**, 53–66.

Smalley, I. J. and Krinsley, D. H. (1979) Eolian sedimentation on Earth and Mars: some comparisons. *Icarus* **40**, 276–288.

Smalley, I. J. and Leach, J. A. (1978) The origin and distribution of the loess in the Danube Basin and associated regions of East–Central Europe. *Sedim. Geol.* **21**, 1–26.

Smalley, I. J. and Smalley, V. (1983) Loess material and loess deposits: formation, distribution and consequences. In M. E. Brookfield and T. S. Ahlbrandt (eds.) *Eolian Sediments and Processes*. Elsevier, Amsterdam, pp.51–68.

Smalley, I. J. and Vita-Finzi, C. (1968) The formation of fine particles in sandy deserts and the nature of 'desert' loess. *J. Sed. Petrol.* **38**, 766–774.

Smalley, I. J., Krinsley, D. H. and Vita-Finzi, C. (1973) Observations on the Kaiserstühl loess. *Geol. Mag.* **110**, 29–36.

Smalley, I. J., Krinsley, D. H., Moon, C. F. and Bentley, S. P. (1978) Processes of quartz fracture in nature and the formation of clastic sediments. In K. E. Easterling (ed.) *Mechanisms of Deformation and Fracture*. Pergamon, Oxford, pp.119–127.

Smith, B. J. (1977) Rock temperature measurements from the northwest Sahara and their implications for rock weathering. *Catena* **4**, 41–63.

Smith, B. J. and Whalley, W. B. (1981) Late Quaternary drift deposits of north central Nigeria examined by scanning electron microscopy. *Catena* **8**, 345–368.

Smith, B. J., McGreevy, J. P. and Whalley, W. B. (1986) Silt production by weathering of sandstone under hot arid conditions: an experimental study. *J. Arid Env.* (in press).

Smith, R. M., Twiss, P. C., Krauss, R. K. and Brown, M. J. (1970) Dust deposition in relation to site, season, and climatic variables. *Soil Sci. Soc. Am. Proc.* **34**, 112–117.

Sneh, A. (1982) Drainage systems of the Quaternary in northern Sinai with emphasis on Wadi el-Arish. *Z. Geomorph.* **NF 26**, 179–195.

Sneh, A. (1983) Redeposited loess from the Quaternary Besor Basin, Israel. *Israel J. Earth Sci.* **32**, 63–69.

Snowden, J. O. and Priddy, R. R. (1968) Geology of Mississippi loess. *Miss. Geol. Surv. Bull.* **111**, 13–203.

Sommeria, G. and Testud, J. (1984) COPT 81: A field experiment designed for the study of dynamics and electrical conductivity of deep convection in continental tropical regions. *Am. Met. Soc. Bull.* **65**, 4–10.

Sparks, R. S. J. (1986) The dimensions and dynamics of volcanic eruption clouds. *Bull. Volcanol.* **48**, 3–15.

Sparks, R. S. J. and Wilson, C. (1976) A model for the formation of ignimbrite by gravitational column collapse. *J. Geol. Soc. Lond.* **132**, 441–451.

Sprunt, E. S. and Brace, W. F. (1974) Direct observations of microcavities in crystalline rocks. *Int. J. Rock Mech. Geomech. Abs.* **11**, 139–150.

Squires, P. and Twomey, S. (1960) The relation between cloud droplet spectra and the spectrum of cloud nucleii. *Geophys. Monog.* **5**, 211–219.

St Arnaud, R. J. and Whiteside, E. P. (1963) Physical breakdown in relation to soil development. *J. Soil Sci.* **14**, 267–281.

Stein, R. and Sarnthein, M. (1984) Late Neogene events of atmospheric and oceanic circulation offshore northwest Africa: high-resolution record from deep-sea sediments. *Paleoecol. Afr. Surround. Is.* **16**, 9–36.

Stevenson, C. M. (1969) The dustfall and severe storms of 1 July 1968. *Weather* **24**, 126–132.

Stix, E. (1975) Pollen and spore content of the air during the autumn above the Atlantic ocean. *Oecologia* **18**, 235–242.

Stow, C. D. (1969) Dust and sand storm electrification. *Weather* **24**, 134–140.

Street, F. A. and Grove, A. T. (1979) Global maps of lake-level fluctuations since 30 000 yr BP. *Quat. Res.* **12**, 83–118.

Street-Perrott, F. A. and Harrison, S. P. (1984) Temporal variations in lake levels since 30 000 yr BP – an index of the global hydrological cycle. In J. E. Hansen and T. Takahashi (eds.) *Climatic Processes and Climate Sensitivity.* American Geophysical Union, Washington DC, pp.118–129.

Street-Perrott, F. A. and Roberts, N. (1983) Fluctuations in closed-basin lakes as an indicator of past atmospheric circulation patterns. In F. A. Street-Perrott, M. Beran and R. A. S. Ratcliffe (eds.) *Variations in the Global Water Budget.* Reidel, Dordrecht, pp.331–345.

Suck, S. H., Thurman, R. E. and Kim, C. H. (1986) Growth of ultrafine particles by Brownian coagulation. *Atmos. Env.* **20**, 773–777.

Summerfield, M. A. (1983) Silcrete. In A. S. Goudie and K. Pye (eds.) *Chemical Sediments and Geomorphology.* Academic Press, London, pp.59–91.

Sundborg, A. (1955) Meteorological and climatological conditions for the genesis of aeolian sediments. *Geog. Ann.* **37**, 94–111.

Sun Tien-ching and Yang Huan-jen (1961) The great Ice Age glaciations in China. *Acta Geol. Sinica* **41**, 234–244 (in Chinese).

Sutton, L. J. (1925) Haboobs. *Q. J. Roy. Met. Soc.* **51**, 25–30.

Sutton, L. J. (1931) Haboobs. *Q. J. Roy. Met. Soc.* **57**, 143–161.

Suzuki, T. and Takahashi, K. (1981) An experimental study of wind abrasion. *J. Geol.* **89**, 23–36.

Svasek, T. N. and Terwindt, J. H. (1974) Measurement of sand transport by wind on a natural beach. *Sedimentology* **21**, 311–322.

Swineford, A. and Frye, J. C. (1945) A mechanical analysis of wind-blown dust compared with analyses of loess. *Am. J. Sci.* **243**, 249–255.

Swineford, A. and Frye, J. C. (1951) Petrography of the Peorian loess in Kansas. *J. Geol.* **59**, 306–322.

Syers, J. K., Jackson, M. L., Berkheiser, V. E., Clayton, R. N. and Rex, R. W. (1969) Eolian sediment influence on pedogenesis during the Quaternary. *Soil Sci.* **107**, 421–427.

Syers, J. K., Mokma, D. L., Jackson, M. L., Dolcater, D. L. and Rex, R. W. (1972) Mineralogical composition and ^{137}C retention properties of continental dusts. *Soil Sci.* **113**, 116–123.

Symmons, P. M. and Hemming, C. F. (1968) A note on wind-stable stone mantles in the southern Sahara. *Geog. J.* **134**, 60–64.

Takayama, Y. and Takashima, T. (1986) Aerosol optical thickness of Yellow Sand over the Yellow Sea derived from NOAA satellite data. *Atmos. Env.* **20**, 631–638.

Talbot, M. R. and Hall, J. B. (1981) Further late Quaternary leaf fossils from Lake Bosumtwi, Ghana. *Paleoecol. Afr. Surround. Is.* **13**, 83–92.

Taylor, G. I. (1915) Eddy motion in the atmosphere. *Phil. Trans. Roy. Soc. Lond.* **A 215**, 1–26.

Taylor, S. R., McLennan, S. M. and McCullough, M. T. (1983) Geochemistry of loess continental crust composition and crustal model ages. *Geochim. Cosmochim. Acta* **47**, 322–332.

Teruggi, M. E. (1957) The nature and origin of Argentine loess. *J. Sed. Petrol.* **27**, 322–332.

Terzaghi, K. (1943) *Theoretical Soil Mechanics*. Wiley, New York, 510pp.

Tetzlaaf, G. and Wolter, K. (1980) Meteorological patterns and the transport of mineral dust from the North African continent. *Paleoecol. Afr. Surround. Is.* **12**, 31–42.

Tetzlaaf, G. and Peters, M. (1986) Deep-sea sediments in the eastern equatorial Atlantic off the African coast and meteorological flow patterns over the Sahel. *Geol. Rund.* **75**, 71–79.

Thiede, J. (1977) Aspects of the variability of the glacial and interglacial North Atlantic boundary current (last 150 000 years). *Meteor Forschung. Reihe* **C28**, 1–36.

Thiede, J. (1979) Wind regimes over the late Quaternary southwest Pacific Ocean. *Geology* **7**, 259–262.

Thiesen, A. A. and Knox, E. G. (1959) Distribution and characteristics of loessial soil parent material in northwestern Oregon. *Soil Sci. Soc. Am. Proc.* **23**, 385–388.

Thom, A. S. (1976) Momentum, mass and heat exchange of plant communities. In J. L. Monteith (ed.) *Vegetation and the Atmosphere*, vol. I, *(Principles)*. Academic Press, London, pp.57–109.

Thompson, L. G. and Mosley-Thompson, E. (1981) Microparticle concentration variation linked with climatic change: evidence from polar ice cores. *Science* **212**, 812–815.

Thompson, R. and Oldfield, F. (1986) *Environmental Magnetism*. Allen and Unwin, London, 227pp.

Thorez, J., Bourguignon, P. and Paepe, R. (1970) Étude preliminaire des associations de minéraux argileux des loess Pleistocènes en Belgique. *Ann. Soc. Géol. Belg.* **593**, 265–285.

Thornthwaite, C. R. (1948) An approach toward a rational classification of climate. *Geog. Rev.* **38**, 55–94.

Thorp, J. and Smith, H. T. U. (1952) *Map of Pleistocene Eolian Deposits of the United States, Alaska and Parts of Canada. Scale 1 : 2 500 000.* Geological Society of America.

Todd, J. E. (1878) Richthofen's theory of the loess, in the light of the deposits of the Missouri. *Proc. Am. Ass. Adv. Sci.* **27**, 231–239.

Todd, J. E. (1897) Degradation of loess. *Proc. Iowa Acad. Sci.* **5**, 46–51.

Tomirdiaro, S. V. (1984) Periglacial landscapes and loess accumulation in the late Pleistocene Arctic and Subarctic. In A. A. Velichko (ed.) *Late Quaternary Environments of the Soviet Union*. Longman, London, pp.141–145.

Tonkin, P. J., Runge, E. C. A. and Ives, D. (1974) A study of Late Pleistocene loess deposits, New Zealand. Part II. Palaeosols and their stratigraphic implications. *Quat. Res.* **4**, 217–231.

Trainer, F. W. (1961) Eolian deposits of the Matanuska Valley agricultural area, Alaska. *US Geol. Surv. Bull.* **1121C**, C1–C35.

Treasher, R. C. (1925) Origin of the loess of the Palouse region, Washington. *Science* **61**, 469.

Tsoar, H. and Møller, J. T. (1986) The role of vegetation in the formation of linear sand dunes (analysis of the case of the Negev–Sinai borderline). In W. G. Nickling (ed.) *Aeolian Geomorphology*. Allen and Unwin, New York, pp.75–95.

Tsoar, H. and Pye, K. (1987) Dust transport and the question of desert loess formation. *Sedimentology* (in press).

Tsoar, H. and Zohar, Y. (1985) Desert dune sand and its potential for modern agricultural development. In Y. Gradus (ed.) *Desert Development*. Reidel, Dordrecht, pp.184–200.

Tucholka, P. (1977) Magnetic polarity events in Polish loess profiles. *Biul. Inst. Geol.* **305**, 117–123.

Tuck, R. (1938) The loess of the Matanuska Valley, Alaska. *J. Geol.* **46**, 647–653.

Tullett, M. T. (1978) A dustfall on 6 March 1977. *Weather* **33**, 48–52.

Tullett, M. T. (1980) A dustfall of Saharan origin on 15 May (1979). *J. Earth Sci. Roy. Dublin Soc.* **3**, 35–39.

Tullett, M. T. (1984) Saharan dustfall in Northern Ireland. *Weather* **38**, 151–152.

Turco, R. P., Toon, O. B., Ackerman, T. P., Pollack, J. B. and Sagan, C. (1983) Nuclear winter: global consequences of multiple nuclear explosions. *Science* **222**, 1282–1292.

Turekian, K. K. (1977) Geochemical distribution of elements. In *The Encyclopedia of Science and Technology*, 4th edn. McGraw-Hill, New York, pp.627–630.

Turekian, K. K. and Cochran, J. K. (1981) ^{210}Pb in surface air at Enewetak and the Asian dust flux to the Pacific. *Nature* **292**, 522–524.

Turner, R. E., Eitner, P. G. and Manning, J. L. (1979) *Analysis of battlefield-induced contaminants for E-O SAEL*. Final Report on Contract DAAD07-79-C-0032. Science Applications Inc., Ann Arbor, Michigan, 166pp.

Tutkovskii, P. A. (1899) The question of the method of loess formation. *Zemlevedenie* **1–2**, 213–311 (in Russian).

Tutkovskii, P. A. (1900) M. Paul Tutkowski on the origin of loess. *Scott. Geog. Mag.* **16**, 171–174.

Tutkovskii, P. A. (1912) Das postglaziale Klima in Europe and Nordamerika, die postglazialen Wüsten und die Lössbildung. *C. R. 11th Int. Geol. Cong. Stockholm, 1910*, pp.359–369.

Twomey, S. (1977) *Atmospheric Aerosols*. Elsevier, Amsterdam.

Udden, J. A. (1894) Erosion, transportation, and sedimentation performed by the wind. *J. Geol.* **2**, 318–331.

Udden, J. A. (1896) Dust and sand storms in the West. *Pop. Sci. Month.* **49**, 655–664.

Udden, J. A. (1897a) Origin of loess. *Am. Geol.* **20**, 274–275.

Udden, J. A. (1897b) Loess as a land deposit. *Geol. Soc. Am. Bull.* **9**, 6–9.

Udden, J. A. (1898) The mechanical composition of wind deposits. *Pub. Augustana Lib.* **1**, 69pp.

Udden, J. A. (1914) Mechanical composition of clastic sediments. *Geol. Soc. Am. Bull.* **25**, 655–744.

Uematsu, M., Duce, R. A., Prospero, J. M., Chen, L., Merrill, J. T. and McDonald, R. L. (1983) Transport of mineral aerosol from Asia over the North Pacific Ocean. *J. Geophys. Res.* **88C**, 5343–5352.

Uematsu, M., Duce, R. A., Nakaya, S. and Tsunogai, S. (1985) Short-term temporal variability of eolian particles in surface waters of the northwestern North Pacific. *J. Geophys. Res.* **90**, 1167–1172.

Van den Ancker, J. A. M., Jungerius, P. D. and Muir, L. R. (1985) The role of algae in the stabilization of coastal dune blowouts. *Earth Surf. Proc. Landf.* **10**, 189–192.

Van der Waals, L. (1967) Morphological phenomena on quartz grains in unconsolidated sands, due to migration of quartz near the Earth's surface. *Meded. Neth. Geol. Sticht.* **18**, 47–51.

Van der Westhuizen, M. (1969) Radioactive nuclear bomb fallout. A relationship between deposition, air concentration and rainfall. *Atmos. Env.* **3**, 241–248.

Van Heuklon, T. K. (1977) Distant source of 1976 dustfall in Illinois and Pleistocene weather models. *Geology* **5**, 693–695.

Vasil'yev, G. I., Bulgakou, D. S., Gaurilenko, L. N. and Kalnichenko, A. S. (1978) Conditions under which dust storms develop in the northern Caucasus. *Soviet Soil Sci.* **2**, 67–77.

Veklich, M. F. (1979) Pleistocene loesses and fossil soils of the Ukraine. *Acta Geol. Acad. Scient. Hung.* **22**, 35–62.

Veklich, M. F. and Sirenko, N. A. (1984) Inter-regional palaeopedological Pleistocene correlation of the USSR loess regions. In M. Pecsi (ed.) *Lithology and Stratigraphy of Loess and Paleosols*. Geographical Research Institute, Hungarian Academy of Sciences, Budapest, pp.249–258.

Vermillion, C. H. (1977) NOAA-5 views dust storm. *Am. Met. Soc. Bull.* **58**, 330.

Vernon, P. D. and Reville, W. J. (1983) The dustfall of November 1979. *J. Earth Sci. Roy. Dublin Soc.* **5**, 135–144.

Virlet-d'Aoust, P. H. (1857) Observations sur un terrain d'origine météorique ou de transport aérien qui existe au Mexique, et sur le phénomène des trombees de poussière auquel il doit principalement son origine. *Geol. Soc. Fr. Bull.* 2nd ser. **16**, 417–431.

Vita-Finzi, C. (1969) *The Mediterranean Valleys*. Cambridge University Press, London.

Vita-Finzi, C. and Smalley, I. J. (1970) Origin of quartz silt: comments on a note by Ph. H. Kuenen. *J. Sedim. Petrol.* **40**, 1367–1369.

Vita-Finzi, C., Smalley, I. J. and Krinsley, D. H. (1973) Crystalline overgrowths on quartz silt particles. *J. Geol.* **81**, 106–108.

Vivian, R. A. (1975) *Les Glaciers des Alpes Occidentales*. Imprimerie Allier, Grenoble.

Volkov, I. A. and Zykina, V. S. (1984) Loess stratigraphy in southwest Siberia. In A. A. Velichko (ed.) *Late Quaternary Environments of the Soviet Union*. Longman, London, pp.119–124.

Volov, M. D. and Orlovskii, O. I. (1977) Drainage of loess lands with a saucer-shaped relief. *Gidroteckh Melior* **8**, 66–68.

Vreeken, W. J. and Mucher, H. J. (1981) Re-deposition of loess in southern Limbourg, The Netherlands. I. Field evidence for conditions of deposition of the Lower Silt Loan Complex. *Earth Surf. Proc. Landf.* **6**, 337–354.

Waggoner, P. E. and Bingham, C. (1961) Depth of loess and distance from source. *Soil Sci.* **92**, 396–401.

Wagner, J. C. (1980) The pneumoconioses due to mineral dusts. *J. Geol. Soc. Lond.* **137**, 537–546.

Walder, J. S. and Hallet, B. (1986) The physical basis of frost weathering: toward a more fundamental and unified perspective. *Arc. Alp. Res.* **18**, 27–32.

Walker, A. S. (1982) Deserts of China. *Am. Scient.* **70**, 366–376.

Walker, P. H. and Costin, A. B. (1971) Atmospheric dust accession in southeastern Australia. *Aust. J. Soil Res.* **9**, 1–5.

Walker, T. R. (1979) Red color in dune sand. *US Geol. Surv. Prof. Pap.* **1052**, 61–81.

Walker, T. R. (1976) Diagenetic origin of continental red beds. In H. Falke (ed.) *The Continental Permian in Central, West and South Europe.* Reidel, Dordrecht, pp.240–282.

Walker, T. R., Waugh, B. and Crone, A. J. (1978) Diagenesis of first-cycle desert alluvium of Cenozoic age, southwestern United States and northwestern Mexico. *Geol. Soc. Am. Bull.* **89**, 19–32.

Walton, W. H. and Woolcock, A. (1960) The suppression of airborne dust by water spray. *J. Air Poll.* **3**, 129–153.

Wang Yong-yan and Zhang Zhong-hu (1980) *Loess in China.* Shansi People's Art Publishing House.

Wang Yong-yan, Teng Zhi-hong and Yue Le-ping (1984) Loess microtextures and the origin of loess in China. In M. Pecsi (ed.) *Lithology and Stratigraphy of Loess and Paleosols.* Geographical Research Institute, Hungarian Academy of Sciences, Budapest, pp.49–58.

Ward, A. W. and Greeley, R. (1984) Evolution of yardangs at Rogers Lake, California. *Geol. Soc. Am. Bull.* **95**, 829–837.

Warn, G. F. and Cox, W. H. (1951) A sedimentary study of dust storms in the vicinity of Lubbock, Texas. *Am. J. Sci.* **249**, 552–568.

Warnke, D. A. (1971) The shape and surface texture of loess particles. *Geol. Soc. Am. Bull.* **82**, 2357–2360.

Warren, G. K. (1878) Valley of the Minnesota and of the Mississippi River to the junction of the Ohio: its origin considered. *Am. J. Sci.* 3rd ser. **16**, 417–431.

Wasson, R. J. (ed.) (1982a) *Quaternary Dust Mantles of China, New Zealand and Australia.* ANU Press, Canberra, 253pp.

Wasson, R. J. (1982b) The contribution of dust to Quaternary valley fills at Belarabon, western NSW. In R. J. Wasson (ed.) *Quaternary Dust Mantles of China, New Zealand and Australia.* ANU Press, Canberra, pp.191–195.

Watkins, W. I. (1945) Observations on the properties of loess in engineering structures. *Am. J. Sci.* **243**, 294–303.

Watson, A. (1985) Structure, chemistry and origins of gypsum crusts in southern Tunisia and the central Namib Desert. *Sedimentology* **32**, 855–876.

Waugh, B. (1970) Petrology, provenance and silica diagenesis of the Penrith Sandstone (Lower Permian) of northwest England. *J. Sed. Petrol.* **40**, 1226–1240.

Weaver, F. M. and Wise, S. R. (1974) Opaline sediments of the southeastern coastal plain and horizon A: biogenic origin. *Science* **184**, 899–901.

Wen Qi-zhong, Diao Gui-yi, Yu Su-hua, Sun Fu-quing, Gu Xiong-fei, Chen Qing-mu and Liu Yu-luan (1982) Some problems of loess geochemistry in China. In R. J. Wasson (ed.) *Quaternary Dust Mantles of China, New Zealand and Australia.* ANU Press, Canberra, pp.69–83.

Wen Qi-zhong, Yang Wei-hua, Diao Gui-yi, Sun Fu-qing, Yu Su-hua and Liu You-mei (1984) The evolution of chemical elements in loess of China and paleoclimatic conditions during loess deposition. In M. Pecsi (ed.) *Lithology and Stratigraphy of Loess and Paleosols.* Geographical Research Institute, Hungarian Academy of Sciences, Budapest, pp.161–169.

Wentworth, C. K. (1922) A scale of grade and class terms for clastic sediments. *J. Geol.* **30**, 377–392.

Whalley, W. B. (1979) Quartz silt production and sand grain surface textures from fluvial and glacial environments. *SEM* 1979/I, 547–554.

Whalley, W. B. (1983) Desert varnish. In A. S. Goudie and K. Pye (eds.) *Chemical Sediments and Geomorphology.* Academic Press, London, pp.197–226.

Whalley, W. B. and Smith, B. J. (1981) Mineral content of Harmattan dust from northern Nigeria examined by scanning electron microscopy. *J. Arid Env.* **4**, 21–30.

Whalley, W. B., Marshall, J. R. and Smith, B. J. (1982) Origin of desert loess from some experimental observations. *Nature* **300**, 433–435.

Whalley, W. B., Smith, B. J., McAlister, J. J. and Edwards, A. (1987) Aeolian abrasion of quartz particles and the production of silt-size fragments, preliminary results and some possible implications for loess and silcrete formation. In I. Reid and L. Frostick (eds.) *Desert Sediments Ancient and Modern.* Blackwell, Oxford (in press).

Wheeler, D. A. (1985) An analysis of the aeolian dustfall on eastern Britain, November 1984. *Proc. Yorks. Geol. Soc.* **45**, 307–310.

Whipple, H. E. (ed.) (1964) Cosmic Dust. *Ann. Am. Acad. Sci.* **119**, 367pp.

White, S. (1976) The role of dislocation processes during tectonic deformations, with particular reference to quartz. In R. G. J. Strens (ed.) *The Physics and Chemistry of Minerals and Rocks.* Wiley, London, pp.75–91.

Whitney, M. I. and Splettstoesser, J. F. (1982) Ventifacts and their formation: Darwin Mountains, Antarctica. *Catena Supp.* **1**, 175–194.

Wilding, L. P. and Drees, L. R. (1968) Distribution and implications of sponge spicules in surficial deposits of Ohio. *Ohio J. Sci.* **68**, 92–99.

Wilding, L. P. and Drees, L. R. (1971) Biogenic opal in Ohio soils. *Soil Sci. Soc. Am. Proc.* **35**, 1004–1010.

Wilding, L. R., Smeck, N. E. and Drees, L. R. (1977) Silica in soils: quartz, cristobalite, tridymite and opal. In J. B. Dixon and S. B. Weed (eds.) *Minerals in Soil Environments.* Soil Science Society of America, Madison, pp.471–552.

Wilke, B. M., Duke, B. J. and Jimoh, W. L. O. (1984) Mineralogy and chemistry of Harmattan dust in northern Nigeria. *Catena* **11**, 91–96.

Williams, G. E. (1971) Flood deposits of the sand bed ephemeral streams of central Australia. *Sedimentology* **17**, 1–40.

Williams, M. A. J. and Clark, M. F. (1984) Late Quaternary environments in north Central India. *Nature* **308**, 633–635.

Williams, R. E. and Allman, D. W. (1969) Factors affecting infiltration and recharge in a loess-covered basin. *J. Hydrol.* **8**, 265–281.

Williams, R. B. G. and Robinson, D. A. (1981) Weathering of sandstone by the combined action of frost and salt. *Earth Surf. Proc. Landf.* **6**, 1–9.

Willman, H. B. and Frye, J. C. (1970) Pleistocene stratigraphy of Illinois. *Illinois State Geol. Surv. Bull.* **94**, 204pp.

Wilshire, H. G. (1980) Human causes of accelerated wind erosion in California's deserts. In D. R. Coates and J. D. Vitek (eds.) *Thresholds in Geomorphology.* Allen and Unwin, London, pp.415–434.

Wilshire, H. G., Nakata, J. K. and Hallet, B. (1981) Field observations of the December 1977 wind storm, San Joaquin Valley, California. *Geol. Soc. Am. Spec. Pap.* **186**, 233–252.

Windom, H. L. (1969) Atmospheric dust records in permanent snowfields: implications to marine sedimentation. *Geol. Soc. Am. Bull.* **80**, 761–782.

Windom, H. L. (1970) Contributions of atmospherically transported trace metals to South Pacific sediments. *Geochim. Cosmochim. Acta* **34**, 509–514.

Windom, H. L. (1975) Eolian contributions to marine sediments. *J. Sed. Petrol.* **45**, 520–529.

Windom, H. L. and Chamberlain, F. G. (1978) Dust storm transport of sediments to the North Atlantic Ocean. *J. Sed. Petrol.* **48**, 385–388.

Windom, H. L., Griffin, J. J. and Goldberg, E. D. (1967) Talc in atmospheric dusts. *Env. Sci. Tech.* **1**, 923–926.

Winkler, E. M. (1977) Insolation of rock and stones a hot item. *Geology* **5**, 188–189.

Winstanley, D. (1972) Sharav. *Weather* **27**, 146–160.

Wintle, A. G. (1981) Thermoluminescence dating of late Devensian loesses in southern England. *Nature* **289**, 479–480.

Wintle, A. G. (1982) Thermoluminescence properties of fine-grained minerals in loess. *Soil Sci.* **134**, 164–170.

Wintle, A. G. (1987) Thermoluminescence dating of loess sections – a reappraisal. *Proc. Int. Conf. Loess Research, Xian, China, November 1985*, (in press).

Wintle, A. G. and Brunnacker, K. (1982) Ages of volcanic tuff in Rheinhessen obtained by thermoluminescence dating of loess. *Naturwissen.* **69**, 181–183.

Wintle, A. G. and Catt, J. A. (1985) Thermoluminscence dating of soils developed in late Devensian loess at Pegwell Bay, Kent. *J. Soil Sci.* **36**, 292–298.

Wintle, A. G. and Proszynska, H. (1983) TL dating of loess in Germany and Poland. *PACT* **9**, 547–554.

Wintle, A. G. and Westgate, J. A. (1986) Thermoluminescence age of Old Crow Tephra in Alaska. *Geology* **14**, 594–597.

Wintle, A. G., Shackleton, N. J. and Lautridou, J. P. (1984) Thermoluminescence dating of periods of loess deposition and soil formation in Normandy. *Nature* **310**, 491–493.

Woodruff, N. P. (1956) Windblown soil abrasive injuries to winter wheat plants. *Agron. J.* **48**, 499–505.

Woodruff, N. P. and Siddoway, F. H. (1965) A wind erosion equation. *Soil Sci. Soc. Am. Proc.* **29**, 602–608.

Worster, D. (1979) *Dust Bowl – The Southern Plains in the 1930s.* Oxford University Press, New York, 277pp.

Wright, W. R. and Foss, J. E. (1968) Movement of silt-sized particles in sand columns. *Soil Sci. Soc. Am. Proc.* **32**, 446–448.

Wu Zi-rhong, Yuan Bao-yin and Gao Fu-qing (1982) Geological environment of loess deposits on Lochuan Yuan, Shaanxi Province. In R. J. Wasson (ed.) *Quaternary Dust Mantles of China, New Zealand and Australia.* ANU Press, Canberra, pp.19–30.

Yaalon, D. H. (1964a) Downward movement and distribution of anions in soil profiles with limited wetting. In E. G. Hallsworth and D. V. Crawford (eds.) *Experimental Pedology.* Butterworth, London, pp.157–164.

Yaalon, D. H. (1964b) Airborne salts as an active agent in pedogenetic processes. *Trans. 8th Int. Cong. Soil Sci., Bucharest*, vol. V, pp.997–1000.

Yaalon, D. H. (1974) Notes on some geomorphic effects of temperature changes on desert surfaces. *Z. Geomorph. Supp. Bd.* **21**, 29–34.

Yaalon, D. H. and Dan, J. (1974) Accumulation and distribution of loess-derived deposits in the semi-desert and desert-fringe areas of Israel. *Z. Geomorph. Supp. Bd.* **20**, 91–105.

Yaalon, D. H. and Ganor, E. (1966) The climatic factor of wind erodibility and dust blowing in Israel. *Israel J. Earth Sci.* **15**, 27–32.

Yaalon, D. H. and Ganor, E. (1973) The influence of dust on soils during the Quaternary. *Soil Sci.* **116**, 146–155.

Yaalon, D. H. and Ganor, E. (1975) Rate of aeolian dust accretion in the Mediterranean and desert fringe environments of Israel. *19th Int. Cong. Sedimentol. Theme 2*, pp.169–174.

Yaalon, D. H. and Ganor, E. (1979) East Mediterranean trajectories of dust-carrying storms from the Sahara and Sinai. In C. Morales (ed.) *Saharan Dust*. Wiley, Chichester, pp.187–193.

Yaalon, D. H. and Ginzbourg, D. (1966) Sedimentary characteristics and climatic analysis of easterly dust storms in the Negev (Israel). *Sedimentology* **6**, 315–332.

Yang Shao-jin, Qian Qin-fen, Zhou Ming-yu, Qu Shoa-hou, Song Xi-ming and Li Yu-yeng (1981) Some properties of the aerosols during the passage of a dust storm over Beijing, China, April 17–20, 1980. *Geol. Soc. Am. Spec. Pap. 186*, pp.159–168.

Yeck, R. D. and Gray, F. (1972) Phytolith-size characteristics between Udolls and Ustolls. *Soil Sci. Soc. Am. Proc.* **36**, 639–641.

Young, D. G. (1964) Stratigraphy and petrography of northeast Otago loess. *New Zealand J. Geol. Geophys.* **7**, 839–863.

Young, D. G. (1967) Loess deposits on the west coast of South Island, New Zealand. *New Zealand J. Geol. Geophys.* **10**, 647–658.

Young, J. A. and Evans, R. A. (1986) Erosion and deposition of fine sediments from playas. *J. Arid. Env.* **10**, 103–115.

Zeuner, F. E. (1949) Frost soils on Mount Kenya. *J. Soil Sci.* **1**, 20–30.

Zhang Zong-hu (1984) Lithological and stratigraphical analysis on the loess profiles of the Loess Plateau in China. In M. Pecsi (ed.) *Lithology and Stratigraphy of Loess and Paleosols*. Geographical Research Institute, Hungarian Academy of Sciences, Budapest, pp.259–270.

Zhao Xi-tao and Qu Yong-xin (1981) Loess of Zhaitang and Yanchi regions, Beijing. *Sci. Geol. Sinica* **1**, 45–54.

Zheng Bex-xing and Li Jijun (1981) Quaternary glaciation of the Qinghai–Xizang Plateau. *Proc. Symp. Qinghai–Zizang (Tibet) Plateau*. Science Press, Beijing, 1631–1640.

Zheng Hong-han (1984) Paleoclimatic events recorded in clay minerals in loess of China. In M. Pecsi (ed.) *Lithology and Stratigraphy of Loess and Paleosols*. Geographical Research Institute, Hungarian Academy of Sciences, Budapest, pp.171–181.

Zhirkhov, K. F. (1964) Dust storms in the steppes of western Siberia and Kazakhstan. *Soviet Geog.* **5**, 33–41.

Zingg, A. W. (1951) A portable wind tunnel and dust collector developed to evaluate the erodibility of field surfaces. *Agron. J.* **43**, 189–191.

Zingg, A. W. and Chepil, W. S. (1950) Aerodynamics of wind erosion. *Agric. Eng.* June 1950, 279–282.

Zurek, R. W. (1982) Martian great dust storms: an update. *Icarus* **50**, 288–310.

Zolotun, V. P. (1974) Origin of loess deposits in the southern part of the Ukraine. *Soviet Soil Sci.* **1**, 1–12.

INDEX